Intraoperative Neurophysiologic Monitoring

Intraoperative Neurophysiologic Monitoring

Gloria M. Galloway
Marc R. Nuwer
Jaime R. López
Khaled M. Zamel

CAMBRIDGE
UNIVERSITY PRESS

CAMBRIDGE UNIVERSITY PRESS
Cambridge, New York, Melbourne, Madrid, Cape Town, Singapore,
São Paulo, Delhi, Dubai, Tokyo, Mexico City

Cambridge University Press
The Edinburgh Building, Cambridge CB2 8RU, UK

Published in the United States of America by Cambridge University Press,
New York

www.cambridge.org
Information on this title: www.cambridge.org/9780521518031

First published 2010

Printed in the United Kingdom at the University Press, Cambridge

A catalog record for this publication is available from the British Library

Library of Congress Cataloging in Publication data
Intraoperative neurophysiologic monitoring / Gloria M. Galloway ... [et al.].
p. ; cm.
Includes bibliographical references and index.
ISBN 978-0-521-51803-1 (hardback)
1. Neurophysiologic monitoring. I. Galloway, Gloria M., 1960– [DNLM: 1. Monitoring, Intraoperative – methods. 2. Intraoperative Complications – prevention & control. 3. Surgical Procedures, Operative – methods. 4. Trauma, Nervous System – prevention & control. WO 181 I609 2010]
RD52.N48I558 2010
616.8′047547–dc22

2010024614

ISBN 978-0-521-51803-1 Hardback

Additional resources for this publication at www.cambridge.org/9780521518031

This book is dedicated to God in whom all is possible; the love and support of my parents; the humor and inspiration of my sons Nadeem and Corey; the encouragement of my darling Bruce; and the tolerance of Jordan my Great Dane who missed many long walks while I spent time writing this book.

Gloria Galloway

Contents

The color plates appear between pages 214 and 215.

Preface

Since its inception clinically in the late 1970s and early 1980s, intraoperative neurophysiologic monitoring has shown a steady increase in use for surgeries in which neural structures may be at risk of injury. The types of and varieties of neurophysiologic techniques available for multimodality monitoring have allowed a patient-centered individualized approach to the practice and planning of these surgical cases. Some neurophysiologic techniques may carry inherent risks. An example of this is the potential risk of induction of seizures with the use of direct cortical and transcranial electric motor stimulation. Other risks involve inadvertent motor movements, tongue and lip lacerations. In addition, the more widespread use of multimodality intraoperative neurophysiologic techniques has allowed surgeons to become somewhat more aggressive and expansive in their surgical approaches. An example of this is the ability to be more expansive during resection of an intramedullary spinal tumor when motor, sensory, and possibly electromyography monitoring indicate that no change in electrophysiologic signals from these pathways has occurred.

The risk of consequential harm as a result of a neurophysiologic technique coupled with the increased ability to be more expansive surgically has changed methods of neurophysiologic monitoring and allowed the field to make an impact on patient safety and quality of care during surgical procedures. Therefore, it is especially important that those performing and interpreting these studies be adequately trained. This has been challenging given the relatively small number of training programs in the field: furthermore, several organizations whose members practice intraoperative neurophysiology have been led to develop guidelines, training courses, and additional certification programs specially geared toward this increasing subspecialty. Fellowship training programs have been around for many years in general neurophysiology as well as subspecialty areas such as epilepsy, including surgical epilepsy and electromyography (EMG). Often interest in intraoperative neurophysiology begins through subspecialty fellowships in EEG, epilepsy, or EMG. Several years ago a separately defined neurologic fellowship track was developed in intraoperative neurophysiology, similar to the tracks that already exist in EEG and EMG.

This book is a compilation of the current trends in intraoperative neurophysiology with chapters on various modalities and clinical uses. Separate chapters devoted to anesthesia, operating-room environment, special considerations in pediatrics and the interpretation and reporting of neurophysiologic data are useful and complementary. This book can be helpful to trainees as well as neurophysiologists already in practice but interested in other approaches to familiar techniques or in reviewing new techniques outside of their typical practice pattern. Questions on the topics covered in the chapters with detailed answers serve as a nice supplement on the accompanying website (www.cambridge.org/9780521518031). In some chapters, illustrative case examples are also included.

Physicians, PhD neurophysiologists, technicians, fellows, and residents can use this book for self-review and preparation and, through improved quality techniques and interpretation, may positively impact patient care.

Gloria Galloway

Acknowledgments

Enormous thanks to Ed Buggie, our neurophysiology laboratory manager, who always supported the need for improved equipment and enhanced training. Our excellent neurophysiologic technologists, Sharon Newell, Judy Brown, Christina Castleberry, and David Brooks, provided excellent technical expertise, were not afraid to give their opinions, and always provided much needed humor and silliness to balance out difficult situations. Much gratitude is given to our orthopedic and neurosurgical colleagues; E. Steve Roach and the neurology division for the support of our neurophysiologic endeavors as we grew together to improve patient care. Lastly, we owe enormous gratitude to those patients who entrusted their care to us.

1 Introduction, history, and staffing for intraoperative monitoring

Marc R. Nuwer

Neurophysiologic intraoperative monitoring (IOM) has grown over three decades into a method widely used to prevent neurologic injury during surgery. Common IOM techniques include electroencephalography (EEG), electromyography (EMG), evoked potentials (EPs), and nerve conduction velocity (NCV).

Intraoperative monitoring can warn the surgeon of changes in time to correct problems and prevent postoperative neurologic deficits. It may also identify some systemic problems. By using IOM to assess a patient's neurologic safety, a surgeon may provide a more thorough procedure or operate on a high-risk patient who might otherwise be turned away. Finally, patients and families can take comfort that neurologic risks are being evaluated during the case.

False alarms do occur, e.g., in 1% of scoliosis spinal-cord monitoring. False alarms are cases when the IOM warns of changes, but the patient awakens from surgery without a new deficit. Some may be false alarms caused by technical failures, difficulty obtaining good quality tracings, or anesthetic changes. In other cases, IOM sometimes raises an alarm, interventions are accomplished, and yet the patient awakens with a neurologic injury. Raising an alarm does not necessarily prevent deficits.

False-negative cases are those in which a patient suffers a postoperative neurologic injury that was not predicted by IOM. In spinal-cord SEP (somatosensory evoked potential) monitoring, the false-negative rate is around 0.1% of SEP-monitored scoliosis procedures (Nuwer *et al.*, 1995). Some false-negatives are due to deterioration soon after surgery. Some injuries are in pathways that were not monitored, e.g., some root lesions. Occasionally, false-negative cases are due to errors by the IOM team, who failed to recognize changes when they occurred. Finally, the existence of some otherwise unexplainable false-negative monitoring cases is a reminder that no technique is 100% accurate in predicting outcomes.

History of monitoring

Early uses of neurophysiologic monitoring during surgery date back to the first half of the twentieth century. Penfield (Penfield and Boldrey, 1937) used direct cortical stimulation in patients undergoing epilepsy surgery to define the locations of motor and sensory cortex. Direct recording of EEG from exposed cerebral cortex (electrocorticography, ECoG) guided the surgeon to resect regions of epileptic discharges, slowing, or lack of fast activity (Jasper, 1949; Marshall and Walker, 1949). Both techniques are still used.

More than a decade later, routine scalp EEG was used during carotid endarterectomy (CEA) (Thompson, 1968; Wylie and Ehrenfeld, 1970; Sharbrough *et al.*, 1973) to assess for cerebral ischemia during carotid clamping. Electroencephalography accurately measures degrees of cerebral ischemia during carotid endarterectomy (CEA) (Sundt *et al.*, 1974). It has provided an excellent substitute for keeping the patient awake during CEA or using other ischemia testing techniques and has become commonly used as a safeguard for CEA patients.

Spinal-cord IOM was first investigated in the early 1970s by Japanese investigators using epidural recordings of spinal potentials evoked by direct spinal stimulation (Shimoji *et al.*, 1971; Imai, 1976). These techniques were validated as accurate methods for monitoring spinal cases (Tamaki *et al.*, 1972, 1981).

Spinal IOM using somatosensory evoked potentials (SEPs) was developed in the mid-1970s. This involved using middle- and long-latency 50–200 ms cortical potentials during orthopedic procedures (Nash *et al.*, 1974, 1977; Nash and Brodkey, 1977), but the attempts were too prone to disruptive signal noise, variability, and sensitivity to anesthesia. Grundy (1982), an anesthesiologist working with Nash, described ways to reduce anesthetic effects and extended the techniques to neurosurgery. These early SEP cortical IOM techniques used filters at 1–100 Hz, and measured middle- and long-latency potentials (e.g., Engler *et al.*, 1978; Speilholz *et al.*, 1979).

Spinal-cord IOM with scalp SEPs remained problematic in the early days because of excessive noise and irreproducible background variability. By the late 1970s, Nuwer and Dawson (1984) studied the variability in these recordings and determined that short-latency SEP techniques improved monitoring, and recorded with restricted filters and other technical modifications, which substantially reduced variability and greatly improved SEP reliability. This SEP method was then widely adopted for spinal-cord IOM

In the UK, Jones (Jones *et al.*, 1982) pioneered epidural spinal recordings with posterior tibial nerve stimulation. This technique was used in response to safety concerns about repeated spinal epidural electrical stimulation.

Spinal-cord IOM with these scalp or epidural SEP methods was the technique of choice for the next two decades. Occasional false-negative monitoring cases were reported. Clinicians noted that SEPs test sensory pathways, whereas the most feared postoperative problems were motor. Burke eventually popularized transcranial electrical stimulation as a practical corticospinal technique for use under anesthesia (Hicks *et al.*, 1991, Burke *et al.*, 1992). The technique has developed through changes in locations of electrodes and stimulus trains.

Methods using auditory EPs and EMG during surgery around cranial nerves were developed in the 1980s. These were used for posterior fossa procedures, initially microvascular decompression and acoustic neuroma resection (Møller and Møller, 1985).

Commercial IOM equipment became available by 1981. Before that, clinical neurophysiologists adapted and customized research or other kinds of clinical equipment for use in specific IOM procedures. Often the early, customized equipment was specific to a particular application. Particular surgeons directed early clinical services, often with a research or technical development purpose. Those techniques were not generally available to other surgeons across their institutions. The first general IOM clinical service, set up at UCLA in 1979, offered a variety of techniques to any surgeon in any surgical discipline. In the early 1980s, annual academic neurophysiology and surgery meetings included IOM research reports. By the mid-1980s, special dedicated symposia taught IOM to clinical neurophysiologists, surgeons, and technologists. Those meetings carried the message of IOM to a much wider audience. The first two IOM textbooks (Nuwer, 1986; Møller, 1988) provided more detail and references to this wider audience. By the late 1980s, IOM was an established technique in general use.

Initially, IOM included EEG, ECoG, EPs, EMG, and NCV techniques. These "first-wave" techniques were familiar to most clinical neurophysiologists. The subsequent "second-wave" techniques expanded the field to include motor EPs and pedicle screw testing, as well as advances in remote supervision, increase channel number, and greater flexibility of commercial equipment. There are now many types of monitoring and surgical neurophysiology testing. Table 1.1 shows a list of the various kinds of monitoring that are or have been used. Not all of these techniques can be recommended for clinical use. Visual EP monitoring has proven too difficult and unreliable so far. The future may bring better, more reliable techniques. Neurogenic EPs are those in which the rostral spinal cord is stimulated and recordings are made at the limb muscles. In theory, the neurogenic EPs are from corticospinal tract conduction, but in reality, they are primarily conducted along other pathways, such as antidromic posterior column transmission with reflex muscle innervation. The visual and neurogenic EPs are no longer used for monitoring at this time. Other techniques described in Table 1.1 are advanced beyond the level of this book. For further information, the reader is referred to Nuwer (2008).

Applications for intraoperative monitoring

There are now many disorders for which intraoperative neurophysiologic monitoring and testing are used. Table 1.2 lists the various clinical disorders for which monitoring and testing are used. Some of these are surgeries commonly encountered in neuromonitoring services. Others are more specialized and advanced surgical situations. This book presents typical neuromonitoring situations. For further discussion of the more advanced and less common techniques, see Nuwer (2008).

In cases involving the cerebral circulation or cardiothoracic cases, EEG and SEPs are often applied. Craniotomies may make use of electrocorticography, EEG, SEPs, and direct cortical stimulation. Deep brain electrode implantation uses microneurography. Posterior fossa cases may use cranial nerve EMG monitoring, SEPs, and BAEPs (brainstem auditory evoked

Table 1.1 Techniques used for intraoperative neurophysiologic monitoring and testing

Electroencephalography (EEG)	
Electrocorticography (ECoG)	
Direct cortical stimulation to localize sensory, motor, and language function	
Deep brain stimulator electrode placement	
Somatosensory evoked potential (SEP)	SEP monitoring with scalp and cervical recording Monitoring spinal epidural potentials with peripheral nerve stimulation SEP for intraoperative mapping of motor cortex
Motor evoked potential (MEP)	Corticospinal tract monitoring with D and I waves from the spinal cord Transcranial electrical stimulation MEPs with recording from limb muscles Mapping the corticospinal tract Neurogenic EPs with rostral spinal cord stimulation and limb recording
Visual, brainstem, and auditory evoked potentials	Visual evoked potentials Brainstem auditory evoked potentials Mapping the brainstem and floor of the fourth ventricle
Electromyographic, reflex, and nerve conduction monitoring	Intraoperative peripheral nerve stimulation and recording Intraoperative facial nerve monitoring Oculomotor and lower cranial nerve monitoring Intraoperative monitoring with free-running EMG Intraoperative EMG during spinal pedicle screw instrumentation Selective dorsal rhizotomy Nerve root assessment with SEPs and MEPs

potentials). Spinal surgery often uses SEPs and MEPs (motor evoked potentials), and may use EMG monitoring for pedicle screw testing. Cauda equina surgery may use SEPs, EMG, and nerve conduction or reflex testing. Surgery in the periphery more often uses nerve conduction techniques, and may also use SEPs.

Table 1.2 Procedures in which intraoperative neurophysiologic monitoring and testing is used

Cerebral, head, and neck surgery	Epilepsy surgery Resection and debulking of cerebral tumors Movement disorders, electrode and ablation placement Resection and debulking of brainstem lesions Cranial base surgery Microvascular cranial nerve decompression Middle ear, mastoid, and parotid surgery Surgery at acoustic and vestibular nerves Surgery at glossopharyngeal, vagus, and laryngeal nerves at thyroid, larynx, carotid, and chest
Spine surgery	Correction and stabilization of scoliosis Spinal surgery for fractures and extramedullary tumors Surgery for intramedullary spinal-cord tumors Surgery for cervical diskectomy and fusion with decompression of myelopathy Lumbar stenosis and fusion surgery Surgery for tethered cord syndrome and other cauda equina lesions Dorsal root entry zone procedures and other surgeries for pain
Peripheral nerve surgery	Brachial plexus surgery Lumbosacral plexus surgery Total hip arthroplasty Pelvic surgery Peripheral nerve surgery
Vascular surgery	Carotid endarterectomy Clipping intracranial aneurysms Descending aortic procedures Cardiac surgery
Interventional radiology procedures	Intracranial endovascular procedures Spinal endovascular procedures Carotid balloon test occlusion

Effective monitoring teams will have many tools available for use as needed. This allows them to vary their techniques or bring in other tools where necessary.

Physician supervision and the monitoring team

Two kinds of IOM are used in surgery. They differ in their goals. *Monitoring* aims to identify any signs of injury and raise an alarm. *Testing* aims to identify neurologic structures. Monitoring occurs over hours, waiting to see if a change occurs from baseline. Testing is performed at a specific discrete time. The two are referred to together as *intraoperative monitoring* (*IOM*), although intraoperative neurophysiology might have been a better general term.

Testing requires personal involvement of a clinical neurophysiologist to identify a neurological structure, e.g., for ECoG or localization of language or motor cortex. Professional judgment is required to recommend what tissue to resect. That professional skill is at a level greater than that demanded of a technologist alone. The clinical neurophysiologist in the room interprets the recordings and discusses recommendations personally with the surgeon. This is referred to as *personal supervision*.

Monitoring does not require in-room personal physician supervision for routine cases. Monitoring requires a team with sufficient expertise and good communication. The IOM team members take on three functions: (a) the hands-on in-room technologist; (b) an expert in IOM technology and application; and (c) a neurophysiologist supervisor. Sometimes two of these functions are combined and performed by one person. For example, an expert monitoring specialist may perform both the in-room hands-on function and serve as the team's IOM technology and application expert. In that case, a physician supervisor relies on the expert for problem solving and related duties as well as hands-on monitoring. In a different scenario, a supervising clinical neurophysiologist is an expert in IOM technology and applications who remotely supervises a registered EEG technologist in the operating room. A third scenario could involve a clinical neurophysiology physician with modest IOM experience who works with an in-room registered EEG technologist, while both use the services of an expert (third person) to help them with monitoring questions, strategies, policies, and problem solving. In this case, the expert serves as a consultant, much like a physician taking advice from a consultant in other areas of practice. In any case, monitoring needs all three functions to be filled: in-room hands-on, clinical neurophysiology interpretation and supervision, and expertise in IOM techniques and problem solving. This team concept for IOM allows for the several kinds of necessary roles and expertise, and preserves the traditional role of the hospital and public agencies for overseeing the health care service.

The supervising physician often monitors in real time at a remote location and is able to intervene in the case as needed. Communication between the physician and in-room technologist may be by phone, paging, or instant messaging. In other situations, the physician enters the operating suite to communicate directly, supervise testing decisions, and resolve problems. The monitoring physician is on the medical staff of the hospital in which the surgery occurs and licensed to practice medicine in that state.

Routine brain or spinal-cord monitoring typically does not require the continuous personal in-room involvement of a clinical neurophysiologist (Nuwer and Nuwer, 1997). Technology now allows for remote on-line real-time monitoring from elsewhere in the building or even off-site. The supervising clinical neurophysiologist is nearby or on-line. The availability to intervene remains, as does the ability to become involved in postoperative care as needed. This is referred to as *direct supervision*.

Some clinical neurophysiologists directly supervise more than one case simultaneously. The literature showing good outcomes for IOM is based on cases in which one to three patients were supervised simultaneously (Nuwer *et al.*, 1995). The original concept of IOM services was based on this direct monitoring of one or two and occasionally three cases. At this time, there is insufficient literature to support monitoring of more than three simultaneous cases. It remains unclear whether the larger number of simultaneous cases could be monitored while still giving sufficient professional attention to each case. The supervising clinical neurophysiologist needs to be ready to be solely involved in a particular case if needed. To pay full attention to one case as needed, there must be a plan for transferring the supervision of any other simultaneous cases to another physician (American Academy of Neurology Professional Association, 2008).

Physician training and certification

Formal physician training in IOM is carried out in clinical neurophysiology fellowships. In the USA these require graduation from a neurology or child neurology residency program. The training in clinical neurophysiology is usually one to two years long. These programs include didactic training broadly in all areas of clinical neurophysiology plus clinical in-depth majors in two of four fields: EEG, EMG, sleep, or intraoperative monitoring. The traditional national Accreditation Council for Graduate Medical Education (ACGME) accredits these training programs.

Two examinations are given to show competence in this field. The American Board of Psychiatry and Neurology (ABPN) administers a clinical neurophysiology examination. The ABPN is a member board of the American Boards of Medical Specialties (ABMS), the national umbrella organization for traditional medical boards. The clinical neurophysiology examination is a written multiple choice examination. It covers broadly the whole discipline of clinical neurophysiology. One can successfully pass that examination without much knowledge about IOM in particular. The other of the two examinations in this field is given by the American Board of Clinical Neurophysiology (ABCN). The board has several examinations, one of which is the examination in intraoperative monitoring. The latter is a comprehensive examination that shows mastery of the IOM field. These two board examinations differ in that the ABCN examination expressly tests for mastery of IOM, whereas the ABPN examination does not.

Continuing medical education (CME) courses for physicians are given through three medical organizations: the American Clinical Neurophysiology Society (ACNS), the American Society of Neuromuscular and Electrodiagnostic Medicine (AANEM), and the American Academy of Neurology (AAN). These formal courses are offered as part of general clinical neurophysiology and neurology educational and scientific meetings for each society. The American Society of NeuroMonitoring (ASNM) is a group composed of IOM technologists, PhD non-physicians who are IOM experts, and some IOM physicians from several specialties. The ASNM annual meetings are another source of continuing education in IOM.

Physicians who do not have the opportunity to take a one–two-year fellowship, and yet still wish to begin providing IOM clinical services, should make use of an expert consultant in IOM who can help define protocols, screen technologist staff, and help with problem solving as needed. That consulting role may be filled by PhD diplomates of the American Board of NeuroMonitoring (DABNM) who are non-physician, or physicians who are experienced in IOM.

Some PhD or MD consultants are available to help with IOM services. Some run independent businesses that can supply traveling technologists as needed. Some provide local physicians with training and consulting. Some are DABNM. That examination is open to interested persons with a master's degree or higher with experience in intraoperative monitoring. It does not require a professional to be licensed. This group of individuals can help with IOM expertise, and are best suited to serve the IOM expert consultant role in an IOM team in which the clinical neurophysiology physician lacks sufficient personal IOM experience.

The standard in the medical community requires involvement of a clinical neurophysiology physician in IOM. The American Medical Association (AMA), in its policy statement on intraoperative neurophysiologic monitoring, states, "Our AMA policy is that supervision and interpretation of intraoperative neurophysiologic monitoring constitutes the practice of medicine . . ." The in-room non-physician personnel work is provided "under the direct or on-line real-time supervision of the ... physician trained in, or who has demonstrated competence in, neurophysiologic techniques and is available to interpret the studies and advise the surgeon during the surgical procedures." (American Medical Association, 2008).

The AMA also has a policy on the role of unlicensed PhD personnel (American Medical Association, 2006):

> It is AMA policy that: (1) the diagnosis of disease and diagnostic interpretation of a study or studies for a specific patient constitutes the practice of medicine; (2) a PhD clinical lab scientist or other non-physician laboratory personnel work under the supervision of a physician under their applicable scopes of work to perform a study or studies that will be the basis of a diagnostic interpretation for a specific patient; and (3) the Medicare physician fee schedule compensates only authorized persons for the diagnostic interpretation of a specific patient and should not provide payments directly to non-physician lab personnel working under the supervision of a physician to perform a laboratory study or studies.

Intraoperative monitoring is dangerous when carried out without the team of IOM expert, clinical neurophysiologist, and trained technologist. The insufficiently prepared team is predisposed to giving overly optimistic or inaccurate information to the surgeon or anesthesiologist. This can promote a false sense of security that the case can proceed without neurologic impairment when the IOM does not show that this is the case.

The team also allows for routine screening by public safety systems in health care. Credentialing and privileging the clinical neurophysiologist by each hospital's neurology service and medical staff office allows for others with expertise in general clinical neurophysiology to evaluate the team's skills, knowledge, abilities, training, and experience in the testing and monitoring requested. Board certification and continuing education can be checked. It also enables the medical staff organization to review the physician's malpractice history, staff and patient complaints, documentation compliance, and other factors that should influence whether privileges and credentials should be granted. Licensing of the physician in that state allows an external review of many of those same items by an independent board less likely to be influenced by internal hospital needs, and answerable to the public for its decisions. Licensing also provides a method for restraining or removing practitioners who fail to meet criteria for safe and effective patient care within the current health care system. Licensing boards can act upon problem practitioners by investigating complaints, bad outcomes, and unprofessional behavior, and can issue public reprimands, demands for further education, restrictions on practice, suspensions and revocations of licenses.

Another point at which IOM falls under the domain of the practice of medicine is the statutory restriction of the practice of medicine to licensed physicians. Diagnosis is a professional opinion about the presence, absence, type, location, or severity of an illness or injury. A diagnostic interpretation is a professional opinion about a diagnosis based upon the findings in a test. Examples of diagnostic interpretations are:

- Normal SEPs,
- Abnormal SEPs due to a central delay in latencies,
- The drop in cortical peak amplitude could be due to the increase in inhalation anesthesia,
- The drop in cortical peak amplitude is severe,
- The most likely location of this impairment is in the spinal cord below the mid-cervical level.

A diagnostic test is one that produces findings suitable for a diagnostic interpretation. Examples of diagnostic tests are: magnetic resonance imaging, transcranial Doppler ultrasound, electrocardiography, and positron emission tomography. Further examples are electroencephalography, electromyography, evoked potentials, and nerve conduction studies. State business and professional statutes and regulations limit the diagnostic interpretation of these diagnostic tests to licensed physicians. The few exceptions are for limited license practitioners, such as the use of X-rays by dentists or podiatrists. These public health care policies are intended to protect the public.

In contrast, technologists and other non-physicians can generally report the findings of a test to a surgeon. Findings are the objective latencies and amplitudes that can be read from cursors, after consideration of where the cursors should be placed and whether tracings are suitable for measurement. Findings can be related to baseline values. Findings are also subject to the diagnostic interpretation of the supervising physician. Examples of test findings are:

- The cortical peaks are at their baseline latencies and amplitudes.
- The cortical peak amplitudes have dropped by 60%.
- The cortical peaks are now absent.

These public processes for the regulation of health care services are handled through the credentialing, privileging, and licensing of physicians. The IOM team needs to stay within the limits of these public health care processes so as to help assure that the public is provided with good quality care and that problem providers are restrained or removed as needed. The clinical neurophysiologist is the key to this, being the only member of the team who is subject to these public health care policy regulations (Nuwer, 2002).

The technologist's role and training

A technologist generally lacks suitable skills, knowledge, abilities, training, and experience to provide the IOM services without supervision by a clinical neurophysiologist. Technologists also lack the medical knowledge and statutory authority to advise a surgeon about clinical options when changes do occur. Most institutions require physician supervision of medical procedures including IOM, much in the same way as for interpretation of MRI (magnetic resonance imaging) or CT (computed tomography). In any case, the monitoring supervisor needs to have medical staff privileges for clinical neurophysiology at the hospital where the monitoring is conducted. The privileging should meet all local public policies to review and approve each individual's suitability to provide these clinical services.

Technologists, a key part of the IOM team, need sufficient experience in neurophysiologic testing, techniques, basic sciences, and relevant clinical sciences before being able to provide remote IOM supervision. Usually, this includes several years of regular inpatient and outpatient experience of conducting EEG and evoked potentials. That experience is an important basis for knowing what signals look like, identifying changes and abnormalities, and dealing with technical problems.

In the USA, formal technologist training programs are usually two years long. Students are trained in didactic fundamentals and clinical practical experience in routine EEG and evoked potentials. Many have additional didactic and clinical experience in IOM. Students are prepared to conduct routine testing, but need additional experience in IOM before being left alone as the in-room member of the IOM team. These training programs are supervised and accredited by a national organization, the Committee on Accreditation for Education in Electroneurodiagnostic Technology (CoA-END).

Each technologist should be proctored and given progressively reduced supervision when introduced to IOM. The technologist's privileging should be specific, i.e., a technologist who knows well how to monitor EEG during a carotid endarterectomy does not necessarily know how to monitor somatosensory EPs in a spinal-cord case. The technologist should also have mastered the outpatient and routine inpatient application of techniques before embarking on the IOM application of that modality. Of course, that cannot be applied to some techniques, such as transcranial electrical stimulation, where there is no outpatient application.

Technologists' competence and knowledge are tested with registration examinations. In the USA, the American Board of Registration of Electroencephalographic and Evoked Potential Technology (ABRET) sponsor these registration examinations. Basic competency examinations are given in EEG and evoked potentials. These require both a written and an oral examination. An additional examination is given in intraoperative monitoring, which grants the successful applicant the Certificate in Neurophysiologic Intraoperative Monitoring (CNIM). The CNIM is obtained by passing a written multiple choice examination without an oral examination component. Well trained IOM technologists have these several registration credentials. Technologists are required to maintain continuing education credits to keep their certificates active. Credits are given through educational offerings in the field, especially those given by the American Society of Electroneurodiagnostic Technologists (ASET).

An alternative track for obtaining the CNIM allows individuals with a bachelor's degree to take the examination despite a lack of skills, knowledge, ability, training, or experience in EEG and evoked potentials. Some have had a small number of cases of actual IOM experience. This leads to the observation that the CNIM credential is a mark of a degree of book knowledge but not necessarily a mark of sufficient practical experience or clinical knowledge. In this situation, the supervising clinical neurophysiologist should assess the actual ability of each technologist, and allow in-room monitoring privileges only to those who have demonstrated actual skills, knowledge, ability, training, or experience in particular IOM techniques. A certificate by itself is not sufficient.

Summary

Intraoperative monitoring applies many useful techniques to various surgical procedures. The monitoring teams encompass several specialists contributing their own expertise. Goals are to enhance patient care, avoid neurological deficits, allow for more complete procedures, allow procedures even on some high-risk patients, and provide feedback to the surgeon about actions that could injure the nervous system.

Questions

1. The false-negative rate for spinal-cord somatosensory evoked potential monitoring is approximately:
   (a) 5%
   (b) 1%
   (c) 0.5%
   (d) 0.1%
2. False-negative cases are those in which:
   (a) The IOM remains stable, but the patient awakens with a new deficit
   (b) The IOM changes, correctly predicting the new deficit
   (c) The IOM remains stable, and the patient awakens without a new deficit
   (d) The IOM changes, but the patient awakens with no new deficit
3. The first kind of intraoperative neurophysiology monitoring used in surgery was:
   (a) Somatosensory evoked potentials
   (b) Electrocorticography
   (c) Electromyography
   (d) Auditory evoked potentials

4. The use of EEG to monitor carotid endarterectomy was popularized around:
   (a) 1960
   (b) 1970
   (c) 1980
   (d) 1990
5. Health care public policy regulations restrict the interpretation of IOM results to:
   (a) Physicians, PhD DABNMs, and CNIM technologists
   (b) Physicians and PhD DABNMs
   (c) Physicians and CNIM technologists
   (d) Physicians
6. A professional opinion about the presence, absence, type, location, or severity of an illness or an injury is a:
   (a) Diagnostic interpretation
   (b) Diagnostic test
   (c) Diagnosis
   (d) Finding
7. Technologists are allowed to report:
   (a) The latency and amplitude values
   (b) That the results are normal
   (c) That the results are abnormal
   (d) That the changes are due to a new spinal-cord level impairment
8. Which organization allows unlicensed non-physicians to take its board examinations?
   (a) American Board of Psychiatry and Neurology (ABPN)
   (b) American Board of Clinical Neurophysiology (ABCN)
   (c) American Board of Electrodiagnostic Medicine (ABEM)
   (d) American Board of NeuroMonitoring (ABNM)
9. The intraoperative monitoring certificate given by the American Board of Registration in Electrodiagnostic Technology to technologists is abbreviated as:
   (a) DABNM
   (b) CNIM
   (c) REEGT
   (d) CLTM
10. Supervision by a physician who is near the operating room and available to enter the room at short notice is referred to as:
   (a) Personal supervision
   (b) Remote supervision
   (c) Direct supervision
   (d) General supervision

References

American Medical Association (2006) *Diagnosis of Disease and Diagnostic Interpretation of Tests Constitutes Practice of Medicine to be Performed by or Under the Supervision of Licensed Physicians*, Policy H-35.971. Available from www.ama-assn.org/go/policyfinder.

American Medical Association (2008). *Intraoperative Neurophysiologic Monitoring*, Policy H-410.957. Available from http://www.ama-assn.org/go/policyfinder.

American Academy of Neurology Professional Association (2008) *Principles of Coding for Intraoperative Neurophysiologic Monitoring (IOM) and Testing: A Model Medical Policy*, Available from http://www.aan.com/globals/axon/assets/4004.pdf.

Burke, D., Hicks, R., Stephen, J., Woodforth, I., and Crawford, M. (1992). Assessment of corticospinal and somatosensory conduction simultaneously during scoliosis surgery. *Electroencephalography and Clinical Neurophysiology*, **85**, 388–396.

Engler, G. L., Speilholz, N. I., Bernhard, W. N., *et al.* (1978). Somatosensory evoked potentials during Harrington instrumentation for scoliosis. *Journal of Bone and Joint Surgery (America)*, **60**, 528–532.

Grundy, B. L. (1982). Monitoring of sensory evoked potentials during neurosurgical operations: methods and applications. *Neurosurgery*, **11**, 556–575.

Hicks, R. G., Burke, D. J., and Stephen, J. P. (1991). Monitoring spinal-cord function during scoliosis surgery with Cotrel–Dubousset instrumentation. *The Medical Journal of Australia*, **154**, 82–86.

Imai, T. (1976). Human electrospinogram evoked by direct stimulation on the spinal cord through epidural space. *Journal of the Japanese Orthopedic Association*, **50**, 1037–1056.

Jasper, H. H. (1949). Electrocorticograms in man. *Electroencephalography and Clinical Neurophysiology. Supplement*, **2**, 16–29.

Jones, S. J., Edgar, M. A., and Ransford, A. O. (1982). Sensory nerve conduction in the human spinal cord: epidural recordings made during scoliosis surgery. *Journal of Neurology, Neurosurgery, and Psychiatry*, **45**, 446–451.

Marshall, C. and Walker, A. E. (1949). Electrocorticography. *Bulletin of the Johns Hopkins Hospital*, **85**, 344–359.

Møller, A. R. (1988). *Evoked Potentials in Intraoperative Monitoring*. Baltimore: Williams & Wilkins.

Møller, M. B. and Møller, A. R. (1985). Loss of auditory function in microvascular decompression for hemifacial spasm. Results in 143 consecutive cases. *Journal of Neurosurgery*, **63**, 17–20.

Nash, C. L., Jr., and Brodkey, J. S. (1977). *Clinical Application of Spinal-Cord Monitoring for Operative Treatment of Spinal Disease*. Cleveland: Case Western Reserve University.

Nash, C. L., Jr., Schatzinger, L., and Lorig, R. (1974). Intraoperative monitoring of spinal-cord function during scoliosis spine surgery. *Journal of Bone and Joint Surgery (America)*, **56**, 1765.

Nash, C. L., Jr., Lorig, R. A., Schatzinger, L. A., and Brown, R. H. (1977). Spinal-cord monitoring during operative treatment of the spine. *Clinical Orthopaedics*, **126**, 100–105.

Nuwer, M. R. (1986). *Evoked Potential Monitoring in the Operating Room*. New York: Raven Press.

Nuwer, M. R. (2002). Regulatory and medical-legal aspects of intraoperative monitoring. *Journal of Clinical Neurophysiology*, **19**, 387–395.

Nuwer, M. R. (2008). *Intraoperative Monitoring of Neural Function*. Amsterdam: Elsevier.

Nuwer, M. R. and Dawson, E. C. (1984). Intraoperative evoked potential monitoring of the spinal cord: enhanced stability of cortical recordings. *Electroencephalography and Clinical Neurophysiology*, **59**, 318–327.

Nuwer, J. M. and Nuwer, M. R. (1997). Neurophysiologic surgical monitoring staffing patterns in the USA. *Electroencephalography and Clinical Neurophysiology*, **103**, 616–620.

Nuwer, M. R., Dawson, E. G., Carlson, L. C., Kanim, L. E., and Sherman, J. E. (1995). Somatosensory evoked potential spinal-cord monitoring reduces neurologic deficits after scoliosis surgery: results of a large multicenter survey. *Electroencephalography and Clinical Neurophysiology*, **96**, 6–11.

Penfield, W. and Boldrey, E. (1937). Somatic motor and sensory representation in the cerebral cortex of man as studied by electrical stimulation. *Brain*, **37**, 389–443.

Sharbrough, F. W, Messick, J. M., Jr., and Sundt, T. M, Jr. (1973). Correlation of continuous electroencephalograms with cerebral blood flow measurements during carotid endarterectomy. *Stroke*, **4**, 674–683.

Shimoji, K., Higashi, H., and Kano T. (1971). Epidural recording of spinal electrogram in man. *Electroencephalography and Clinical Neurophysiology*, **30**, 236–239.

Speilholz, N. I., Benjamin, M. V., Engler, G. L., and Ransohoff, J. (1979). Somatosensory evoked potentials during decompression and stabilization of the spine: methods and findings. *Spine*, **4**, 500–505.

Sundt, T. M., Jr., Sharbrough, F. W., Anderson, R. E., and Michenfelder, J. D. (1974). Cerebral blood flow measurements and electroencephalograms during carotid endarterectomy. *Journal of Neurosurgery*, **41**, 310–320.

Tamaki, T., Yamashita, T., Kobayashi, H., and Hirayama, H. (1972). Spinal-cord monitoring. *Journal of Japanese Electroencephalography and Electromyography*, **1**, 196.

Tamaki, T., Tsuji, H., Inoue, S., and Kobayashi, H. (1981). The prevention of iatrogenic spinal-cord injury utilizing the evoked spinal-cord potential. *International Orthopaedics*, **4**, 313–317.

Thompson, J. E. (1968). *Surgery for Cerebrovascular Insufficiency (Stroke) With Special Emphasis on Carotid Endarterectomy*. Springfield: Charles C Thomas.

Wylie, E. J. and Ehrenfeld, W. K. (1970). *Extracranial Occlusive Cerebrovascular Disease: Diagnosis and Management*. Philadelphia: WB Saunders Company.

2

The preoperative assessment

Gloria M. Galloway

It is extremely helpful for patients undergoing surgery with intraoperative monitoring to have a preoperative assessment by the neurophysiologic intraoperative monitoring (NIOM) team prior to the surgical case. A preoperative assessment is important to allow the monitoring team to gather additional information on the patient that is not always readily available in the surgical team's clinic notes (Table 2.1). This includes current medications, allergies, prior surgical and medical history and an understanding of what risks may be involved, given any prior anesthetic or surgical complication. It allows a discussion of expectations by the patients or their parents for the day and days leading up to the surgical procedure and those days needed for recovery after the procedure.

It has been well established that collaborative communication can reduce the number of communication failures and subsequent adverse events in surgical procedures (Lingard *et al.*, 2004). Surgical teams may follow preoperative team checklists to enhance communication and avoid mishaps. These anesthetic and surgical checklists typically focus on operative medications, anesthesia requirements, the use of special instruments, and a description of the operative plan, which includes the estimated duration of the surgery, description of procedure, and the use and choice of blood products (Lingard *et al.*, 2008). Similarly, it is important that the neurodiagnostic monitoring team be able to evaluate the patient prior to the day of surgery, since their focus will be different from what is evaluated for in most surgical and anesthetic preoperative checklists and evaluations.

The NIOM preassessment will include a history and in some cases possibly a focused physical examination as well. The history will be focused on key features, which will be pertinent in the results obtained by monitoring particularly neurological, orthopedic, and neurosurgical history and the existence of medical, anesthetic, and surgical complications. Studies have shown that perioperative complications affect patient outcome (Glassman *et al.*, 2007) and can significantly affect the results of NIOM (Albert *et al.* 1995).

This assessment by the NIOM group will help determine the choice of correct monitoring modalities and the exclusion of inappropriate ones. It will allow for a better understanding of abnormalities, including conduction delays that may be seen in the intraoperative baseline study, and correlate these with the patient's clinical scenario. The NIOM group preoperative assessment will also allow the acquisition of clinical information from the physical examination and history, in order to select appropriate muscle groups for recording. In addition, information that may exclude the patient from certain monitoring modalities, such as transcranial electric stimulation of motor tracts, will be obtained.

Much collaboration and communication occurs between the NIOM and anesthesia teams during the procedure and having the staff and equipment in close proximity can facilitate this communication. Figure 2.1 demonstrates a common set-up of equipment for NIOM and anesthesia monitoring. One can see from this picture that the proximity of the staff may enhance communication between them.

Figure 2.2 shows a preoperative SEP, obtained by recording from the ulnar nerve on a patient with a deformed left upper extremity. No responses could be obtained from the deformed side, eliminating the need to troubleshoot this in the OR case (Thuet *et al.*, 2005). Recording had also been attempted on the median nerve but without success in this patient.

Figure 2.3 shows additional equipment, which has been set out in the operation room for a complicated spine surgery. The stacks of additional trays of instrumentation equipment that are required can often impact the space available in the operation room for monitoring equipment and personnel. This limitation of space can make it difficult for the technical staff to get to the patient during times of troubleshooting and

Table 2.1 Information obtained during preoperative assessments

Contraindication to a monitoring modality
Choice of modalities to include in NIOM
Muscle selection for recording
Choice of stimulation sites in limbs
Preoperative baseline monitoring studies
Limitations to NIOM techniques based on preoperative assessments
Review of comorbidities

Figure 2.1 A typical set-up in the operating room, showing the proximity of the NIOM equipment and technical staff to the surgery site, as well as to the anesthesia monitoring equipment and staff.

supports the need for good communication between the NIOM, surgeon, and anesthesia teams to enhance patient care. In some cases a modification of monitoring modalities, such as described above in the case of the deformed extremity, may preclude the ability to monitor over certain peripheral nerve pathways. If this important clinical and electrophysiological information can be obtained during the preoperative assessment it may avoid difficulties in the set-up of the case on the day of the surgery. In this way, a preoperative assessment can be a very useful, efficient, and, ultimately, cost-saving approach to enhance patient care and safety in preparation for NIOM during complicated surgical procedures.

The preoperative assessment should include at least four key components: a detailed history should be obtained, the modalities to use should be selected, muscle groups for recording should be selected, and contraindications to the use of TES (transcranial electrical motor stimulation) should be determined.

Medical, surgical, and anesthetic history

This information should include:

1. The clinical history and indications for surgery
2. List of outpatient medications and allergies
3. Prior surgical history, including prior intracranial surgery
4. Significant past medical history
5. Anesthetic complications, and significant anesthetic history
6. Possible relative contraindications to the use of standard monitoring techniques

It is well recognized that the baseline health status of a patient is a significant factor in the surgical outcome in that patient. The ability to include NIOM during these complicated surgeries (Spivak *et al.*, 1994) has allowed newer surgical techniques to be employed in patients with other significant medical conditions. Preparation for surgery in these patients includes paying attention to these medical conditions in order to identify potential unstable situations and the potential for improvements in treatment and care (Hu and Berven, 2006).

Complications following adult spine deformity surgery have been reported to be as high as 40% (Baron and Albert, 2006). The risk of complications increases with expansion of the surgery, increased patient age, and significant medical comorbidities (Faciszewski *et al.*, 1995). This risk of complication is lessened with optimum nutritional state and pulmonary status as well as multimodality NIOM (Thuet *et al.*, 2005).

The medical and neurological condition of the patient is a major factor in the success of obtaining

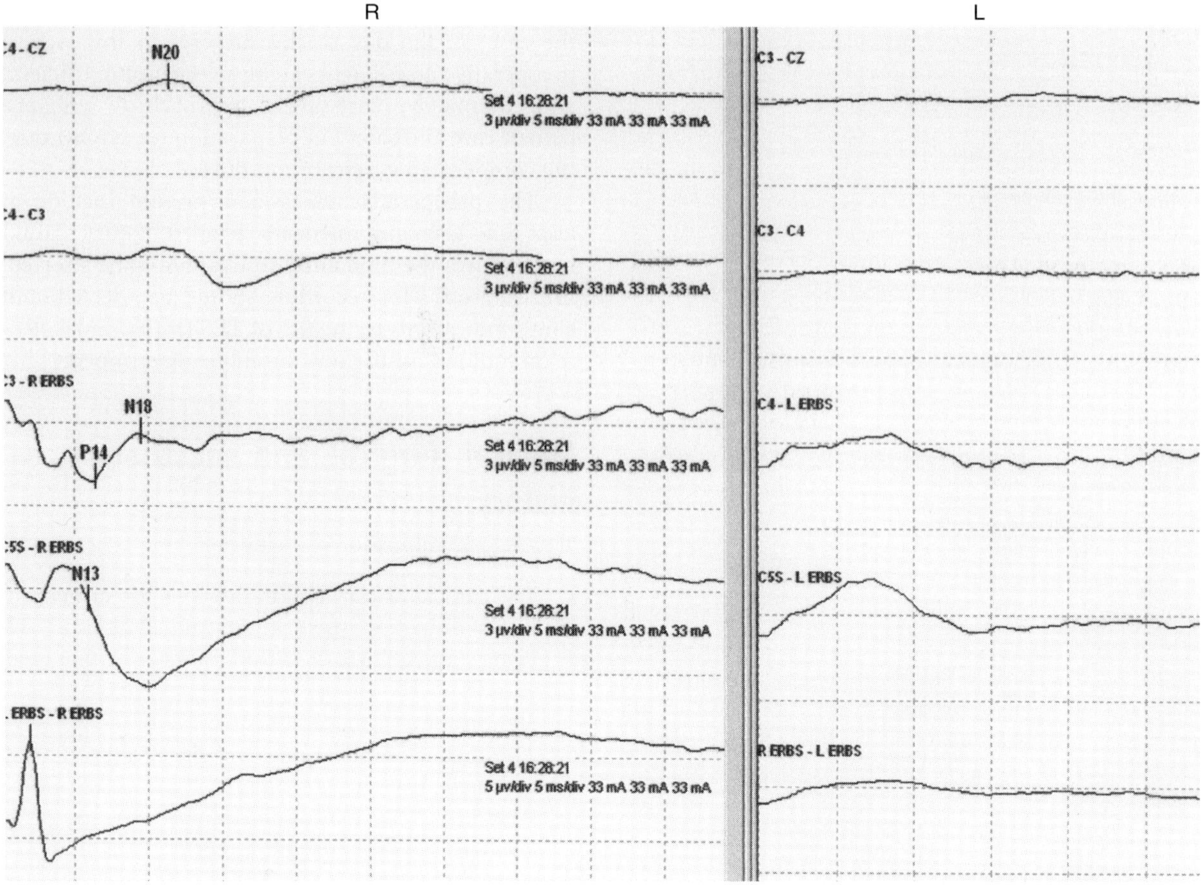

Figure 2.2 Normal right-sided SEP responses with recording from the ulnar nerve but no reproducible responses on the left.

Figure 2.3 A significant amount of instrumentation is typically placed near the surgical site, as noted here during a complicated spine case, showing additional equipment and discard basin. These provide a reduced amount of space to maneuver near the surgical site during technical troubleshooting.

satisfactory intraoperative monitoring of evoked potentials. In a patient with neurological deficits, the likelihood of obtaining these evoked potentials is decreased (Fisher *et al.*, 1995). Pulmonary function is often tested prior to surgical correction for scoliosis because the majority of these patients have significant potential for restrictive pulmonary disease. Recovery from complicated spinal procedures may involve the development of significant pulmonary compromise if care is not taken. When the scoliosis is properly treated, however, the result is an improvement in the pulmonary vital capacity and a decreased potential for pulmonary infections. A preoperative assessment of pulmonary function is warranted for other spine conditions prior to surgical repair. Clinical studies and experience have shown that the premorbid pulmonary status of the patient correlates with the surgical outcome (Vedantam and Crawford, 1997).

During the preoperative assessment, unexpected situations may arise. At our institution, patients of all ages may come for evaluation and it is not uncommon to discover lice infestation when setting a child up for a preoperative study. This may mean a cancellation or delay of a surgical case so that the child can be appropriately and adequately treated prior to surgery. Although this same patient may have been seen for clinical evaluation numerous times beforehand and may have undergone other testing (Kober *et al.*, 2001), an evaluation of the patient's head and hair may not have been thoroughly made, and the lice infestation would not have been discovered until then.

Selection of modalities to be used during the operative procedure

Unfortunately, in some cases insurance may not cover payment of this important preoperative assessment and patient care may suffer as a consequence. Not only might the preoperative assessment uncover details of the clinical history that pose surgical risk, it can also lessen the risk of the monitoring procedure.

In many surgical cases, multimodality monitoring occurs (DiCindio *et al.*, 2003) and therefore preparation and knowledge of the patient's history and what the patient's monitoring study demonstrates outside of the operating room is very useful. In this way a strategy for monitoring based on the type of surgery, and those structures most at risk for neurological injury, the preoperative neurological status, and neurological findings will provide the most helpful information in minimizing troubleshooting and surprises on the day of surgery (Chu *et al.*, 2006).

With monitoring modalities that allow for transcranial stimulation, an evaluation of prior intracranial history should include a history of intracranial surgeries, intracranial hardware placement, seizures, and prior ischemic cerebrovascular disease (Lieb and Selim, 2008). Although in most cases, the referring physician is typically aware of significant complicating premorbid conditions, most physicians have come across scenarios where in some cases a portion of the history may go undiscovered. This may alert the other members of the monitoring team, including the surgeon and anesthesiologist, to clinical disorders that may place the patient at a higher surgical risk (McGarvey, 2007). Having the support of the NIOM team provides a type of check and balance in this important system.

Monitoring techniques pose a risk, although small, of potentially inducing seizures through transcranial electrical or magnetic stimulation in a small subset of patients. The TMS-associated seizure risk has been reported to range up to 2.8% for single-pulse and 3.6% for paired-pulse stimulation. Medically intractable epilepsy and a lowering of antiepileptic medications appear to be associated with an increase in this risk (Schrader *et al.*, 2004).

In addition, the placement of electrodes used in monitoring requires abrasion of the skin for surface electrodes or puncturing of the skin for needle electrodes, causing some degree of risk, albeit small. Patients who have been anticoagulated may require additional precautions, particularly with the placement of needle electrodes over the face, paraspinal muscles, and abdominal muscles.

Patients who have had a prior history of convulsive seizures, particularly in their recent history, and are planned for transcranial electrical stimulation, may require some additional monitoring techniques during the surgery to evaluate for potential epileptiform activity. Therefore, a proper preoperative assessment is strongly recommended for any patient undergoing a planned surgical procedure with intraoperative monitoring, to enhance the quality of patient care (Hausmann *et al.*, 2003). Patients will typically have undergone imaging with MRI of the spine at some point during their work-up. Studies have shown a good correlation between functional magnetic resonance imaging (fMRI), magnetoencephalography (MEG) and sensorimotor activation (Kober *et al.*, 2001).

Monitoring of the brainstem

The choice of monitoring modalities is determined by the type of surgery and patient factors. Therefore, for brainstem surgical lesions somatosensory evoked potentials (SEPs), brainstem auditory evoked potentials (BAEPs), motor evoked potentials (MEPs), and cranial nerve monitoring may be needed. These modalities allow for prompt and reliable evaluation of the functional integrity of brainstem structures. In addition to these monitoring modalities, mapping of the cortical spinal tract and motor nuclei on the floor of the fourth ventricle are recognized techniques, which can be utilized during monitoring in surgeries involving intrinsic brainstem tumors.

Pedicle screw stimulation

Stimulation of pedicle screws is often done during select spinal repair surgeries with the use of free run and stimulated electromyography recording of potentials in the limb, abdominal, and paraspinal muscles (if accessible). This type of stimulation and recording may be added to lessen the risk of radicular injury after pedicle screw placement. Pedicle screw stimulation is complementary to spinal-cord somatosensory and motor tract monitoring, which would also be done in these patients. The decision over which surgeries will utilize hooks, pedicle screws, or both is made by the operating surgeon based on the clinical case.

Upper extremity SEP (median or ulnar)

It is important to know the position of the patient during surgery and what support is being provided to limbs during the surgical procedure. Although the surgical field may be small or large, attention to details of patient positioning is important in regard to monitoring. This is because during surgeries, particularly those lasting several hours, limbs can be prone to nerve compression or traction injuries due to maintenance of a prolonged posture. An example of this is seen with brachial plexus palsies during prolonged spine surgeries in which the patient is placed in a prone position and arms are resting alongside the head. Therefore monitoring somatosensory tracks of peripheral nerves and attention to any change explainable by patient position may avoid additional injury.

Stimulating, for example, the median nerve in the upper extremity and recording somatosensory evoked potentials through the brachial plexus area may assist in detecting this type of injury. The patient's arms can then be repositioned during surgery or arm rolls for cushioning them can be used if a neurodiagnostic change occurs that is thought to be due to compression of the peripheral nerve. The location of the conduction delay can be seen at the brachial plexus level using somatosensory evoked potentials identifying the source. In this scenario, all waveforms proximal to the brachial plexus will be delayed on the affected side. So the choice of monitoring modalities that will be useful intraoperatively will determine the choice of neurodiagnostic modalities that are evaluated in the preoperative assessment (Banoub *et al.*, 2003).

Transcranial electrical motor stimulation (TES)

Unfortunately, the use of transcranial electrical motor stimulation, which is done in the operating room, cannot be assessed during the preoperative assessment, because of the inability of the majority of awake patients to tolerate transcranial electrical stimulation. Although transcranial magnetic (TMS) and transcranial electrical motor stimulation (TES) are not directly equal modalities they are similar and one can argue for using transcranial magnetic stimulation in the preoperative assessment of patients who are to undergo transcranial electrical stimulation intraoperatively.

The use of TMS can provide information on the status of the motor tracts prior to the intraoperative procedure. Although using transcranial magnetic stimulation preoperatively as a diagnostic test is not in routine use at this time, it has been shown in numerous studies to be a safe and reliable means of stimulating the motor cortex to elicit a motor response (Malcolm *et al.*, 2006).

Selective dorsal rhizotomy

The preoperative assessment for selective dorsal rhizotomy (SDR) requires special considerations. A thoughtful approach to preoperative assessment in these patients utilizes a team that includes a neurologist, neurosurgeon, rehabilitative therapist or therapists, and psychologist. This multidisciplinary team approach to patient care and treatment, which may ultimately involve many additional types of medical caregivers allows caregivers with different skills and approaches to patient care to provide input.

The indications for SDR in spastic cerebral palsy are well described by Albert *et al.* (1995). These indications are based on degree of ambulation or dependence and can be divided into six categories:

1. Patient ambulating independently without assistive devices. These patients may have early contractures or need tendon releases.
2. Patients ambulating with assistive devices, most often due to trunk instability.
3. Quadruped crawlers with impaired postural control of standing.
4. Commando (belly) crawlers with severe postural impairments.
5. Totally dependent – require complete assistance with ADLs (assisted devices for living).
6. Spasticity that is asymmetric.

To assess the benefit of selective dorsal rhizotomy in a given patient, quantifiable measures of functional ability should be used. Some of these measures include the gross motor function measure, the pediatric evaluation of disability inventory, the Edinburgh visual gait score, and passive range of motion scores.

There has been significant debate as to whether there is greater potential benefit in using selective dorsal rhizotomy (SDR) earlier in the course of disease or waiting until later. This has become particularly significant given the greater realization that serious complications can occur in these patients with the use of implantable pumps that administer antispasmodics.

When a selective dorsal rhizotomy procedure is being planned, preoperative assessment of the patient by the involved neurophysiologist allows for useful identification of the particular muscle groups most useful to monitor during the procedure as each patient may present unique features and particular characteristics. Therefore, a good working relationship with the surgeon and an assessment of the patient prior to the day of surgery allows for better coordination of care and more useful information from the monitoring procedure. This preoperative assessment is often made several days prior to the procedure or it may need to be made on the day of the actual surgery. Being able to plan for the procedure and to plan which muscle groups should be monitored in a given patient is the key to a successful result (Turner, 2009).

The goal of selective dorsal rhizotomy is to reduce the functionally limiting spasticity without abolishing the spasticity of certain muscles in a given patient that are needed for stability. This is true in the ambulatory patient requiring a degree of spasticity to stand but also in the non-ambulatory patient who still needs assistance in transfers.

The procedure of selective dorsal rhizotomy does not remove all sensory input but is instead selective in partially removing sensory input, so that the degree of motor tone is reduced but the patient is not left hypotonic. This is based on the fact that in spastic cerebral palsy there is reduction in the spinal inhibitory influence on the alpha motor neuron pool so that increased tone with spasticity and clonus result. Spasticity occurs as a result of impairment of the normally inhibitory descending corticospinal tracts. Sensory information mediated through the posterior or dorsal roots is relayed to the anterior horn cells with reduced inhibition from the corticospinal tracts. This results in an imbalance with increased tone, spasticity, and clonus.

Selection of recording muscle groups

Preoperative planning in patients who are to undergo selective dorsal rhizotomy allows for the neurophysiologist to examine the patient clinically and assess and select the most appropriate muscle groups to record from during the surgical procedure (Tables 2.2 and 2.3). This ensures that those muscles most affected will be the ones examined during the procedure, allows the clinical history to be reviewed and allows the patient and family to better understand what will be taking place from a neurophysiologic standpoint.

A review of the history will include evaluations from physical therapists as well as the neurosurgical staff in regard to ambulation, bladder and bowel function. Careful examination includes assessment of muscle power and muscle tone, including evaluation of the presence of clonus and the degree and intensity of the clonus, which will again be evaluated during the intraoperative monitoring procedure (Turner, 2009).

Table 2.2 Muscle selection for SDR

Tibialis anterior	L4–5
Medial gastrocnemius	S1–2
Vastus lateralis	L2–4
Gluteus maximus	L5–S2
Anal sphincter	S1–2

Table 2.3 Muscle selection in motor potential monitoring

Hypothenar muscle group	C8–T1
Thenar muscle group	C8–T1
Brachioradialis, or biceps	C5–6
Abductor hallicus	S1–S2
Gastrocnemius	S1–2
Tibialis anterior	L4–5

This ultimately ensures successful neurophysiologic monitoring. This is an opportunity to evaluate the patient who is still awake. Once under anesthesia, recording electrodes will be placed in the appropriate muscles to be used for recording.

In the pediatric population, preoperative assessment needs to be well-planned (McGarvey, 2007). To maximize the interpretation of neurodiagnostic monitoring preoperatively, a mild degree of sedation can be helpful. The use of sedation poses its own particular risks, which need to be explained to the patient (in an age appropriate way, using language that can be easily understood when possible) and to the family. It can be very helpful to utilize a sedation team in very young patients to optimize the ability to obtain a preoperative neurodiagnostic study.

In older patients, the technique of distraction can be very useful. The use of a family member to distract a patient can help. Alternatively, a skilled provider trained in the technique of distraction may optimize the performance of a neurodiagnostic study preoperatively, and avoid the use of sedating medications. At our hospital a skilled provider referred to as a *child life provider* will usually attend sedated procedures to assist in patient as well as parent comfort when indicated. These valuable providers often help reduce patient anxiety, so that I have been able to perform electromyography procedures on these young patients with greater ease.

Determination of relative contraindications to the use of TES

Transcranial electrical motor stimulation, as mentioned, has the theoretical risk of inducing seizures, although there have been only a few reports focusing on patients with a prior history of seizures rather than new seizure induction. One can use a screening survey during the preoperative assessment to aid in determining the risk of potential seizures occurring intraoperatively. At our institution, we have developed a screening questionnaire to be used when the patient is brought in for preoperative studies. This questionnaire is used in addition to reviewing the patient records in the preoperative assessment. This screening questionnaire includes questions that are asked to the parents and patient during the preoperative assessment at our institution.

Screening questionnaire for TES (developed at Nationwide Children's Hospital, Columbus, Ohio)

1. Do you have any implanted devices, such as a pacemaker or a pump for medications?
2. Have you had a history of non-febrile seizures, epilepsy, or recurrent seizures, including absence seizures (not including only simple febrile seizures)?
3. Any history of a stroke?
4. Any known history of brain structural abnormality, presence of vascular clips, shunts, or prior history of intracranial surgery or skull defects? (autism, ADHD, and mental retardation by themselves are NOT contraindications for monitoring at our institution).
5. History of raised intracranial pressure or what may have been referred to as "benign intracranial hypertension"?
6. Is there a known cardiac disease, such as cardiomyopathy or abnormal rhythm? Solely, mitral valve prolapse is not a contraindication. However, although significant heart history is not an absolute contraindication at our institution, it needs to be considered in the overall risk assessment for a given patient.

If the answer to any one of these questions raises suspicion that there may be a risk to the patient of inducing seizures intraoperatively then the neurologist trained in and responsible for interpreting the neurodiagnostic testing preoperatively and intraoperatively discusses the most appropriate course of action with the referring physician and surgeon. This may include limiting the neurodiagnostic monitoring during the course of the surgical case to the use of certain modalities only and, in particular, typically avoiding the use of transcranial electrical motor stimulation to lessen the seizure risk. Another option might be the assessment and discussion with the surgeon and referring physician over use of antiepileptic medications in the patient, to limit the risk of seizures while allowing for appropriate intraoperative neurophysiologic monitoring modalities to occur. In this way communication prior to the day of surgery ensures patient safety and optimizes care.

Questions

1. Information found on anesthetic and surgical checklists includes all of the following except?
   (a) Operative medications
   (b) Anesthesia requirements
   (c) Estimated duration of the surgery
   (d) Description of procedure
   (e) Hepatitis status
2. Preoperative NIOM assessments are helpful in gaining information on which of the following?
   (a) Choice of monitoring modalities
   (b) Muscle selection for recording
   (c) Preoperative baseline monitoring studies
   (d) Review of comorbidities
   (e) All of the above
3. The TMS associated seizure risk has been reported to range up to what percent for single- and paired-pulse stimulation?
   (a) 5.2% for single-pulse and 10 % for paired-pulse stimulation
   (b) 2.8% for single-pulse and 3.6% for paired-pulse stimulation
   (c) 1.2% for single-pulse and 2.1% for paired-pulse stimulation
4. The aim of selective dorsal rhizotomy is to do which of the following?
   (a) Remove all sensory input and decrease motor tone so that the patient is hypotonic
   (b) Remove all sensory input and increase motor tone to leave the patient hypertonic
   (c) Partially remove sensory input so that the degree of motor tone is reduced but the patient is not left hypotonic
5. A deformity of the left upper extremity at the wrist and elbow is found on a patient during a preoperative assessment for spine surgery. Which of the following statements is true?
   (a) Because of the deformity, electrode placement for a typical median SEP will need to be modified
   (b) Modification of electrode placement will not be needed intraoperatively because stimulation at the Erb's point is routinely done anyway
   (c) This information is not important because the upper extremities are typically not monitored during spine surgery cases
6. Which of the following statements is true?
   (a) There is currently not a method available to evaluate motor pathways preoperatively in an awake and non-sedated patient
   (b) Routine monitoring of SEPs has eliminated the need to assess the motor tracts preoperatively as this would not give any additional information
   (c) Single-pulse transcranial magnetic stimulation can be done in an awake and non-sedated patient and will give information on motor pathways
7. Which of the following perioperative conditions has been shown to affect surgical outcome?
   (a) Presence of comorbidities
   (b) Patient age
   (c) Pulmonary status
   (d) All of the above
8. Which monitoring modality is typically utilized to reduce the risk of radicular injury during surgery?
   (a) SEP
   (b) Monitoring cranial nerve 7
   (c) BAER
   (d) Pedicle screw stimulation
9. Recording from the anal sphincter is important during which type of surgery?
   (a) Posterior fossa decompression
   (b) Cervical diskectomy
   (c) SDR
   (d) Epilepsy surgery
10. During a preoperative NIOM assessment for scoliosis surgery scheduled for the next day, the patient is discovered to have lice. Anesthetic and surgical assessments have already been made and the patient has been cleared for surgery the next day. What is the most appropriate action for the NIOM staff to take?
    (a) Proceed with the preoperative monitoring study, avoiding contact with the lice
    (b) Contact the surgeon's office and inform them of the presence of lice on the patient. This will likely require treatment and postponing surgery until resolved

(c) Request that the anesthesia staff increase the inhalational anesthetic on the surgery day so that the lice will be sedated and monitoring can proceed

References

Albert, T. J., Purtill, J., Mesa. J., McIntosh, T., and Balderston, R. A. (1995). Health outcome assessment before and after adult deformity surgery: a prospective study. *Spine*, **20**, 2002–2004.

Banoub, M., Tetzlaff, J. E., and Schubert, A. (2003). Pharmacologic and physiologic influences affecting sensory evoked potentials: implications for perioperative monitoring. *Anesthesiology*, **99**(3), 716–737.

Baron, E. M. and Albert, T. J. (2006). Medical complications of surgical treatment of adult spinal deformity and how to avoid them. *Spine*, **31**(19), 106–118.

Chu, W., Lam, W., Chan, Y., *et al.* (2006). Relative shortening and functional tethering of spinal cord in adolescent idiopathic scoliosis? Study with multiplanar reformation magnetic resonance imaging and somatosensory evoked potential. *Spine*, **31**(1), E19–E25.

DiCindio, S., Theroux, M., Shah, S., *et al.* (2003). Multimodality monitoring of transcranial electric motor and somatosensory evoked potentials during surgical correction of spinal deformity in patients with cerebral palsy and other neuromuscular disorders. *Spine*, **28**(16), 1851–1856.

Faciszewski, T., Winter, R. B., Lonstein, J. E., Denis, F., and Johnson, L. (1995). The surgical and medical perioperative complications of anterior spinal fusion surgery in the thoracic and lumbar spine in adults: a review of 1223 procedures. *Spine*, **2**, 1592–1599.

Fisher, R. S., Raudzens, P., and Nunemacher, M. (1995). Efficacy of intraoperative neurophysiological monitoring. *Journal of Clinical Neurophysiology*, **12**(1), 97–109.

Glassman, S., Hamill, C., Bridwell, K., *et al.* (2007). The impact of perioperative complications on clinical outcome in adult deformity surgery. *Spine*, **32**(24), 2764–2770.

Hausmann, O., Boni, T., Pfirrmann, C., Curt, A., and Min, K. (2003). Preoperative radiological and electrophysiological evaluation in 100 adolescent idiopathic scoliosis patients. *European Spine Journal*, **12**, 501–506.

Hu, S. and Berven, S. (2006). Preparing the adult deformity patient for spinal surgery. *Spine*, **31**(195), S126–S131.

Kober, H., Nimsky, C., Moller, M., *et al.* (2001). Correlation of sensorimotor activation with functional magnetic resonance imaging and magnetoencephalography in presurgical functional imaging: a spatial analysis. *Neuroimage*, **14**(5), 1214–1228.

Lieb, K. and Selim, M. (2008). Preoperative evaluation of patients with neurological disorders. *Seminars in Neurology*, **28**, 603–610.

Lingard, L., Espin, S., Whyte, S., *et al.* (2004). Communication failures in the operating room: an observational classification of recurrent types and effects. *Quality and Safety in Health Care*, **13**(5), 330–334.

Lingard, L., Regehr, G., Orser, B., *et al.* (2008). Evaluation of a preoperative checklist and team briefing among surgeons, nurses and anesthesiologists to reduce failure in communication. *Archives of Surgery*, **143**(1), 12–17. Discussion, 18.

Malcolm, M. P., Triggs, W. J., Light, K. E., *et al.* (2006). Reliability of motor cortex transcranial magnetic stimulation in four muscle representations. *Clinical Neurophysiology*, **117**, 1037–1046.

McGarvey, M. L. (2007). Perioperative management. *Journal of Clinical Neurophysiology*, **24**(6), 467.

Schrader, L. M., Stern, J. M., Koski, L., Nuwer, M. R., and Engel, J. Jr., (2004). Seizure incidence during single and paired-pulse transcranial magnetic stimulation (TMS) in individuals with epilepsy. *Clinical Neurophysiology*, **115**, 2728–2737.

Spivak, J. M., Neuwirth, M. G., Giordano, C. P., and Bloom, N. (1994). The perioperative course of combined anterior and posterior spinal fusion. *Spine*, **19**, 520–525.

Thuet, E. K., Padberg, A., Raynor, B. L., *et al.* (2005). Increased risk of postoperative neurologic deficit for spinal surgery patients with unobtainable intraoperative evoked potential data. *Spine*, **30**, 2094–2103.

Turner, R. (2009). Neurophysiologic intraoperative monitoring during selective dorsal rhizotomy. *Journal of Clinical Neurophysiology*, **26**(2), 82–84.

Vedantam, R. and Crawford, A. (1997). The role of preoperative pulmonary function tests in patients with adolescent idiopathic scoliosis undergoing posterior spinal fusion. *Spine*, **22**(23), 2731–2734.

3 Magnetic stimulation

Khaled M. Zamel

Introduction

The idea of inducing electrical current using a changing magnetic field began with the discovery of Michael Faraday, in 1831. He found that when an electric current is repeatedly turned on or off in a (primary) coil of wire, another electric current is induced in a nearby (secondary) coil by the changing magnetic field around the primary coil.

Another milestone was achieved in 1896 when d'Arsonval found that the brain can be magnetically stimulated by placing the head inside a large coil carrying high electrical current. This stimulation produces flashing lights (magnetophosphenes), which are seen by the subject and are presumably caused by the secondary induced current in the visual cortex.

It took almost 90 more years before technical developments made it possible for Anthony Barker and his associates in 1985 to use stimulating coils that were compact. These coils could produce a magnetic field of up to 2 Tesla, a pulse with a rise time of 100 ms and a total duration of 400 ms (Barker *et al.*, 1985). With such a simple round coil, they were able to stimulate the motor cortex and elicit muscle contractions. Since then, rapid advances ensued and many investigators discovered that they could use this new and exciting tool to stimulate visual, somatosensory, and auditory cortex selectively and produce illusory sensory perceptions.

Using the same basic principle of Michael Faraday, transcranial magnetic stimulation (TMS) is essentially achieved by a sudden change in electrical current in the primary coil that induces a changing magnetic field (Figure 3.1), subsequently inducing an electrical current in the brain, which acts as the secondary coil in Faraday's experiment.

Therefore, TMS is a form of brain stimulation that is based on a time-varying magnetic field, which penetrates unimpeded through skin and bones, generating an electric current in focused and selected areas of the brain.

Moreover, rather than directly inducing instant changes by applying single TMS pulses to the classical sensory or motor cortex, it has been found that if applied repeatedly, TMS could modulate cerebral cortical activity for periods that lasted beyond the time of stimulation. Repetitive TMS (rTMS) at different frequencies can either facilitate or inhibit the targeted cortical area.

Although the induced electrical current is both weak and brief, it can depolarize nerve cell membranes when the direction of the current is across the membrane. If the depolarization exceeds some level, the nerve cell will discharge and, as the propagated action potential greatly outlasts the electrical pulse, the effect of one TMS pulse can last tens of milliseconds. The effectiveness of the stimulation depends on the strength of the magnetic field produced by the stimulating coil (typically 1–2 T), the shape of the pulse (monophasic or biphasic), its rate of change, and the frequency of the pulses either a single pulse or repetitive or rapid pulses, typically around 10 Hz. By altering the configuration of

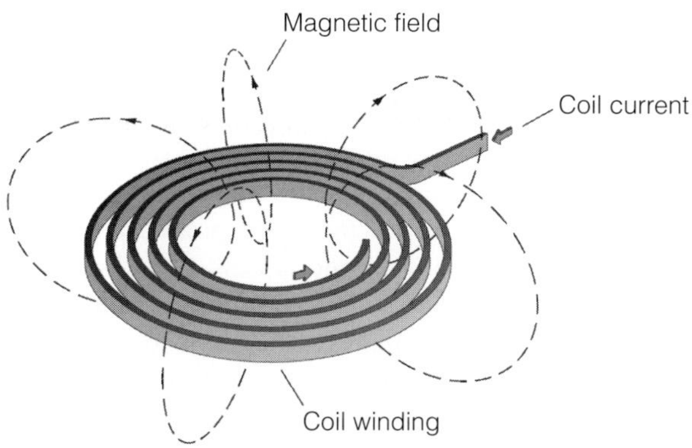

Figure 3.1 The direction of current flow in a circular magnetic coil and the induced magnetic field (from Jalinous, 1995 with permission from Magstim Company Ltd).

the TMS coil on the scalp, the direction of the induced electrical field can be changed. This is particularly important when stimulating white matter tracts since current flowing parallel to the tracts and therefore not crossing the cell membrane, has negligible effects. The effect of coil orientation can even be seen when stimulating cortical gray matter, where the excitable membranes are less ordered.

Intraoperative monitoring of the corticospinal tracts

Neurophysiologic intraoperative monitoring (NIOM) of the spinal cord during neurosurgical and orthopedic operations has traditionally relied on the assessment of the ascending (sensory) pathways by means of monitoring the somatosensory evoked potentials (SEPs). While it does not directly assess motor pathways, SEP monitoring acts as a good surrogate for monitoring the integrity of the corticospinal tracts, as most intraoperative spinal cord injuries would involve both sensory and motor tracts. Nuwer and colleagues (Nuwer *et al.*, 1995) found that when used in NIOM for scoliosis surgeries, monitoring SEPs alone reduced postoperative major neurological deficits by 50%. The SEPs identified insults with a sensitivity of 92% and a specificity of 98.9%.

However, while SEPs can reliably detect lesions that involve the dorsal column, lesions to the motor pathways can occasionally occur in isolation, sparing the sensory pathways with no changes to the SEP waveforms. This is because of the distinct anatomic territory and vascular supplies of both the anterior cord, where most of the motor pathways reside, and the dorsal columns that run in the posterior cord. Several cases have been reported of patients who suffered injury to their motor pathways with no concurrent changes to their SEPs (Ginsburg *et al.*, 1985; Minahan *et al.*, 2001; Hilibrand *et al.*, 2004). Therefore, the introduction of a second NIOM modality specifically for testing the motor pathways provides an essential technique for more comprehensive monitoring of the spinal cord and is now considered the standard of care. In addition to testing the integrity of different neurologic pathways independently, the combined use of these different modalities provides a layer of redundancy that could be very useful should one of them fail during a critical part of the surgery for any physiologic, surgical, anesthetic, or technical reason.

Assessing the integrity of the motor pathways can be accomplished by triggering motor-evoked potentials (MEPs) through two different techniques, which use either transcranial electrical stimulation (TcES) or transcranial magnetic stimulation (TMS) of the motor cortex.

For several reasons, eliciting MEPs intraoperatively in the anesthetized patient is best achieved by TcES, which is now considered the standard and optimal technique for use in the operating room. On the other hand, extraoperative testing of the motor pathways by stimulating the corticospinal tracts is best accomplished by TMS.

Chapter 7 will review in detail the basics of using TcES. This chapter focuses on the principles and clinical applications of TMS, particularly in preoperative and intraoperative testing of the corticospinal pathways in patients undergoing spinal cord surgeries.

Basic principles of magnetic stimulation

The development of the first TMS device by Barker in 1985 (Barker *et al.*, 1985) led to a new era of research in neuroscience as well as a growing number of applications in neurology, psychiatry, and clinical neurophysiology. Based on Michael Faraday's theory of electromagnetic induction, this new device provided the first painless method of stimulating the human motor cortex and triggering myogenic responses, allowing the integrity of the central and peripheral motor pathways to be evaluated.

When a sufficiently high intensity focused magnetic field is delivered to the primary motor cortex, efferent volleys along the corticospinal pathway are induced, thus producing myogenic evoked responses in the muscles. The TMS device is capable of delivering a magnetic pulse that is intense enough to induce electric current in the brain tissue.

The basic TMS power electronics consists of three essential parts: a storage capacitor that can be charged to hold up to several kilovolts of electrical energy; a thyristor switch, which acts as a gate to allow the capacitor to fully discharge in a few microseconds; and, finally, a stimulating coil to conduct this large stored charge. When this intense current flows through the coil, it results in a rapidly fluctuating magnetic field. If the coil is placed above the head, these rapidly changing magnetic pulses penetrate the scalp and the skull with very little attenuation. The brain then acts as the secondary circuit for these magnetic pulses, inducing an ionic current and consequently an electric current flow (eddy current) in the brain.

Types of magnetic stimulators

Single-pulse TMS

This is widely available and very safe to use. Single-pulse TMS can depolarize neurons and produce measurable effects virtually from any part of the peripheral or central nervous system; therefore, it is used in various research and clinical applications. It is most useful in assessing the descending motor pathways and is frequently used to measure delays in the central motor conduction time.

Based on the TMS device capacitor discharge and the directions of the current flow, as well as how the thyristor gating is terminated, different types of magnetic coil current pulse can be generated by the single-pulse TMS (Figure 3.2):

1. Monophasic pulses have a very rapid rise time to a peak and a slower linear return to zero. The stimulation occurs during the initial rapid phase. This provides a stable and well-defined stimulus.
2. Biphasic pulse: this consists of one damped sine pulse that is shorter and more efficient than the monophasic pulse in stimulating the brain, even if the initial phase is similar in intensity. Therefore, it is well suited for bilateral cortical stimulation but is less precise than the monophasic pulse and produces a higher noise level.
3. Polyphasic pulse: this consists of a few cycles of exponentially decaying sinusoidal waves. This is also more efficient in eliciting MEPs, which can be achieved at a lower peak magnetic field than would be required using a monophasic pulse (Maccabee *et al.*, 1998). However, it produces the highest heat and noise of the three pulse types.

Paired-pulse TMS device

Paired-pulse TMS is achieved by connecting two single-pulse systems, so as to deliver two consecutive magnetic pulses through the same coil. The inter-pulse interval can be adjusted so that it can selectively excite either excitatory or inhibitory neurons. Accordingly, this is a very attractive tool in assessing inhibitory or excitatory mechanisms of the brain or spinal cord.

Two single-pulses can also be used synchronously but discharged through two independent coils to stimulate different regions at the same time or in succession. This is labeled a double-pulse TMS system.

Triple and quadruple-pulse TMS devices

Like the paired-pulse TMS, a set of three or four magnetic stimulators can be connected together that

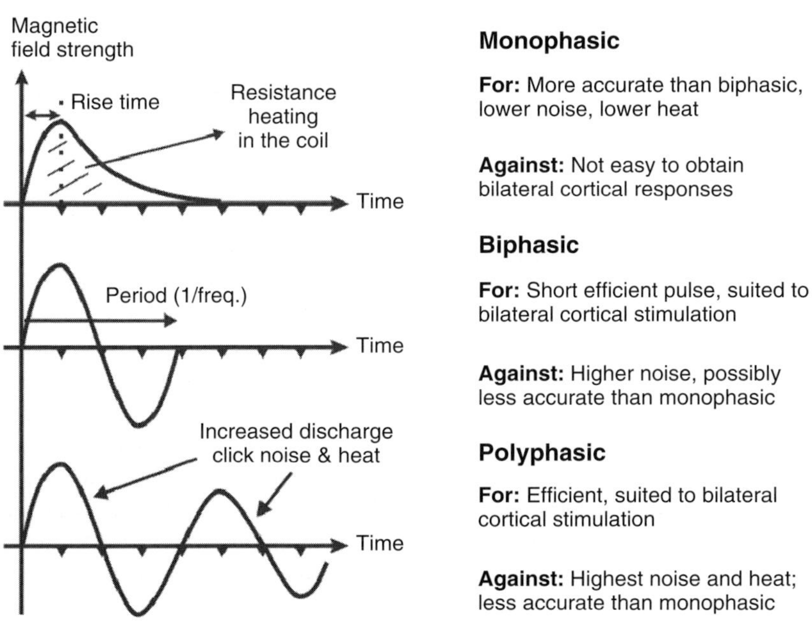

Figure 3.2 The different magnetic field output waveforms produced by the three types of single-pulse TMS devices (from Jalinous, 1995 with permission from Magstim Company Ltd).

is capable of delivering trains of three or four stimuli with an adjustable inter-stimulus interval. This is particularly useful in eliciting MEPs in the anesthetized patient that can't otherwise be generated by single-pulse devices (Hargreaves and Watt, 2005; Rohde *et al.*, 2003).

Rapid-rate TMS (rTMS)

Long trains of stimuli at different frequencies ranging from 1 Hz to 50 Hz can be achieved by rTMS and lead to a transient modification of the brain cortical excitability. Low frequencies, less than 2 Hz, are typically inhibitory, while faster frequencies are excitatory. This ability to temporarily alter the physiology of the neural circuitry has made rTMS a very important tool in various areas of neuroscience research.

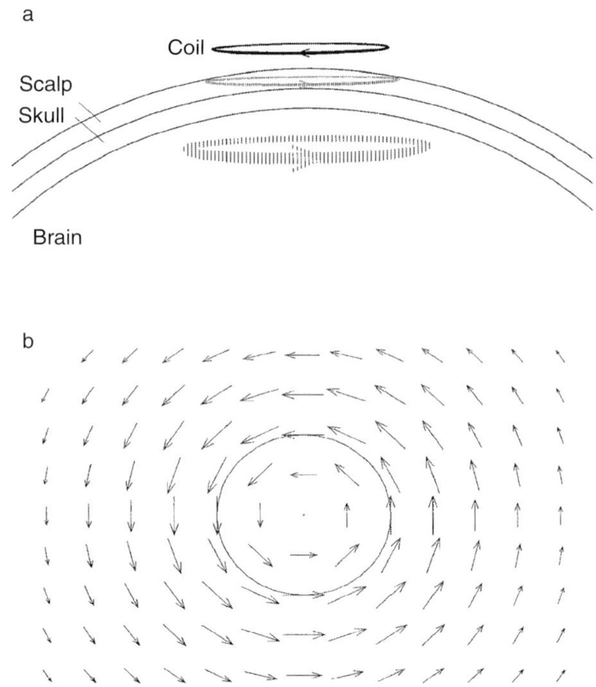

Figure 3.3 (a) A spherical model of the head with a single circular coil positioned parallel to the surface of the scalp. Note that the induced current direction is reversed and the width of the induced current is wider with increasing distance from the stimulating coil. (b) The distribution and direction of the induced current in the cortex; the density of the current is indicated by the length of each arrow. The magnitude of the current is minimal in the center and periphery of the current field and maximal in a circular distribution that correlates with the winding of the circular coil. (Reprinted with permission from Cohen and Cuffin, 1991, with permission from Lippincott, Wolters Kluwer Health.)

Types and characteristics of TMS coils

The basic structure of a typical magnetic coil consists of a set number of copper windings that are tightly wound together and well covered by a rubber compound for electrical insulation. The magnetic field generated by TMS devices and their induced intracranial electrical current is largely determined by the shape, size as well as the direction of electric current of the magnetic coil used. The first transcranial magnetic stimulator used a circular coil (Barker *et al.*, 1985). The circular coil remains widely used in both research and clinical applications, in particular when the goal is to stimulate a large area of the cortex. Later, a double coil (also called a figure-of-eight coil) was developed, which is able to provide a more focused and smaller area of stimulation. Several other magnetic coils with different shapes, sizes and current features were also developed. In this section, focus will be on the commonly used coils which are the circular and figure-of-eight, in addition to the cone or cap coils that are relevant to TMS use in intraoperative and perioperative testing.

Circular coils

Circular coils are relatively powerful. When held parallel to the surface of the skull, an electric current is induced in the brain that runs in the reverse direction to that of the electric current in the coil. The induced electric current is mostly under the entire windings of the coil (Figure 3.3); hence the area of the cerebral cortex that is stimulated is rather large and non-focal. Therefore, the circular coil is particularly useful when the targeted area of stimulation is uncertain. Better focused stimulation can be achieved by tilting the coil so that only the edge is on the head (i.e., the coil is perpendicular to the surface of the skull Cohen and Cuffin, 1991 Figure 3.4). The intensity and the depth of the induced electric current increases with the diameter of the stimulating coil (Cohen and Cuffin, 1991).

Figure-of-eight coils

The development of figure-of-eight-shaped coils was an important technical advance. Two circular coils are combined so that the current flows in the same direction at the junction of the two coils. This provides the most focused TMS stimulation with the maximum activation occurring directly under the junction of the two circular components (Figure 3.5). Therefore,

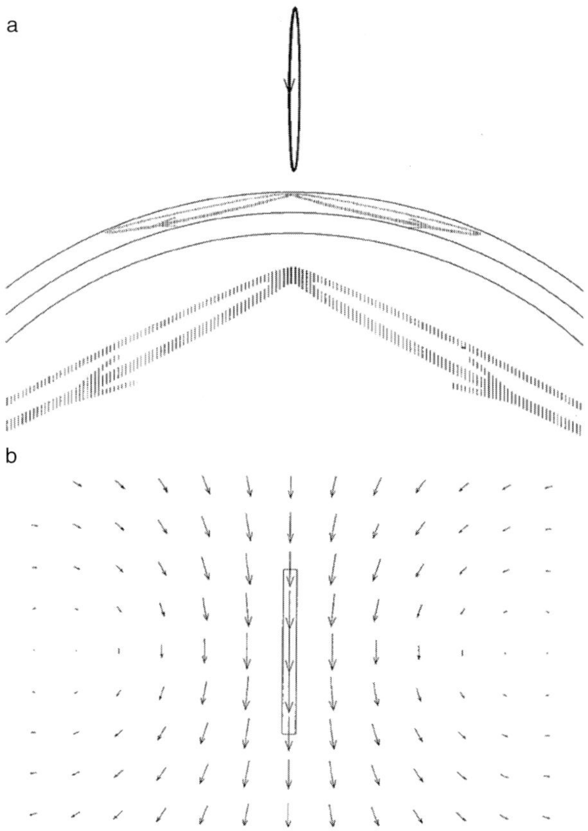

Figure 3.4 (a) The circular coil is repositioned to be perpendicular to the surface of the scalp, resulting in two circular bands of induced current. (b) The computed cortical current is more focal than in Figure 3.3 with the maximal current magnitude in the area under the edge of the stimulating coil. (Reprinted with permission from Cohen and Cuffin, 1991, with permission from Lippincott, Wolters Kluwer Health.)

this is a particularly useful coil for mapping the cerebral cortex.

Cap-shaped and cone coils

These deviate from the typical planar (i.e., flat) coils, having a more curved or bent shape. This better fits with the spherical shape of the human skull and provides a magnetic field that is able to generate the greatest electrical fields for depths comparable with the circular and figure-of-eight coil designs (Kraus *et al.*, 1993 Figure 3.6). The magnetic field of the cap-shaped coil is larger, more intense, and can simultaneously and bilaterally stimulate the motor cortex for both upper and lower extremities. The cap-shaped coil is shaped like a saddle and was designed to trigger

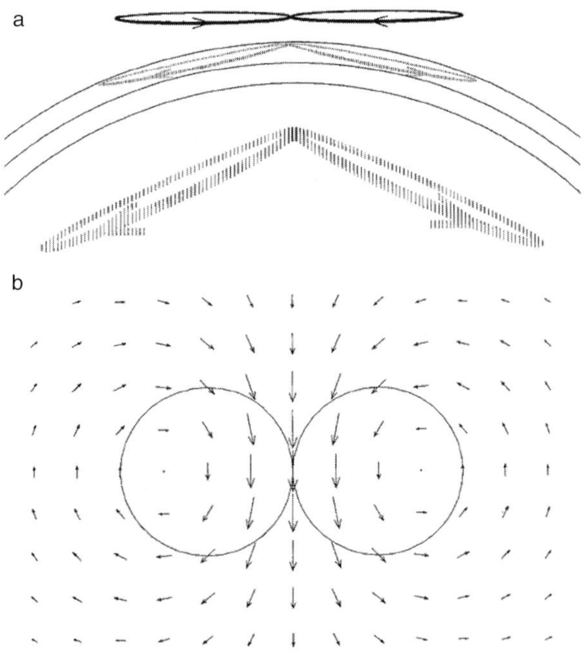

Figure 3.5 (a) Stimulation provided using a figure-of-eight coil. (b) Note that the length of the arrows is maximal in the center of the induced field, indicating an induced current in the cortex that is more focal and greater than that which can be achieved using the circular coil. (Reprinted with permission from Cohen and Cuffin, 1991, with permission from Lippincott, Wolters Kluwer Health.)

motor-evoked potentials for spinal-cord monitoring (Kraus *et al.*, 1993; Aglio *et al.*, 2002a).

The cone-shaped coil (Figure 3.7) consists of an over-sized figure-of-eight coil with the two wings of the coil bent at an angle. Like the cap-shaped coil, the cone-shaped coil allows the generated magnetic field to better follow the curvature of the head, producing a more intense and deeper induced electric current.

TcES versus TMS

Five years before the development of the first TMS device, Merton and Morton built a high-voltage stimulator that could directly stimulate muscles, and found that the same technique could be effectively applied to the head to stimulate the brain through the intact scalp and skull and synchronously trigger an MEP response (Merton and Morton, 1980). Since then, transcranial electrical stimulation has been selectively used for several clinical purposes.

Transcranial electrical stimulation directly produces a flow of ions between the stimulating electrodes

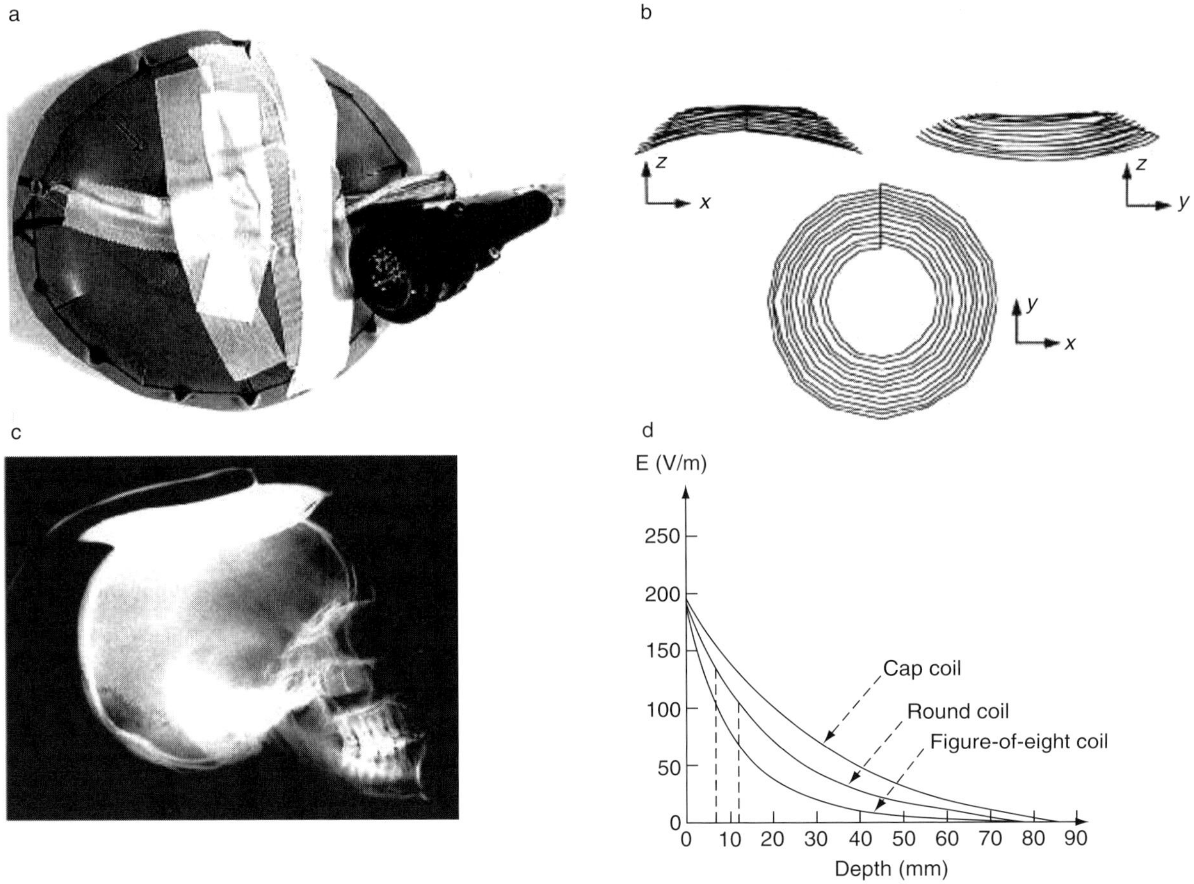

Figure 3.6 (a) The cap coil as designed by Cadwell and colleagues, consisting of 12 turns of copper wire encased in a rubber insulator. The outer diameter of the coil is 17 cm and the inner diameter is 7 cm. (b) Computer reconstruction of the cap coil with the side view showing the saddle-shape design. (c) Radiograph of the cap coil positioned on a subject's head. (d) The diagram illustrates the greatest electric field strength being induced at all depths when the cap coil is used, in comparison with the circular or figure-of-eight coils. (Reprinted from Kraus *et al.*, 1992 with permission from Lippincott, Williams, and Wilkins.)

(Figure 3.8). Compared with TMS, transcranial electrical stimulation is fairly precise and localized because there is no doubt about the position of the electrodes. On the other hand, TMS works by a brief magnetic field that consecutively induces an electric field, which in turn causes ionic current flow; therefore, pinpointing the effective locus of action of TMS is far more difficult. However, TMS has major advantages in that indwelling electrodes are avoided and, with minor exceptions, the discomfort or pain produced by electrical stimulation from electrodes on the scalp is absent.

As a result, in the unconscious or anesthetized patient, such as in the case of intraoperative monitoring, TMS offers no further advantage over transcranial electrical stimulation. On the other hand, due to avoidance of pain in the awake human, TMS has greatly expanded the opportunities for probing and examining the motor pathways.

Intraoperative monitoring using TMS has been successively achieved by multiple investigators (Ubags *et al.*, 1999; Sihle-Wissel *et al.*, 2000; Hargreaves and Watt, 2005). In the operating room the use of TMS has many challenges. The stimulating coil is typically heavy and bulky, and so it is difficult to maneuver and hold in place over the patient's head. Slight changes in the coil position could lead to large variations in its efficacy in transcranial stimulation. This remains a challenge even with the use of the modified skull cap or double-cone coils. Additionally, the size of the coil could hinder the surgical or anesthetic fields, making it impractical to use. With repetitive

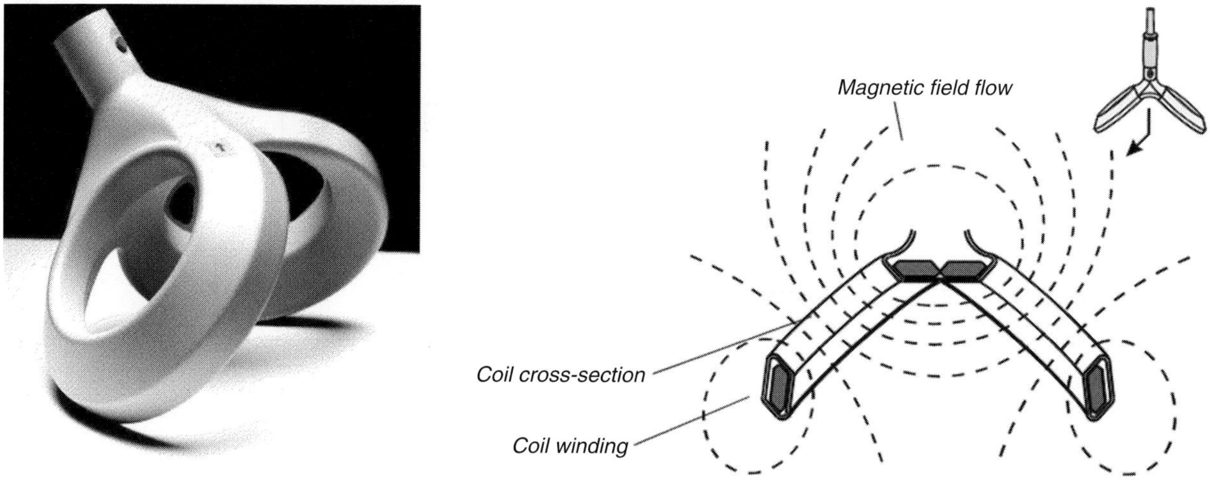

Figure 3.7 The double-cone coil has two wings bent at an angle, which allows for a better fit with the curvature of the head and therefore produces a more intense and deeper induced electric current. This type of coil is especially useful for stimulating the leg areas (from Jalinous, 1995, with permission from Magstim Company Ltd).

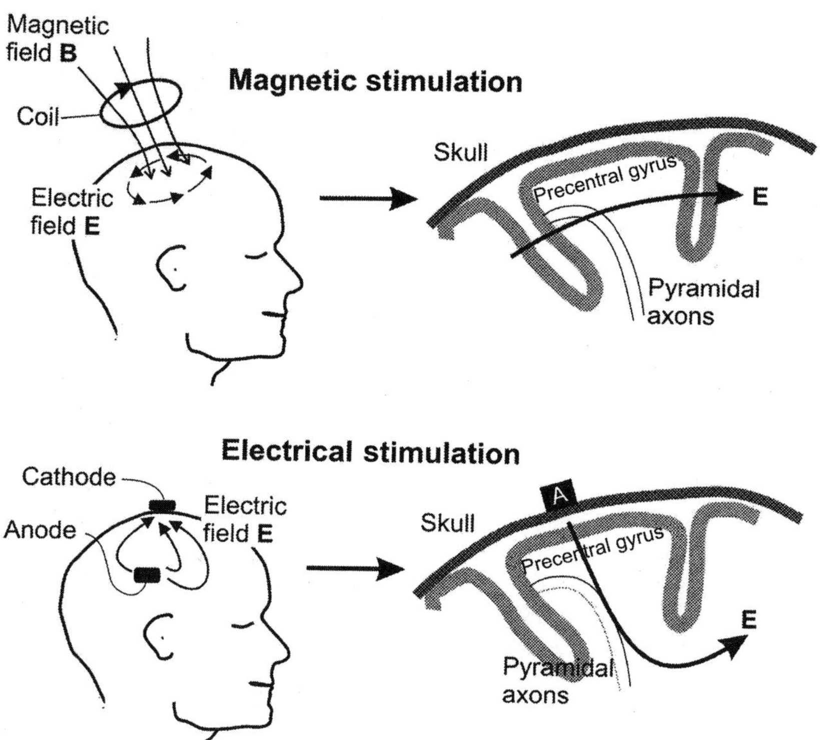

Figure 3.8 Transcranial electric stimulation (TES) and transcranial magnetic stimulation (TMS). The top figure shows an induced electric current in the brain triggered by a magnetic coil. In this case the electric current is parallel to the scalp, most probably stimulating those pyramidal axons near bends or other fibers parallel to the cortex. The bottom figure shows TES with the more localized current flowing from the anode to the cathode probably running radial then parallel to the scalp, thereby stimulating the pyramidal axons deeper in the white matter. (Reprinted with permission from Ruohonen and Ilmoniemi, 2005, with permission from Elsevier.)

stimulations, the coil is subject to overheating, which could lead to interruption of the monitoring and increase the operative risk. The myogenic responses (MEPs) obtained with TMS are typically smaller in amplitude than TcES (Figure 3.9). Ubags compared myogenic MEP responses from both TMS and TES in 11 patients and reported that MEPs were not obtainable by TMS in two patients, while all had good reproducible MEPs using TES (Ubags *et al.*, 1999).

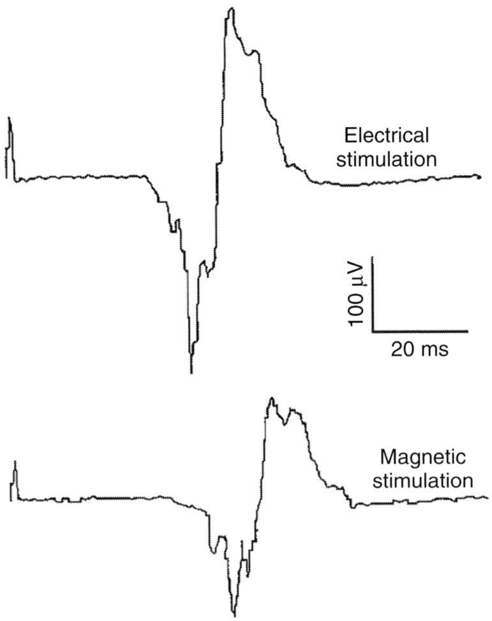

Figure 3.9 Comparison of compound muscle action potentials (CMAPs) in an anesthetized patient elicited with both electrical and magnetic stimulation, demonstrating that the myogenic responses obtained with TMS are typically smaller in amplitude than with TES. (Reprinted with permission from Ubags *et al.*, 1999, with permission from Lippincott, Wolters Kluwer Health.)

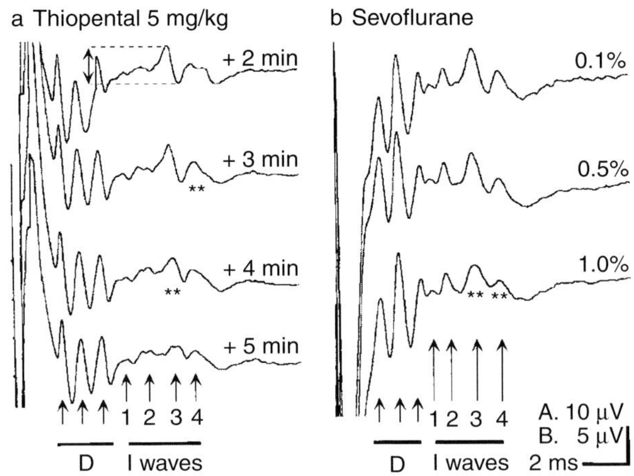

Figure 3.10 The effect of two different anesthetic medications (thiopental and sevoflurane) showing similar impacts on the recordings of the descending corticospinal volleys in response to transcranial electric stimulation (TcES). Both drugs caused attenuation of the late I waves (indicated by asterisks) while no change occurred in the complex D wave. (Reprinted with permission from Woodforth *et al.*, 1999, with permission from Lippincott, Wolters Kluwer Health.)

Differences in the response to transcranial electrical or magnetic stimulation of the motor cortex

Transcranial electrical and magnetic stimuli act differently on the human motor cortex, although the generated responses are similar. After applying a single electrical pulse stimulus to the motor cortex, two main types of response can be recorded from the epidural surface of the spinal cord. An earlier response, called the direct (D) wave, is believed to be generated by direct activation of the fast fibers in the corticospinal tracts; therefore, this D wave has a short latency and relatively large amplitude (Figure 3.10). The D wave is followed by a series of smaller-amplitude and less well-defined waves that are also generated in the cortex but are thought to reflect activation of the corticospinal fibers, predominantly through transsynaptic transmission in the deeper cortical interneurons; these responses are termed the indirect (I) waves (Amassian *et al.*, 1987).

D waves are much more resistant to the effect of anesthetic medications, but I waves in contrast can be abolished by a synaptic block of the vertically oriented excitatory chains of neurons terminating on the cortical motor neuron's body (Woodforth *et al.*, 1999) (see Figure 3.10). However, the orientation of the anode–cathode pair does play an important role in preferentially eliciting these responses. A coronally oriented anode–cathode pair will enhance D waves, while a midline posterior-to-anterior orientation enhances I waves (Maccabee, 2000) (Figure 3.11).

When applied to a spherical conductor, such as the human head, TMS induces an electrical field that is parallel to the inner surface of the volume conductor being stimulated (Cohen and Cuffin, 1991). The induced intracranial electric current in general mimics the electric current that runs in the coil but is opposite in direction (Figure 3.8). On the contrary, direct transcranial electrical stimulation applied to the cortical surface generates a field that is more radially directed toward the center of the head. This fact defines the fundamental differences in the characteristics of the electric field generated by either electrical or magnetic stimulation.

In contrast to transcranial electrical stimulation TMS results only in transsynaptic excitation of corticospinal tracts through cortical interneurons (Day *et al.*, 1987a,b). Another significant difference stems from the fact that direct electrical stimulation excites neural elements deeper in the brain than magnetic stimulation (Hess *et al.*, 1986b). These differences were proposed to give an explanation of why the latencies of the induced MEPs with TMS were consistently found to be delayed by about 2 ms when compared with responses triggered by transcranial electrical stimulation (Amassian *et al.*, 1989).

Amassian and colleagues later demonstrated that changing the orientation of the magnetic coil to a more lateral–sagittal position could stimulate the corticospinal tracts with a latency similar to that produced by TcES. In fact, both direct and indirect (D and I) waves can be preferentially excited by altering the orientation of the coil, so that when a figure-of-eight coil is oriented in the posterior to anterior direction, the I waves are selectively activated, while reorienting the coil to the side of the head (lateral–medial direction) primarily activated the D wave with a latency similar to the effect of the anodal electrical stimulation (Nakamura *et al.*, 1996) (Figure 3.12). Furthermore, it has been found that these different responses to orientation apply more to stimulation of the upper extremities. In the lower extremities, I waves are more prominent than D waves, regardless of the direction of the magnetic coil (Nakamura, 1998).

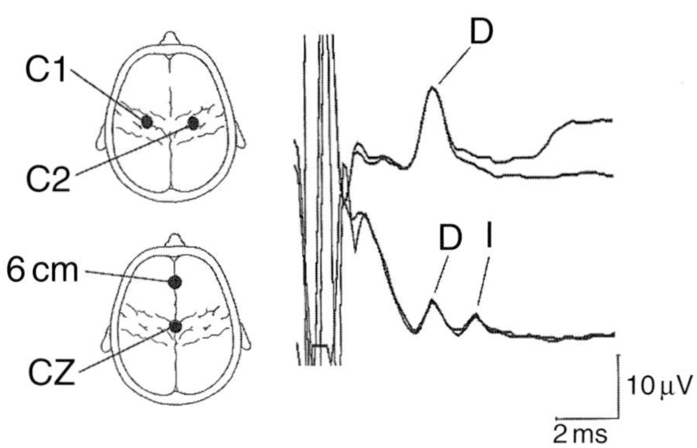

Figure 3.11 Epidural recordings of D and I waves, demonstrating an enhancement of the D wave when the anode and cathode (C1 and C2) are oriented in the coronal plane, while positioning the anode midline at CZ with the cathode 6 cm more anteriorly, led to more recruitment of the I waves (from Maccabee *et al.*, 2000, with permission from Elsevier).

In addition to the epidural spinal cord recording, motor evoked potentials (MEPs) can also be recorded from muscles (compound muscle action potentials, CMAPs) using transcranial electrical or magnetic stimulation. Recording CMAPs is the more widely used technique in the USA, although spinal cord recording is more widely used in other parts of the world. Many also advocate dual recording of the D and I waves from the spinal cord as well as CMAPs from the muscles. Recording CMAPs from the muscles, while easier and less invasive, can be abolished by muscle relaxants and is more sensitive to most anesthetic medications. Nevertheless, it is usually successful by adjusting stimulation and recording parameters.

The utility of TMS in NIOM

As discussed before, in the awake subject, TMS is a more desirable method of motor cortex stimulation

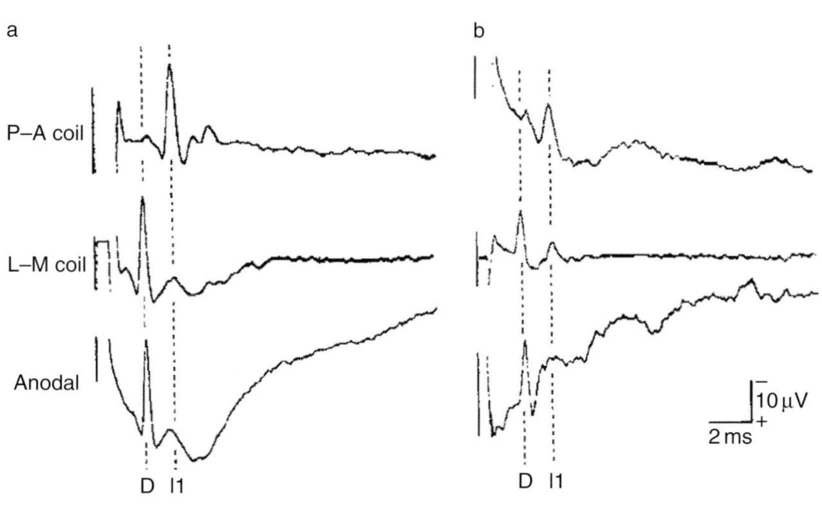

Figure 3.12 (a)(b) Comparison of the corticospinal responses obtained in two patients with both TMS and TES. The TMS was delivered using a figure-of-eight coil held over the motor area of the right hemisphere in two different orientations with the current flow in either posterior–anterior (P–A stimulation) or in the lateral–medial (L–M stimulation) direction. The TES was delivered with the anode placed 7 cm lateral to the vertex and cathode at the vertex. The corticospinal responses were recorded using a bipolar catheter epidural electrode placed between the third and fourth cervical vertebrae. Note that the P–A stimulation selectively activated the I waves while L–M stimulation activated the D wave similar to the response obtained with anodal stimulation. (Reprinted from Nakamura *et al.*, 1996 with permission from Elsevier.)

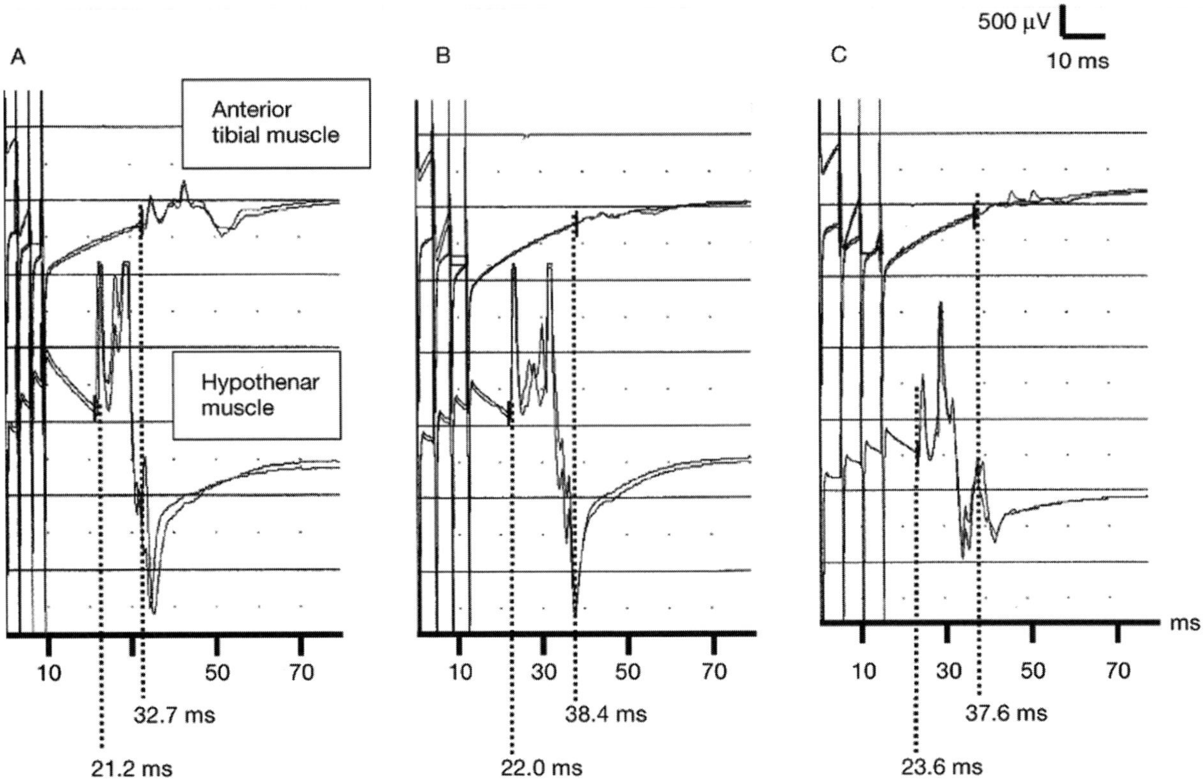

Figure 3.13 Intraoperative monitoring using quadruple TMS stimulation of the left motor cortex with three different stimulation frequencies (333, 250, and 200 Hz). Compound muscle action potential (CMAP) responses were obtained from the right hypothenar and the anterior tibial muscle. Propofol and nitrous oxide were used for anaesthesia. Note the increase of the CMAP mean latency when the frequency of stimulation is reduced without significant decrease in the mean amplitude. (Reprinted from Rohde *et al.*, 2003, with permission from Oxford University Press.)

than direct transcranial electrical stimulation, with less discomfort and pain. On the contrary, in anesthetized patients, pain is controlled by anesthetic agents and direct electrical stimulation produces more robust corticospinal volleys which can be recorded over the spinal cord, as well as more reliable compound muscle action potentials from muscle recording (Amassian *et al.*, 1989; Burke and Hicks, 1998). Additionally, as mentioned before, the bulky magnetic coil and it's potential for overheating pose technical challenges in the operating room. The intensity of cortical stimulation using TMS is also highly sensitive to minimal changes in the orientation of the magnetic coil. Finally, blocking of I waves with anesthetic medications is more likely to occur with TMS, which is more dependent on interneuronal synaptic transmission. Therefore intraoperative monitoring using magnetic stimulation offers no advantage over TcES.

Nevertheless, several groups of investigators have used TMS for intraoperative monitoring with either single-pulse TMS (Amassian *et al.*, 1989; Edmonds *et al.*, 1989; Schmid *et al.*, 1992; Kraus *et al.*, 1993; Herdmann *et al.*, 1993; Tabaraud *et al.*, 1993; Glassman *et al.*, 1995; Lee *et al.*, 1995; Kaneko *et al.*, 1997; Burke and Hicks, 1998; Ubags *et al.*, 1999; Sihle-Wissel *et al.*, 2000; Aglio *et al.*, 2002b), paired-pulse TMS (Kitagawa *et al.*, 1995), triple-pulse stimuli (Hargreaves and Watt, 2005), or quadruple-pulse TMS (Rohde *et al.*, 2003; Figure 3.13). In most of these studies, total intravenous anesthesia (TIVA) led to more successful monitoring than the use of conventional anesthetic regimens. Special coils, such as the cap-shaped coil or the cone coil, were also reported to enhance the cortical induced electrical currents and thus enhance the recorded evoked potentials.

Extraoperative uses of single-pulse TMS

Transcranial magnetic stimulation is an attractive method of painlessly studying the central and peripheral nervous system. The potential clinical and research applications of TMS are enormous and the literature has grown rapidly on the uses of TMS in vast areas of neuroscience and psychiatry.

Single-pulse TMS has been used in many conditions where assessment of the central motor conduction time could be helpful to indicate an injury to the corticospinal cord. For example, TMS has been helpful in detecting subclinical spinal cord abnormalities in patients with multiple sclerosis (Hess *et al.*, 1986a; Schriefer *et al.*, 1989; Humm *et al.*, 2003) and central motor conduction time using TMS has been used to distinguish between the various types of multiple sclerosis.

Magnetic stimulation has also been helpful in other spinal cord disorders, such as traumatic, neoplastic, and compressive myelopathy, as well as motor neuron disease, including amyotrophic lateral sclerosis (ALS). Similarly, TMS can be used in the preoperative assessment of patients undergoing spinal cord surgery. This could potentially be helpful in screening and obtaining baselines in patients who are considered for intraoperative transcranial electrical stimulation.

At our center, we applied TMS to 36 children and young adults as part of their preoperative assessment prior to scoliosis surgery. The double-cone magnetic coil (Figure 3.6) was used to deliver single-pulse TMS, starting at 50% of the stimulator output (MagStim model 200 stimulator; MagStim Company Ltd, Whitland, Wales, UK). Compound muscle action potentials (CMAPs) were recorded from selected muscles of both upper and lower extremities, the brachioradialis and tibialis anterior muscles. Thirty-four patients had reproducible CMAPs in all four extremities (Figure 3.14) while two patients had triggered CMAPs from the tibialis anterior muscle only on one side. Intraoperatively, all patients had CMAPs recorded from all four extremities using transcranial electrical stimulation (Figure 3.15 data not yet published).

In a patient with spinal cord cavernous angioma (Figure 3.16) preoperative TMS revealed absent CMAPs from either leg. During surgery, monitoring using transcranial electrical stimulation also revealed absent CMAPs responses from either leg. The clinical implications of this screening is not yet clear, given the limited number of patients, but it appears that the responses from magnetic and electrical stimulation are concordant in the majority of patients.

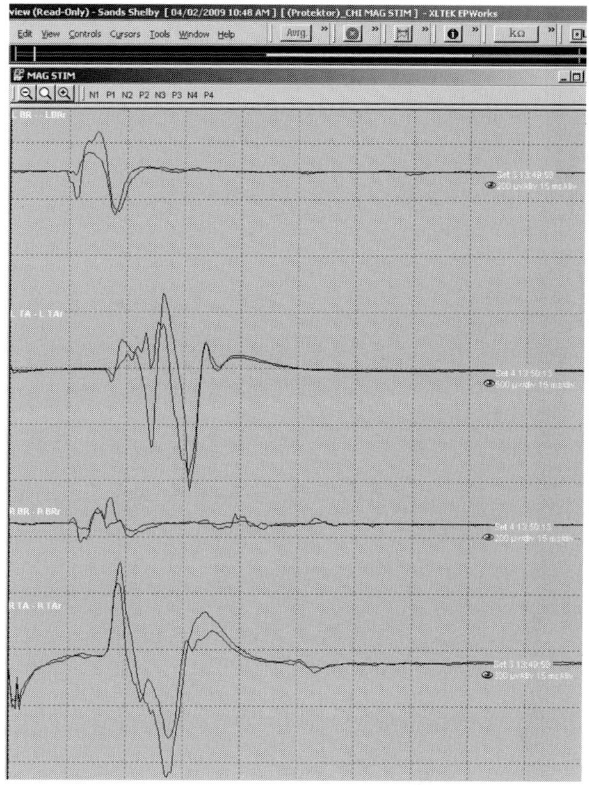

Figure 3.14 Example of surface-recorded CMAPs resulting from single-pulse TMS stimulations at 50% full stimulator strength using a double-cone coil in a representative patient. The recordings are obtained from the brachioradialis and tibialis anterior muscles on both sides with the gain set at 200 μV/division and time base at 15 ms/division.

Questions

1. Which of the following is not a component of the magnetic stimulator?
   (a) Capacitor
   (b) Thyristor switch
   (c) Amplifier
   (d) Stimulating coil
2. Which of the following is true about the magnetic coils used in transcranial magnetic stimulation (TMS)?
   (a) A circular coil provides the best focused stimulation

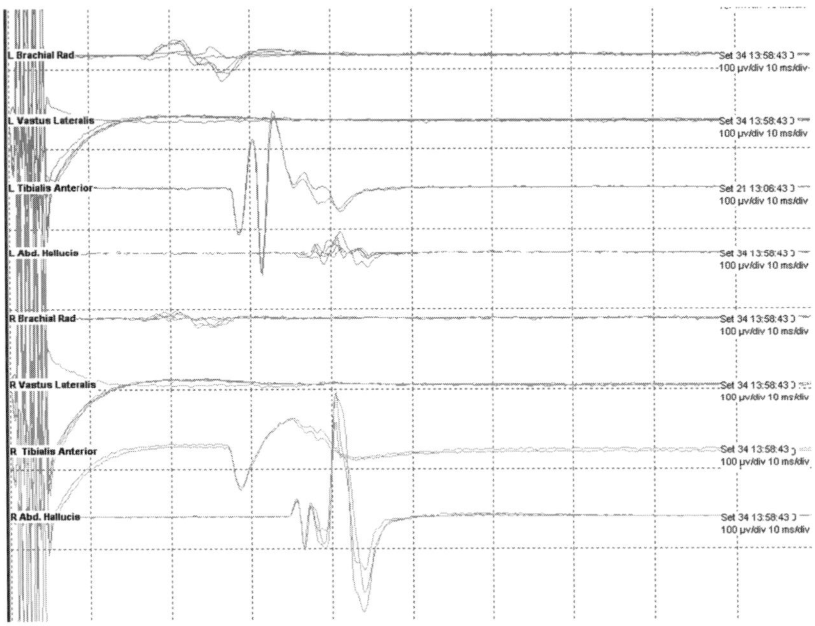

Figure 3.15 Motor evoked potentials (MEPs) after transcranial electric stimulation recorded in the same patient shown in Figure 3.14. The CMAPs are recorded from left and right brachioradialis, vastus lateralis, tibialis anterior, and abductor hallucis muscles.

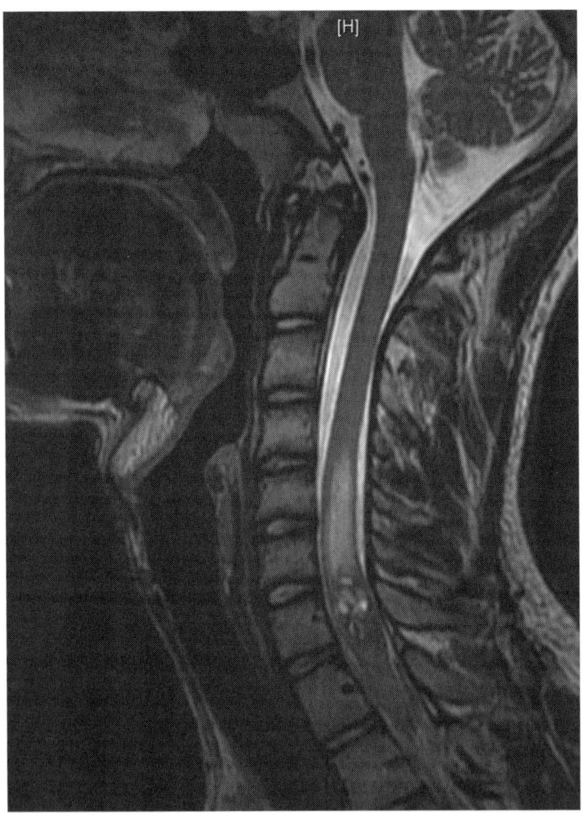

Figure 3.16 A T2-weighted MRI showing a venous angioma in the cervical spinal cord.

(b) The figure-of-eight coil is capable of stimulating a larger cortical area than the circular coil
(c) The cone-shaped coil generates a more intense and deeper electric current and can potentially stimulate both upper and lower extremities simultaneously
(d) None of the above

3. Which of the following is true with regard to achieving transcranial stimulation in the anesthetized patient?
   (a) Transcranial electrical stimulation (TES) is more practical as it involves less bulky equipment and there is less interference with the surgical and anesthesia field
   (b) Small changes in the orientation of the TMS coil can lead to large variations in the MEP responses
   (c) Trains of TMS stimulation can be successful in eliciting MEP responses
   (d) All of the above
4. Which of the following statements about responses recorded at the spinal cord after electrical stimulation of the motor cortex is false?
   (a) There is a series of smaller amplitude waves (indirect waves) that reflect activation of the corticospinal fibers via transsynaptic transmission

   (b) A short-latency, large amplitude single wave (direct wave) is generated by direct activation of the fast fibers in the corticospinal tracts
   (c) The indirect waves are much more resistant to the effect of anesthetic medications than the direct wave
   (d) The orientation of the stimulating anode–cathode pair can preferentially enhance either the direct wave or the indirect waves
5. Assessment of the central motor conduction time is best obtained by:
   (a) Single-pulse TMS
   (b) Paired-pulse TMS
   (c) Repetitive TMS
   (d) Quadruple-pulse TMS
6. Obtaining reliable motor evoked potentials in the anesthetized patient is best obtained by:
   (a) Single-pulse TMS
   (b) Paired-pulse TMS
   (c) Repetitive TMS
   (d) Quadruple-pulse TMS

References

Aglio, L. S., Kraus, K. H., Desai, S., *et al.* (2002a). Efficacious use of a cap-shaped coil for transcranial magnetic stimulation of descending motor paths. *Clinical Electroencephalography*, **33**, 21–29.

Aglio, L. S., Romero, R., Desai, S., *et al.* (2002b). The use of transcranial magnetic stimulation for monitoring descending spinal cord motor function. *Clinical Electroencephalography*, **33**, 30–41.

Amassian, V. E., Cracco, R. Q., and Maccabee, P. J. (1989). Focal stimulation of human cerebral cortex with the magnetic coil – a comparison with electrical stimulation. *Electroencephalography and Clinical Neurophysiology*, **74**, 401–416.

Amassian, V. E., Stewart, M., Quirk, G. J., and Rosenthal, J. L. (1987). Physiological basis of motor effects of a transient stimulus to cerebral cortex. *Neurosurgery*, **20**, 74–93.

Barker, A. T., Freeston, I. L., Jalinous, R., and Jarratt, J. A. (1985). Non-invasive stimulation of motor pathways within the brain using time-varying magnetic-fields. *Electroencephalography and Clinical Neurophysiology*, **61**, S245–S246.

Burke, D. and Hicks, R. G. (1998). Surgical monitoring of motor pathways. *Journal of Clinical Neurophysiology*, **15**, 194–205.

Cohen, D. and Cuffin, B. N. (1991). Developing a more focal magnetic stimulator. 1. Some basic principles. *Journal of Clinical Neurophysiology*, **8**, 102–111.

Day, B. L., Rothwell, J. C., Thompson, P. D., *et al.* (1987a). Motor cortex stimulation in intact man. 2. Multiple descending volleys. *Brain*, **110**, 1191–1209.

Day, B. L., Thompson, P. D., Dick, J. P., Nakashima, K., and Marsden, C. D. (1987b). Different sites of action of electrical and magnetic stimulation of the human brain. *Neuroscience Letters*, **75**, 101–106.

Edmonds, H. L., Paloheimo, M. P. J., Backman, M. H., *et al.* (1989). Transcranial magnetic motor-evoked potentials (TCMMEP) for functional monitoring of motor pathways during scoliosis surgery. *Spine*, **14**, 683–686.

Ginsburg, H. H., Shetter, A. G., and Raudzens, P. A. (1985). Postoperative paraplegia with preserved intraoperative somatosensory evoked-potentials. *Journal of Neurosurgery*, **63**, 296–300.

Glassman, S. D., Zhang, Y. P., Shields, C. B., Johnson, J. R., and Linden, R. D. (1995). Transcranial magnetic motor-evoked potentials in scoliosis surgery. *Orthopedics*, **18**, 1017–1023.

Hargreaves, S. J., and Watt, J. W. H. (2005). Intravenous anaesthesia and repetitive transcranial magnetic stimulation monitoring in spinal column surgery. *British Journal of Anaesthesia*, **94**, 70–73.

Herdmann, J., Lumenta, C. B., and Huse, K. O. W. (1993). Magnetic stimulation for monitoring of motor pathways in spinal procedures. *Spine*, **18**, 551–559.

Hess, C. W., Mills, K. R., and Murray, N. M. F. (1986a). Measurement of central motor conduction in multiple sclerosis by magnetic brain stimulation. *Lancet*, **2**, 355–358.

Hess, C. W., Mills, K. R., and Murray, N. M. F. (1986b). Percutaneous stimulation of the human brain – a comparison of electrical and magnetic stimuli. *Journal of Physiology – London*, **378**, 35.

Hilibrand, A. S., Schwartz, D. M., Sethuraman, V., Vaccaro, A. R., and Albert, T. J. (2004). Comparison of transcranial electric motor and somatosensory evoked potential monitoring during cervical spine surgery. *Journal of Bone and Joint Surgery – American Volume*, **86**A, 1248–1253.

Humm, A. M., Magistris, M. R., Truffert, A., Hess, C. W., and Rosler, K. M. (2003). Central motor conduction differs between acute relapsing-remitting and chronic progressive multiple sclerosis. *Clinical Neurophysiology*, **114**, 2196–2203.

Jalinous, R. (1995). *Guide to Magnetic Stimulation*. Whitland (Wales): The MagStim Company.

Kaneko, K., Kawai, S., Taguchi, T., *et al.* (1997). Spatial distribution of corticospinal potentials following transcranial electric and magnetic stimulation in human spinal cord. *Journal of the Neurological Sciences*, **151**, 217–221.

Kitagawa, H., Nakamura, H., Kawaguchi, Y., *et al.* (1995). Magnetic-evoked compound muscle action potential neuromonitoring in spine surgery. *Spine*, **20**, 2233–2239.

Kraus, K. H., Gugino, L. D., Levy, W. J., Cadwell, J., and Roth, B. J. (1993). The use of a cap-shaped coil for transcranial magnetic stimulation of the motor cortex. *Journal of Clinical Neurophysiology*, **10**, 353–362.

Lee, W. Y., Hou, W. Y., Yang, L. H., and Lin, S. M. (1995). Intraoperative monitoring of motor function by magnetic motor evoked-potentials. *Neurosurgery*, **36**, 493–500.

Maccabee, P. J., Nagarajan, S. S., Amassian, V. E., *et al.* (1998). Influence of pulse sequence, polarity and amplitude on magnetic stimulation of human and porcine peripheral nerve. *Journal of Physiology – London*, **513**, 571–585.

Maccabee, P. J., Amassian, V. E., Ziemann, U., Wassermann, E., and Deletis, V. (2000). Emerging applications in neuromagnetic stimulation. In *Comprehensive Clinical Neurophysiology*, ed. K. H. Levin and H. O. Luders. Philadelphia: WB Saunders.

Merton, P. A., and Morton, H. B. (1980). Stimulation of the cerebral cortex in the intact human subject. *Nature*, **285**, 227.

Minahan, R. E., Sepkuty, J. P., Lesser, R. P., Sponseller, P. D., and Kostuik, J. P. (2001). Anterior spinal cord injury with preserved neurogenic 'motor' evoked potentials. *Clinical Neurophysiology*, **112**, 1442–1450.

Nakamura, H. (1998). D- and I-waves in humans. *Electroencephalography and Clinical Neurophysiology*, **107**(3), 76P.

Nakamura, H., Kitagawa, H., Kawaguchi, Y., and Tsuji, H. (1996). Direct and indirect activation of human corticospinal neurons by transcranial magnetic and electrical stimulation. *Neuroscience Letters*, **210**, 45–48.

Nuwer, M. R., Dawson, E. G., Carlson, L. G., Kanim, L. E., and Sherman, J. E. (1995). Somatosensory evoked potential spinal cord monitoring reduces neurologic deficits after scoliosis surgery: results of a large multicenter survey. *Electroencephalography and Clinical Neurophysiology*, **96**, 6–11.

Rohde, V., Krombach, G. A., Baumert, J. H., *et al.* (2003). Measurement of motor-evoked potentials following repetitive magnetic motor cortex stimulation during isoflurane or propofol anaesthesia. *British Journal of Anaesthesia*, **91**, 487–492.

Ruohonen, J. and Ilmoniemi, R. J. (2005). Basic physics and design of transcranial magnetic stimulation devices and coils. In *Magnetic Stimulation in Clinical Neurophysiology*, ed. M. Hallett and S. Chokroverty. Philadelphia: Butterworth-Heinemann, pp. 17–30.

Schmid, U. D., Boll, J., Liechti, S., Schmid, J., and Hess, C. W. (1992). Influence of some anesthetic agents on muscle responses to transcranial magnetic cortex stimulation – a pilot study in humans. *Neurosurgery*, **30**, 85–92.

Schriefer, T. N., Hess, C. W., Mills, K. R., and Murray, N. M. F. (1989). Central motor conduction studies in motor neuron disease using magnetic brain stimulation. *Electroencephalography and Clinical Neurophysiology*, **74**, 431–437.

Sihle-Wissel, M., Scholz, M., and Cunitz, G. (2000). Transcranial magnetic-evoked potentials under total intravenous anaesthesia and nitrous oxide. *British Journal of Anaesthesia*, **85**, 465–467.

Tabaraud, F., Boulesteix, J. M., Moulies, D., *et al.* (1993). Monitoring of the motor pathway during spinal surgery. *Spine*, **18**, 546–550.

Ubags, L. H., Kalkman, C. J., Been, H. D., Koelman, J. H., and de Visser, B. W. O. (1999). A comparison of myogenic motor evoked responses to electrical and magnetic transcranial stimulation during nitrous oxide opioid anesthesia. *Anesthesia and Analgesia*, **88**, 568–572.

Woodforth, I. J., Hicks, R. G., Crawford, M. R., Stephen, J. P. H., and Burke, D. (1999). Depression of I waves in corticospinal volleys by sevoflurane, thiopental, and propofol. *Anesthesia and Analgesia*, **89**, 1182–1187.

4 The operating-room environment and team approach: pitfalls and technical factors

Khaled M. Zamel

The history of neurophysiologic intraoperative monitoring (NIOM) in the operating room dates back to the mid-1970s. Since then the field has flourished with remarkable advances in both the number and the quality of modalities available. This has been achieved by the rapid technological advancements in computers and electronics coupled with an extensive and growing clinical experience over the past 25 years. While this has tremendously improved the ability of the NIOM team to monitor different components of the nervous system extensively, successful NIOM faces many challenges in the operating room.

The ultimate goal of NIOM is to improve the outcome of surgeries that involve risk of injury to the peripheral or central nervous system. Achieving this goal depends on the coordination of efforts by the surgeon, anesthesiologist, and NIOM team.

Many different NIOM techniques have been developed, as discussed in other chapters of this book, with the goal of providing the surgeon with continuous feedback on the at-risk area. While many of the techniques used in NIOM have initially been studied and mastered in the outpatient setting, challenges occur intraoperatively because of anesthesia, and intervening artifacts in the operating room environment.

For the NIOM to be successful, therefore, it is essential to build a team that is knowledgeable and experienced, in the preoperative and intraoperative interpretation and analysis. This is necessary to differentiate clinically important signal changes from false-positive physiologic or non-physiologic changes.

The NIOM team

Tremendous technological advancements have improved the quality and availability of NIOM, nevertheless successful NIOM continues to depend largely upon a well-skilled, experienced, and dedicated team of clinical neurophysiologists and electroneurodiagnostic technologists.

Although it has become a standard of care for many surgical procedures, the degree of involvement of the clinical neurophysiologist in NIOM has remained controversial and continues to vary among different institutions (Nuwer and Nuwer, 1997). Experienced and certified neurophysiology technologists are indispensable members of the NIOM team.

In addition to obtaining proper certification from the American Board of Electroencephalographic and Evoked Potentials Technologists (ABRET), technologists need to develop adequate skills and knowledge to provide NIOM properly and safely.

The American Society of Electroneurodiagnostic Technologists (ASET) has stated (American Society of Electroneurodiagnostic Technologists, 1998) that after graduating from an accredited electroneurodiagnostic technology program, technologists require 3–5 years of supervised training experience in routine diagnostic electroneurodiagnostic testing and a minimum of one year of experience in the operating room under the supervision of a qualified neurophysiologist.

In addition to acquiring broad knowledge about a range of NIOM equipment and techniques, the NIOM technologist must develop important organizational skills and adhere to strict checklists to avoid errors which could potentially lead to major consequences. The NIOM technologists develop skills in applying a variety of recording and stimulating electrodes not typically used in the outpatient setting (Figure 4.1).

The technologist should provide integrated, computer time-locked, detailed documentation of the course of surgical events, anesthetic medications used, hemodynamic changes, and monitoring data

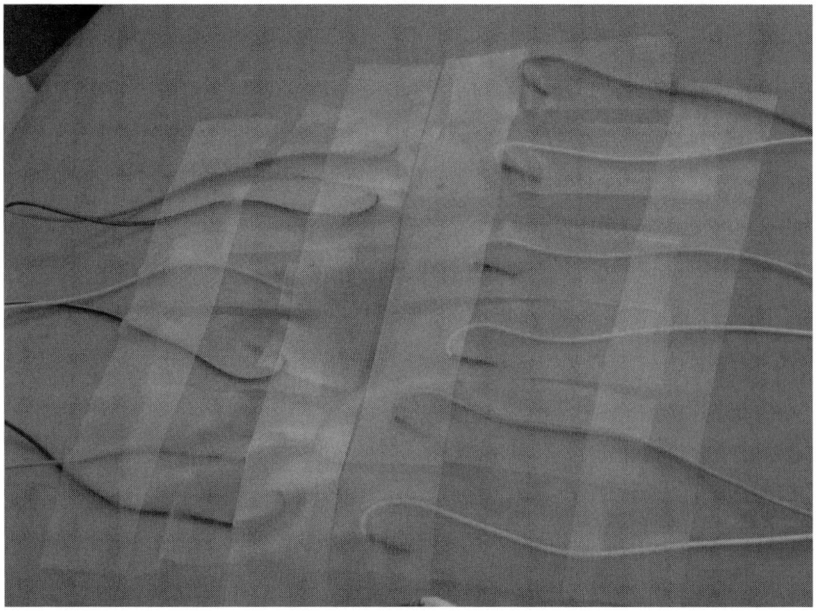

Figure 4.1 A set of needle electrodes placed by the NIOM technologist in the abdominal muscles to record triggered and spontaneous electromyographic responses during scoliosis surgery. The electrode wires are color coded to identify and label the different muscle groups easily. For color version, see plate section.

throughout the surgery. Technologists can collect, measure and identify the recorded signals.

Technologists should not monitor more than one surgery at the same time, to assure adequate attention. The availability of more than one technologist is advisable, to allow for cross coverage during prolonged surgeries. Also, the availability of additional technologists can significantly reduce the time needed to prepare the patient for the NIOM and avoid excessive delays after the induction of anesthesia and before the surgery starts (Figure 4.2).

The essential training, skills, and experience of the technologist complements the expertise of the clinical neurophysiologist, who is sufficiently experienced and knowledgeable in neuroanatomy, neurophysiology, and effect of anesthesia and other physiological factors on the neural structures being studied. The practice of NIOM has changed and become more complex considerably over the past few decades. Individual hospitals set practice guidelines and credentialing for NIOM. However, most hospitals require a licensed clinical neurophysiologist to supervise NIOM as part of patient care.

This supervision has been facilitated by the development of instantaneous communication between the technologists in the operating room and the neurophysiologist, through continuous remote online monitoring. It is recommended that the monitoring clinical neurophysiologist be available nearby at all times to intervene if needed and be able to come to the operating room at short notice if necessary.

An example is during scoliosis surgeries, in which the neurophysiologist may be present in the operating room during certain steps, to provide immediate feedback and interpretation directly to the surgeon. The American Society of Electroneurodiagnostic Technologists recommends that a clinical neurophysiologist must always be physically present in the operating room when electrical activity is recorded directly from the surface of the brain and that in the absence of a neurophysiologist, the surgeon or anesthesiologist should take responsibility in interpreting the waveform description provided by the technologist (American Society of Electroneurodiagnostic Technologists, 1998).

Common current practice dictates that no more than three cases are monitored simultaneously by a single neurophysiologist, to assure adequate attention to each individual case (Nuwer, 2008). When critical changes occur in two or more surgeries simultaneously, a feasible back-up plan is important, to allow involvement of more than one neurophysiologist.

These recommendations are arguable and are not universally applied mainly due to the limited number of neurophysiologists and NIOM teams that are available.

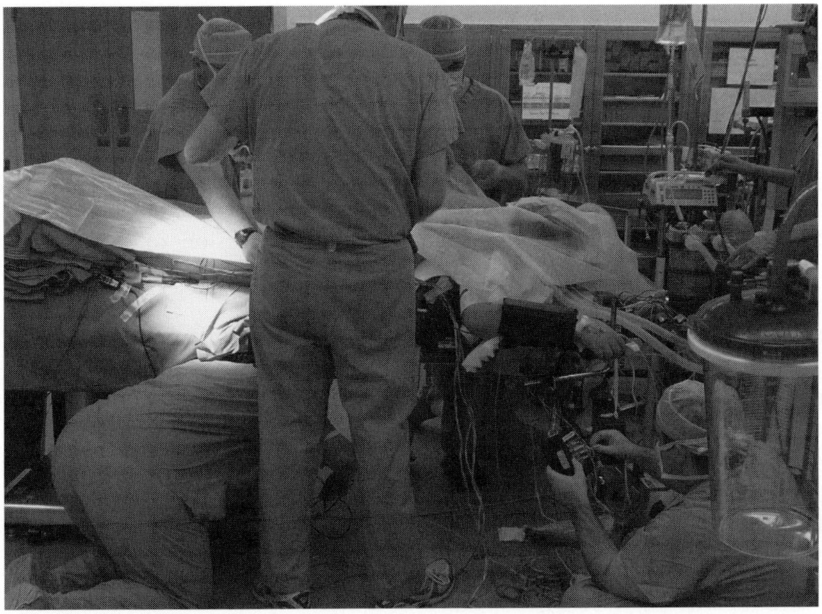

Figure 4.2 The availability of more than one technologist allows for better efficiency in setting up the patient more quickly while the surgeon prepares the patient for the surgery, therefore reducing any delay after the induction of anesthesia.

Technological advances in the quality, speed, and bandwidth of the Internet allow Web-based telemedicine to be used by some NIOM programs to provide services to off-site facilities.

While not optimal, this may allow availability of essential NIOM service (Kinney and Slimp, 2007). Experience, level of training, and certification are all important requirements for the clinical neurophysiologist interpreting NIOM. In a recent survey conducted at the American Society of Neurophysiological Monitoring meeting (Stecker and Robertshaw, 2006), it was found that surgeons were more likely to respond to warnings issued by a neurophysiologist with a higher degree of experience.

While in general the degree of agreement between neurophysiologists was at most moderate, respondents with the higher levels of experience and certification were more likely to agree and were more likely to provide correct interpretations. An earlier study by Nuwer (Nuwer *et al.*, 1995) surveyed 173 orthopedic spine surgeons and found that the rate of neurologic injury was more than twice as high when the NIOM was provided by teams with lesser experience (100 or fewer cases) as opposed to teams with higher experience (having performed more than 300 cases).

Practical experience helps in avoiding rigid application of commonly accepted or published criteria. In general, more than 50% reduction in the amplitude, or more than 10% latency prolongation of somatosensory evoked potentials (SEPs) is significant. However, other factors, such as the initial quality of the waveforms, the rate and speed of change in these signals, and changes in anesthetic medications and hemodynamic status of the patient need to be considered.

In such cases, the use of a different modality, such as transcranial electric stimulation (TES) to record motor-evoked potentials (MEPs) can be very helpful in establishing whether the SEP changes reflect an actual tissue injury. In general, the use of multimodal NIOM is invaluable and should be attempted in all cases. For example, when monitoring spinal cord surgery cases, monitoring SEPs alone is not reliable in detecting injury to the descending motor pathways (Lesser *et al.*, 1986).

It is now clear that MEP changes may precede SEP changes or may occur without any change in SEPs (Machida *et al.*, 1988; Aminoff, 1989; Zornow *et al.*, 1990). Therefore, NIOM for spinal cord surgeries should include monitoring MEPs and SEPs.

Anesthesia

Our anesthesia colleagues are often asked to make appropriate changes in order to optimize the recorded waveforms for the different monitoring modalities. The anesthesiology team should be made aware of the NIOM modalities used during a surgery. For example, knowing that a modality such as transcranial electrical stimulation (MEPs) is highly sensitive to the effect of anesthetic agents is essential. Motor evoked potentials can be abolished with partial neuromuscular blockade and halogenated anesthetics. The use of intravenous agents, such as propofol, can be very helpful in maintaining stable MEPs throughout surgery (Sekimoto *et al.*, 2006).

Special anesthetic protocols using total intravenous anesthesia (TIVA) have been widely adopted and have reduced the chances of unsuccessful NIOM. Variations in the amplitudes of recorded MEPs can be minimized with tight control of the neuromuscular blockade with steady continuous infusion rather than bolus administration (Sloan and Heyer, 2002). Similarly, somatosensory evoked potentials (SEPs) are usually affected by most anesthetics, certain agents, such as etomidate and ketamine, can actually enhance the amplitudes of the recorded SEPs.

Knowledge of the different types of anesthetic agents, as well as their effect on the nervous system, is obviously essential for NIOM. The neurophysiologist performing NIOM should work with the anesthesiologist to optimize anesthetic medication selection.

Communication with the anesthesiologist is critical to correlate any interval change in the neuroelectric data with interventions by the anesthesia team. This includes feedback on the use of drugs, as well as feedback on the hemodynamic status of the patient, which can affect perfusion to the nervous system tissues and lead to delayed conduction in neuroanatomic pathways.

It is well understood that the operating room environment is busy and it is helpful for the monitoring team to independently obtain feedback. For example, most technologists performing NIOM learn how to read the bedside monitors, to note and log the patient's vital signs periodically throughout the surgery. The use of closed-circuit TV monitors recording from the surgical field allow the team to follow the different surgical steps and anticipate critical interventions, such as distraction or rod placement in scoliosis surgeries (Figure 4.3).

Overall, the monitoring team should develop a method to facilitate communication with the anesthesia team. All members involved in the care of the NIOM case should have a mutual understanding of each other's activities so that false assumptions and subsequently wrong interpretations and warnings are avoided.

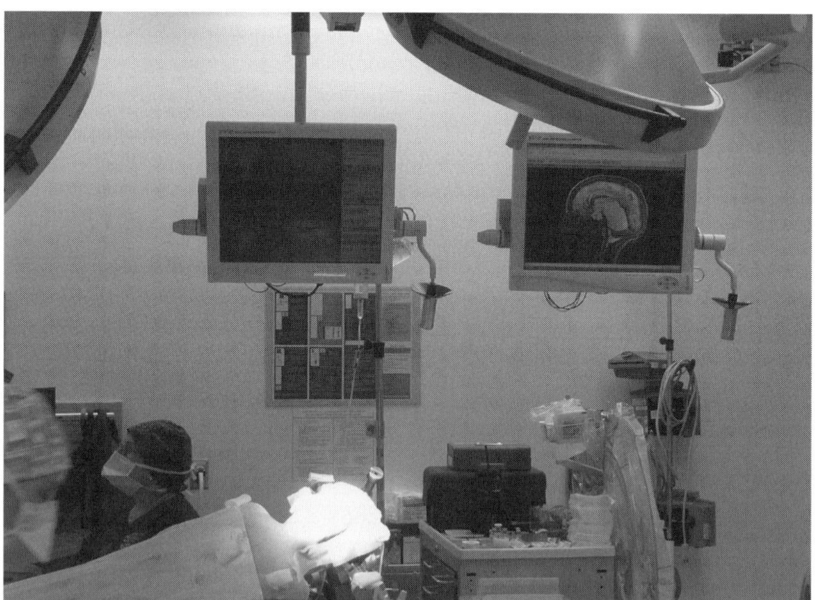

Figure 4.3 Overhead closed-circuit TV monitors recording directly from the surgical field are helpful to display and follow the different surgical steps.

A detailed discussion of the basic pharmacology of anesthetic agents and their effects on neurophysiological monitoring is provided in Chapter 5.

NIOM equipment

Over the past two decades, cumulative experience and advancement in clinical neurophysiology, paralleled by major and rapid advances in computer technology and electronics, have largely influenced the production of NIOM equipment, which is remarkably better in quality, reliability, versatility, and storage capacity. Earlier, most centers that provided NIOM depended on equipment often built by the centers' own engineers and neurophysiologists. Today, all NIOM equipment sold uses computer-based programs and is thus capable of automatically storing as well as displaying the data in many different formats. Better filters and more stable amplifiers have now improved the quality of the recorded potentials. Data can also be obtained and displayed simultaneously from several stimulators, allowing for interleaved responses rather than waiting for an entire set of averaging to be completed before the next stimulator is activated. All these features have therefore reduced the time needed to obtain an interpretable record to allow more rapid feedback to the surgeon. The minimum technical requirements of NIOM equipment have recently been published by the ACNS (see https://www.acns.org/guidelines.cfm). Today's machines are also more compact in size and many are laptop-based making it easier to fit them into a crowded operative suite. This reduction in size did not come at the expense of efficiency, capacity, or speed. In contrast, modern NIOM equipment has faster computer processors with signal processors that are able to display up to 64 channels. This allows for simultaneous stimulation and recording of several modalities. Special attention should be given to the amplifiers, which can pick up noises from numerous sources of electrical output in the operating room. Too much noise directly translates into excessive time needed for averaging responses (i.e., a reduced signal-to-noise ratio), which in turn prolongs the feedback time to the surgeon. If the source of the electrical interference cannot be eliminated, moving the amplifier to a different location may eliminate the noise. In addition to knowledge regarding the equipment, the NIOM team must be knowledgeable about other equipment in the operating room which may produce artifacts and interfere with recorded signals. It is also important to plan, prior to surgery, the location of the NIOM equipment in the operating room. Depending on the type of surgery, the NIOM technologist will select an optimum location to minimize interference with the surgeon's and anesthesiologist's space without compromising the service provided by the NIOM team. In some surgeries a position close to the surgeon is essential for certain procedures, as with direct stimulation of neural structures, such as in selective dorsal rhizotomy or release of tethered cord surgeries (Figure 4.4). On the other hand, monitoring other surgeries may not require direct interaction with the surgeon and the NIOM equipment can be far from the operating table, allowing more space for the surgeon and the anesthesiologist (Figure 4.5). Most manufacturers of NIOM equipment also provide software that allows for real-time on-line communication between the technologists and the supervising neurophysiologist, thus improving the ability to respond in a timely fashion. Standards for transferring digital neurophysiological data between computer systems have recently been published by the ACNS (2008) (https://www.acns.org/pdfs/3.03_acns-ts1_2008-0306.pdf).

The surgeon

The monitoring procedure should be selected in advance, in consultation with the surgeon. Neurophysiologic monitoring is considered based on the surgeon's assessment of the risk of neurologic injury from a given operative procedure. The NIOM team can then decide if the neurologic structure at risk can be monitored using one or more of the known NIOM modalities.

There is no one technique that is optimal for every operation, therefore it is crucial to have communication with the surgeon, regarding anatomic sites involved and the extent of potential damage to any neurologic pathways. It is also important to discuss in advance any limitations to the monitoring techniques, so that the surgeon's expectations from the NIOM are realistic and feasible. On the other hand, learning in advance about the surgical technique to be used allows the NIOM team to understand the limitations in accessing parts of the neural system and to plan in advance which modalities are best suited.

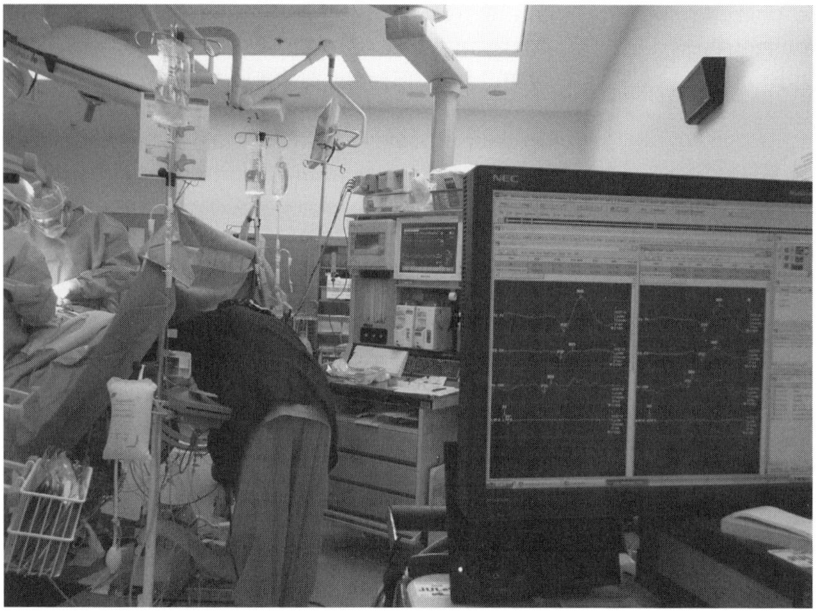

Figure 4.4 The neurophysiologic monitoring equipment is placed near the operating table. This facilitates handling of stimulating electrodes and direct interaction with the surgeon during scoliosis surgery when monitoring of stimulus-triggered electromyography for pedicle screws placement.

Figure 4.5 Monitoring equipment placed away from the operating table in an attached room, eliminating any interference with the surgeon's field. This facilitates a posterior fossa surgery, involving monitoring of brainstem auditory and somatosensory evoked potentials.

It is important to recognize that experience with using NIOM varies widely among surgeons. Occasionally senior surgeons have had little exposure to NIOM and might seldom request IOM for their patients. On the other hand, other surgeons introduced to IOM should be well informed that false-positives are common and that not all neurophysiologic changes are due to the surgical intervention.

Despite the multitude of studies that have demonstrated the value of different NIOM modalities in reducing operative morbidity, evidence-based guidelines are lacking. There is debate as to what value should be used as the warning threshold above which the surgeon needs to be notified.

In most cases, the interpretation criteria should vary, depending on the type of surgery and the underlying neurologic function of the patient. The decision to alert the surgeon should be balanced between avoiding false-positives and delaying a warning. If gradual deterioration of the recorded responses is noted but has not yet reached an already set warning level, it is unacceptable to wait until the threshold value has been reached before informing the surgeon.

A surgeon who understands the basics and limitations of NIOM techniques will be more tolerant and understanding of such warnings even if they are repeated during the surgery. Such a surgeon will appreciate the early warning, as it would provide an opportunity to correlate the changes with an alteration in the operative procedure. On the other hand, experienced NIOM teams are able to reduce the number of false-positives by eliminating technical factors.

The operating room

Recording evoked potentials or electromyography in the outpatient setting often involves shielded rooms and careful planning for the installation and operation of the equipment. It is very challenging to obtain these recordings in the operating room, using portable equipment in an environment that is electrically hostile and crowded, which can interfere with the EP equipment. In addition to the careful selection of optimal parameters for each intended EP modality, sources of electrical or magnetic interference need to be eliminated. The numerous sources of electrical interference in the operating room include but are not limited to electrocautery, surgical drills, cavitron ultrasonic surgical aspirator (CUSA), cardiopulmonary bypass machines, electric operating room table, C-arm fluoroscopes, overhead lighting, surgical microscope, monitors, intravenous pumps, blood warmers, and patient warmers.

The EP equipment has to be checked for proper grounding. Electrical interference can be removed either by replacing faulty equipment or moving it away from the EP equipment. If the interference persists, it might be helpful to braid both inputs to the differential amplifier so that the interfering artifacts would be canceled out by common-mode rejection of the differential amplifier.

Magnetic fields in the operating room, such as those produced by intraoperative MRI devices, can interact with the recording or stimulating metal electrodes, potentially inducing motion, heat, or electric current. This can lead to significant artifacts as well as skin burns from overheating. The availability of MRI-compatible-electrodes could now permit the combined use of intraoperative low-field open MRI and NIOM (Szelenyi *et al.*, 2008).

Safety in the operating room

The importance of safety in the operating room cannot be overemphasized. Members of the NIOM team should follow safe measures to prevent the risk of electrical shock to the patient as well as to themselves and other personnel in the operating room. Electrical shock can lead to burns, localized damage to the affected tissues, or even death from cardiac arrest.

All NIOM equipment should comply with the International Electrotechnical Commission standards (International Electrotechnical Commission, 2009). After placement of all electrodes, the technologist has to screen and secure the usually large number of wires and cables to prevent physical injury. Wires should be taped together under or on the side of the bed and checked to ensure that they are not in the path of the surgical or anesthesia teams or other movable equipment during the surgery. Careful attention should be paid to the electric power cord, so that it is not eroded or damaged.

Protocols should be developed to control the amount of electrical stimulation given to the patient's peripheral or central nervous system. Care should be given not to use excessive electric current, which could lead to thermal injury. The technologist should assist the surgical nurse in assuring that the ground electrode of the electric cautery is properly placed and secured. Improper grounding leads to shunting of the electrical current through a different route, usually the recording electrodes. This can lead to overheating and burns in the skin or other tissues. Exit points or electrodes with a smaller surface area are subjected to more heating and burns. This is more likely to occur when using needle electrodes.

The use of transcranial electric stimulation in the operating room demands further consideration regarding patient safety due to the risk of scalp burns (Parikh *et al.*, 2003), bite injuries, or the possibility of causing seizures (MacDonald, 2002). Finally, special care of the NIOM equipment is necessary, if exposed to excessive use. Inspection and maintenance should be performed at least every six months or as directed by the manufacturer (International Electrotechnical Commission, 2009).

Questions

1. According to the American Society of Electroneurodiagnostic Technologists (ASET), it is highly recommended that all neurophysiologic intraoperative monitoring (NIOM) be attended by a certified technologist who should not monitor more than one surgery at the same time.
   (a) True
   (b) False
2. When recording brain activity directly from the cortical surface, the American Society of Electroneurodiagnostic Technologists (ASET) recommends that:
   (a) The monitoring technologist monitors and interprets the results to the surgeon
   (b) A clinical neurophysiologist does not necessarily need to be present in the operating room as the responses can be monitored remotely
   (c) If the clinical neurophysiologist cannot be present then the surgeon or the anesthesiologist assume responsibility in interpreting the waveforms
   (d) None of the above

References

Aminoff, M. J. (1989). Intraoperative monitoring by evoked potentials for spinal-cord surgery – the cons. *Electroencephalography and Clinical Neurophysiology*, **73**, 378–380.

American Society of Electroneurodiagnostic Technologists (ASET) (1998). *Electroneurodiagnostic Technologists in the Operating Room*, Position statement. www.aset.org.

International Electrotechnical Commission (IEC) (2009). *Medical Electrical Equipment: Particular Requirements for the Basic Safety and Essential Performance of High Frequency Surgical Equipment and High Frequency Surgical Accessories*. International Electrotechnical Commission [IEC].

Kinney, G. A. and Slimp, J. C. (2007). Intraoperative neurophysiological monitoring technology: recent advances and evolving uses. *Expert Review of Medical Devices*, **4**(1), 7, 33–41.

Lesser, R. P., Raudzens, P., Luders, H., *et al.* (1986). Postoperative neurological deficits may occur despite unchanged intraoperative somatosensory evoked potentials. *Annals of Neurology*, **19**, 22–25.

MacDonald, D. B. (2002). Safety of intraoperative transcranial electrical stimulation motor evoked potential monitoring. *Journal of Clinical Neurophysiology*, **19**, 416–429.

Machida, M., Weinstein, S. L., Yamada, T., Kimura, J., and Toriyama, S. (1988). Dissociation of muscle action potentials and spinal somatosensory evoked potentials after ischemic damage of spinal-cord. *Spine*, **13**, 1119–1124.

Nuwer, M. R. (2008). Intraoperative monitoring of the spinal cord. *Clinical Neurophysiology*, **119**, 247.

Nuwer, J. M. and Nuwer, M. R. (1997). Neurophysiologic surgical monitoring staffing patterns in the USA. *Electroencephalography and Clinical Neurophysiology*, **103**, 616–620.

Nuwer, M. R., Dawson, E. G., Carlson, L. G., Kanim, L. E., and Sherman, J. E. (1995) Somatosensory evoked potential spinal cord monitoring reduces neurologic deficits after scoliosis surgery: results of a large multicenter survey. *Electroencephalography and Clinical Neurophysiology*, **96**, 6–11.

Parikh, S. N., Mehlman, C. T., and Keith, R. W. (2003). A third-degree burn caused by a neurogenic motor-evoked potential monitoring electrode during spinal surgery: a case report. *Spine*, **28**(1), E21–24.

Sekimoto, K., Nishikawa, K., Ishizeki, J., Kubo, K., Saito, S., and Goto, F. (2006). The effects of volatile anesthetics on intraoperative monitoring of myogenic motor-evoked potentials to transcranial electrical stimulation and on partial neuromuscular blockade during propofol/fentanyl/nitrous oxide anesthesia in humans. *Journal of Neurosurgical Anesthesiology*, **18**, 106–111.

Sloan, T. B. and Heyer, E. J. (2002). Anesthesia for intraoperative neurophysiologic monitoring of the spinal cord. *Journal of Clinical Neurophysiology*, **19**, 430–443.

Stecker, M. M. and Robertshaw, J. (2006). Factors affecting reliability of interpretations of intra-operative evoked potentials. *Journal of Clinical Monitoring and Computing*, **20**, 47–55.

Szelenyi, A., Gasser, T., and Seifert, V. (2008). Intraoperative neurophysiological monitoring in an open low-field magnetic resonance imaging system: clinical experience and technical considerations. *Neurosurgery*, **63**, 268–275.

Zornow, M. H., Grafe, M. R., Tybor, C., and Swenson, M. R. (1990). Preservation of evoked potentials in a case of anterior spinal artery syndrome. *Electroencephalography and Clinical Neurophysiology*, **77**, 137–139.

5 Basic pharmacology of anesthetic agents and their effects on neurophysiological monitoring

Gloria M. Galloway

The role of the anesthesiology staff during surgical cases in which neurophysiologic monitoring (NIOM) occurs is complex. This includes the ability to administer anesthesia effectively and safely while maintaining and manipulating blood pressure, minimizing increases in intracranial pressure and facilitating rapid emergence from anesthesia, and appropriate neurological assessment. At the same time, careful attention must be paid to the anesthetic agents used so that stability and reliability of continuous NIOM can occur.

It is not surprising then that this balance is difficult at times to maintain and is a source of difficulty in the operating room. Therefore, successful neurophysiologic intraoperative monitoring requires a team approach, which includes use of those general anesthetic agents that do not significantly affect the electrophysiological signals (Iwasaki *et al.*, 2003) so that accurate and reliable analysis can occur. The team approach includes regular constructive communication between members of the anesthesiology, surgical, and NIOM teams.

The appropriate choice of anesthetic agent depends on many factors, including the patient's underlying medical condition, the type of surgical procedure, and subsequent NIOM needed. Anesthetic agents, either intravenous or halogenated, are thought to have their primary mode of action at the γ-amino butyric acid and *N*-methyl-D-aspartic acid receptors. These receptors mediate electrolyte channels, affecting synaptic transmission involved in anesthesia.

The effect of anesthetic agents on the motor pathway is the cause of much consideration and active involvement during NIOM. Transcortical motor stimulation involves direct activation of pyramidal cells, producing D or direct waves and a series of I, or indirect, waves from interneuronal connections, probably in lamina V, and the premotor cortex. Once a descending motor volley occurs, stimulation of the alpha motor neuron causes a peripheral response to occur. Anesthetic agents primarily express their effects on the alpha motor neuron through these synaptic connections and can have a significant inhibitory impact on the corticospinal tract alpha motor neuron.

Anesthetic agents exert their significant effects on the motor pathway through several different synaptic connections, including interneuronal synapses, synapses at the alpha motor neuron, and synapses at the neuromuscular junction. The D response, which is recorded from the epidural space, is resistant to the effects of most anesthetic agents, including the halogenated agents even at high concentrations, and thus can be recorded easily. Magnetic stimulation of the motor cortex involves transsynaptic activation and is especially sensitive to the effects of inhalational anesthetics, making their use in the operative setting difficult. Using higher magnetic stimulation field strengths or avoiding inhalational anesthetics, particularly at high concentrations, may allow this type of stimulation to take place.

Since, in the operating room, the conditions are not routinely present for reliable and effective magnetic stimulation to occur, for practical purposes electrical stimulation is the preferred method of transcranial stimulation used intraoperatively. The effects of anesthetic agents on cortical stimulation are lessened by using multipulse motor stimulation of the motor cortex. At very high anesthetic concentrations, an anesthetic depressive effect may be more obvious and adjustments in stimulation parameters, including inter-stimulus interval (ISI) and the number of stimulations that are in the train, may need to be made to produce the desired effect. The inter-stimulus interval should be set so that the cortical neuronal refractory period is completed and is not so long as to negate the effect of summation of the train of stimuli.

Table 5.1 Agents used in the stages of anesthesia

Stage	Premedication	Induction	Maintenance	Emergence
Agent	Intravenous benzodiazepine	Intravenous sedative hypnotic	Intravenous sedative hypnotic	Oxygenation, hyperventilation
	Narcotic	Intravenous opioid	Intravenous opioid	
	+/− anti-emetic, +/− antihypertensive agent	Paralytic agent	Paralytic agent	Reversal of paralytic agent

Studies have shown, particularly with inhalational agents, that the ISI requires much adjustment to enhance recorded responses during monitoring (Hu *et al.*, 2001).

Understanding the anesthetic experience of the patient during a given procedure is important in gaining insight into the effects of surgical manipulations and anesthetic changes on systemic variables and NIOM data. A patient under general anesthesia is described as passing through four primary anesthetic stages. Various anesthetic agents guide the smooth transition from one stage to the next. The first stage involves *premedication*, which allows for the relief of anxiety. The second stage is *induction* of anesthesia and this is done with the use of a bolus dose of intravenous sedative and hypnotic agent, and often an intravenous opioid and neuromuscular relaxant or paralytic agent. The third stage is *maintenance* of anesthesia and typically constitutes the anesthetic state during the majority of the surgical procedure. The last stage is that of *emergence* from anesthesia, and the goal of this state is that it occurs as an easy transition without complication and is typically preceded by limitation of the volatile agent used through breathing and wearing off or reversing of the paralytic agent.

NIOM is typically done after induction of anesthesia and completed prior to emergence of anesthesia. Table 5.1 outlines the various anesthetic agents typically utilized during various stages of anesthesia.

Premedication in the preoperative period typically includes agents such as intravenous benzodiazepines, including midazolam at 1–2 mg, which can lessen anxiety and allow preparation for endotracheal intubation and reduction of pain. Alternatively, premedication can include a narcotic agent such as morphine, at 1–4 mg, or fentanyl, at 25–50 micrograms. In addition, in some cases antihypertensive agents and anti-emetics are also used, depending on the particular patient during this preoperative period.

After peripheral intravenous catheter insertion and preoxygenation has been completed, induction of anesthesia occurs and requires administration of an intravenous sedative hypnotic, often along with a paralytic agent and opioid to allow endotracheal intubation. Sometimes preoxygenation is followed by intravenous induction anesthesia with a sedative hypnotic, such as a barbiturate, and a muscle relaxant for intubation, such as rocuronium. Rocuronium is both fast acting and non-depolarizing in nature. Control of ventilation occurs to maintain oxygenation and normocapnia often using 30% oxygen in an air mixture.

Pulse oximetry, ECG, arterial BP, capnography, nasopharyngeal temperature, and muscle twitch responses are routinely monitored during these surgeries. If potentials are monitored during the period of induction this would reveal smaller amplitude cortical potentials than those seen later in the case and therefore it is best to wait to obtain baselines after induction is complete. Scouting of potentials is routinely done prior to obtaining baselines quickly, often during induction. It is also during this period of induction that temperature decreases often occur again, contributing to the decrease in EP amplitude that would be seen. Although core body temperature is being monitored, the distal extremity will be lower in temperature than the core temperature readout indicates. Warming devices are often used called Baer huggers. Warmed intravenous fluids may also be used to warm the patient.

To maintain anesthesia, ongoing sedative hypnotic agents along with a volatile anesthetic or nitrous oxide may be used in combination with an opioid agent. Controlled infusions of these agents, rather than a bolus, will cause less anesthetic effect on evoked potential signals and are preferred. Either no or limited amounts of paralytic agents should be used while motor potentials are being evaluated. After the surgical procedure, emergence from anesthesia occurs with reversal of the neuromuscular paralytic agent and an

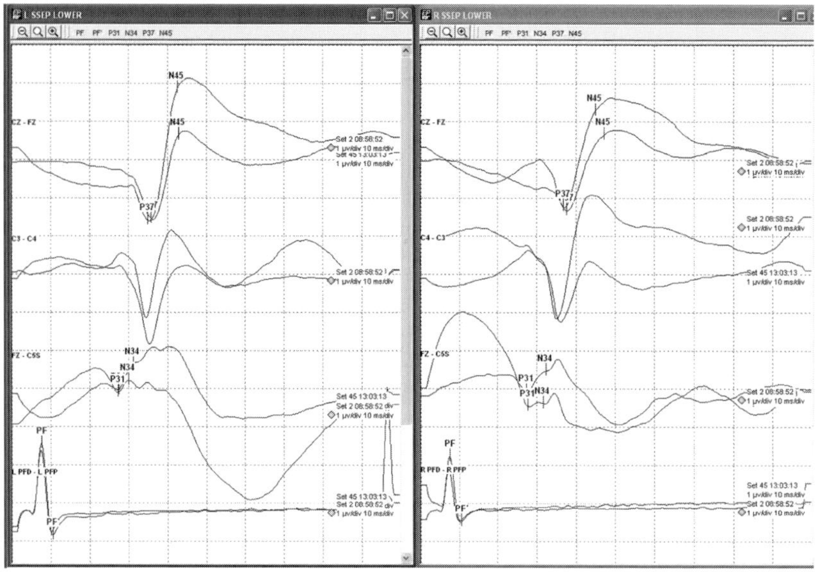

Figure 5.1 Decrease in SEP amplitude in subcortical (P31–N34) and cortical (P37–N45) potentials as sevoflurane anesthetic was increased. This improved when anesthetic levels were reduced.

increase in ventilation, which allows for removal of the inhalational agent.

Inhalational agents commonly used include halogenated agents and nitrous oxide. The common halogenated agents include halothane, isoflurane, nitrous oxide, desflurane, and sevoflurane, and their level of administration is described as a percentage of the mean alveolar concentration (MAC). This is defined as the alveolar partial pressure of a gas at which 50% of patients will not move with a 1 cm surgical incision to the abdomen. Both halogenated agents and nitrous oxide cause decreases in the amplitude of somatosensory (SEP) as well as transcranial electrical motor (MEP) waveforms and prolongation of latency of SEP signals, which is greatest at the cortical level and least at the subcortical or peripheral levels.

In Figure 5.1, anesthetic effects of sevoflurane are shown on subcortical and cortical potentials in a patient undergoing scoliosis repair, illustrating a decrease in the amplitude of the SEPs. Figure 5.2 shows that as the sevoflurane level is further increased the MEPs are lost from all limbs. In direct correlation, halogenated anesthetic agents cause a decrease in cerebral perfusion pressure. When cerebral blood flow falls below 15 (ml/min)/100 g, SEP potentials may be lost altogether. Figure 5.3 illustrates the effect of a reduction in mean arterial pressure on the TcMEPs, with a loss of MEPs in the lower extremities during a spine surgery. Improvement in these potentials is seen in Figure 5.4, with return of the lower extremity MEPS after the arterial pressure is corrected. Therefore, changes in MAP (mean arterial pressure) may enhance the negative effect on evoked potential responses seen with halogenated anesthetic agents.

These halogenated anesthetic agents may also cause prolongation of latencies of brainstem auditory evoked potential (BAEP) signals at higher concentrations. With these agents, the EEG predominant rhythm is shifted frontally and the amplitude and frequency of the rhythm is decreased. If possible, attempts should be made to keep the MAC below 0.5 for successful NIOM, although absolute numbers are not as significant as the effect of that agent in the particular patient being monitored. In this regard, if the use of a halogenated agent cannot be avoided during surgeries in which NIOM is undertaken, then the level of agent used needs to be titrated to the patient need and its effect on the ability to interpret continuous electrophysiological data effectively and reliably with NIOM. A combination of halogenated anesthetic agents and nitrous oxide would be expected to have a summated attenuating effect on evoked potential amplitudes and latencies.

Isoflurane, *sevoflurane*, and *desflurane* appear to be more potent anesthetic depressants of the electrophysiological signal than halothane. Isoflurane increases cerebral blood flow to a minor degree but increases intracranial pressure and may need to be avoided in patients with intracranial lesions and reduced cerebral vascular compliance. Sevoflurane and desflurane may have a more rapid onset and offset of action than isoflurane but may be more potent during increasing

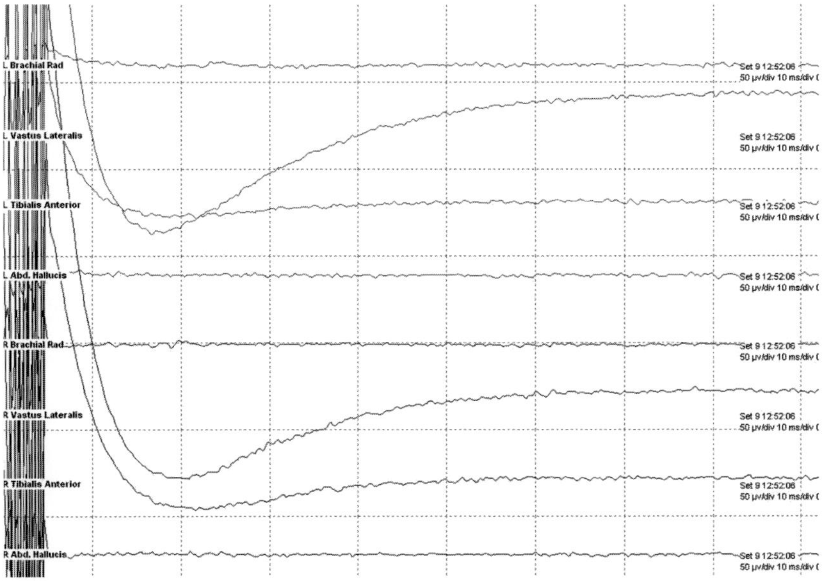

Figure 5.2 Effect of increasing the sevoflurane level to 2.5%, showing loss of all previously recorded MEPs. For color version, see plate section.

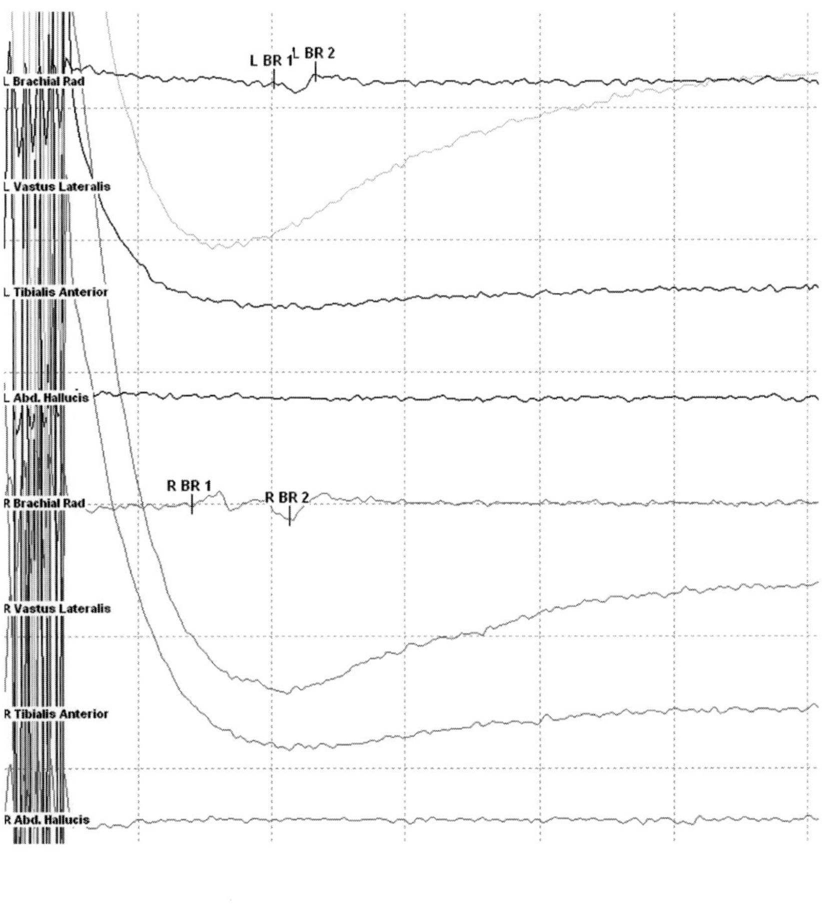

Figure 5.3 Loss of lower extremity MEPs with decrease of MAP during scoliosis repair. For color version, see plate section.

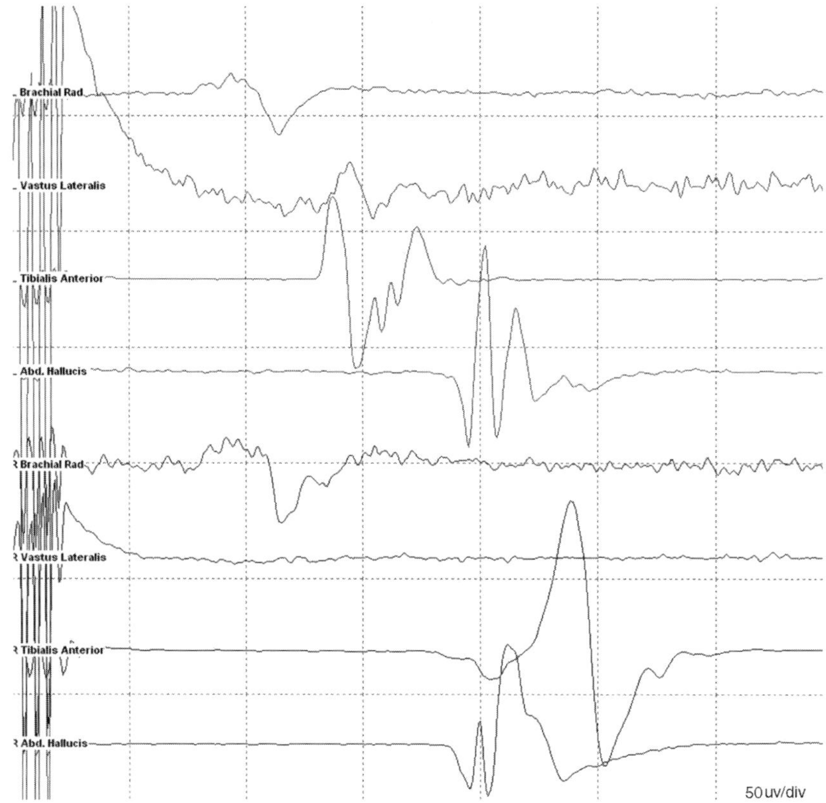

Figure 5.4 Return of lower extremity MEPs, although small in amplitude, after the MAP was increased in the patient in Figure 5.3. For color version, see plate section.

concentration periods, such as during titration, and can therefore confuse interpretation in NIOM.

In a study by Boisseau and others (2002) the effects of sevoflurane and propofol on somatosensory evoked potentials were compared in 24 patients by random allocation during spine surgery. Sevoflurane demonstrated a dose-dependent depression on the SEP amplitude and latency measures. Propofol did not have a significant effect on the recording signals. Figure 5.2 illustrates the effect of sevoflurane on motor-evoked potentials, with the loss of all previously recorded potentials.

In contrast, ketamine and etomidate cause the amplitude of SEP signals to increase and therefore in some patients may be useful to enhance those potentials, which are difficult to record. However, the use of ketamine or etomidate can confuse the interpretation of neurophysiologic monitoring signals if it is added later on during the operation. A recent review of ketamine in transcranial electrical motor stimulation in children suggested that it may be a very useful agent (Zaarour *et al.*, 2007). This study suggested that the use of ketamine allowed reliable motor-evoked potentials to transcranial electrical stimulation to be obtained in children without significant side effect or complication.

Although the ability to obtain adequate cortical motor potentials with single-pulse stimulation in the presence of halogenated anesthetic agents is problematic, this can often be overcome with multipulse cortical stimulation and the prudent use of anesthetic agents (Nathan *et al.*, 2003). Figure 5.5 shows an example of an anesthetic monitoring screen used by the anesthesiology team to review systemic and anesthetic variables and data trends during surgery. Following the data trends will allow one to see the effects on systemic variables that occur with changes to the anesthetic regimen. Using blood pressure agents to control blood pressure would be recommended over using increased doses of anesthetic agent, owing to the effect on NIOM.

Some studies have suggested that the use of remifentanyl may infrequently create an acute tolerance to opioids postoperatively (Anschel *et al.*, 2008) in both children and adults. More work on this is needed, to know how frequently or how consistently this occurs. Ketamine has a significant analgesic effect that may be

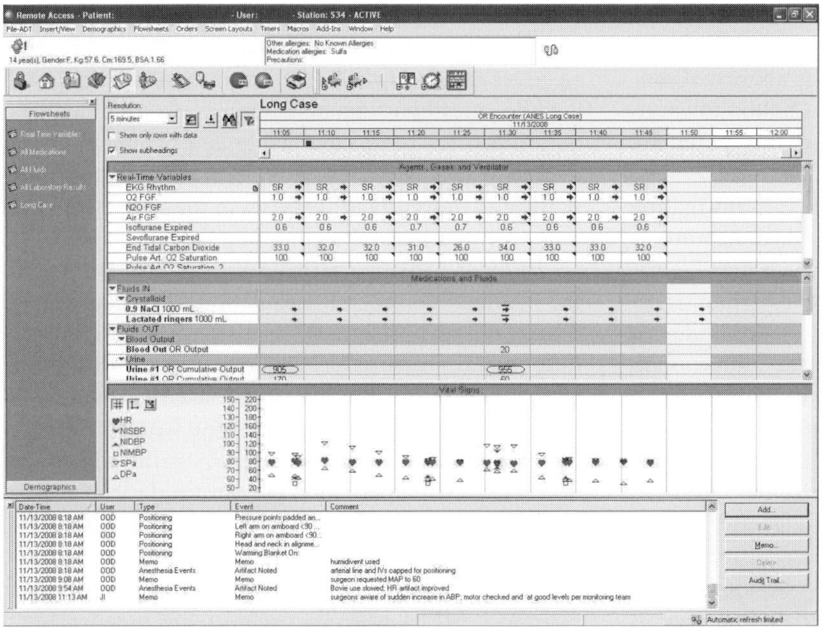

Figure 5.5 Example of an equipment screen utilized by the anesthesia team to review systemic and anesthetic variables intraoperatively.

useful in the postoperative setting. It is used for acute postoperative pain after other surgical procedures. It does, however, cause dose-related sympathetomimetic and cardiovascular excitatory effects.

Neuromuscular blocking paralytic agents are used in limited doses during NIOM that involves MEPs or electromyographic monitoring. Typically these agents are not used at all or in small amounts to allow for two to four twitches in a train of four (TOF). There is much patient variability in the amount of muscle activity needed to record adequate MEPs and so titration during the operative procedure is done by measuring the muscle responses after the patient is given four 2 Hz peripheral stimuli or trains of four. Vecuronium is a non-depolarizing muscle relaxant and does not increase intracranial pressure as opposed to succinylcholine. In addition, vecuronium has less effect on blood pressure, heart rate and ultimately on NIOM and is therefore the muscle relaxant most often used in small doses when monitoring is needed (Ben-David *et al.*, 1987).

In selective dorsal rhizotomies (SDRs), muscle relaxants need to be avoided so that clinical and electrophysiological muscle responses occurring as a result of dorsal root stimulation are accurately interpreted. These surgeries typically involve children with severe spasticity and an exaggerated muscle response, referred to as a brisk hyperactive response (BHR), may occur. Anesthetic agents used in these surgeries must control for the BHR response while not reducing or limiting the muscle response to dorsal root stimulation.

A small study by Konya and colleagues (2009) suggests that the use of sevoflurane may have benefit in selective dorsal rhizotomies over the use of isoflurane in its influence on the BHR response. Premedication with an anxiolytic agent, such as midazolam, along with atropine to control secretions, is typically done. The anesthesia is induced with either total intravenous anesthesia (TIVA) or nitrous oxide in combination with a small dose of muscle relaxant, such as vecuronium, prior to the rhizotomy procedure. The use of muscle relaxant is typically avoided after induction of anesthesia. As might be expected from its effect on enhancing MEP potential amplitudes, the use of ketamine during rhizotomies, particularly prior to dorsal root stimulation, may be associated with an unfavorable excitatory effect, confusing interpretation.

A study by Frigon and coworkers (2008) looked at the use of low-dose ketamine in selective dorsal rhizotomy procedures and found that its use was associated with a lower maximal threshold for stimulation of the dorsal root and also with a more sustained response to a train of stimulation. The threshold of the response and the response to a train of stimulation can be used in determining whether muscle responses to dorsal root stimulation are normal or not. Therefore, the use of ketamine in this situation might conceivably allow misinterpretation of the responses and potentially inadvertent severing of dorsal rootlets, which may not have been classified as abnormal outside of ketamine use. Frigon's study (2008), however, was a retrospective cohort study so this potential, although conceivable, was not determined. The authors noted a greater difficulty in discriminating contralateral EMG patterns with the use of ketamine.

Magnetic cortical stimulation is much more sensitive to the effects of inhalational anesthesia than transcranial electrical stimulation. This is probably due to transsynaptic activation of the cortex rather than direct stimulation and production of indirect or I waves, which are much more susceptible to anesthetic depression. In children, inhalational anesthetic agents affect mean arterial pressure, intracranial pressure, and cerebral perfusion pressure to a greater degree and in children with known increased intracranial pressure, intravenous anesthesia may be more desirable (Gibson, 2004). In all patients, inhalational agents can increase intracranial pressure through vasodilatation of the cerebral vasculature. All inhalational agents except nitrous oxide will depress cerebral metabolism.

In the neonate (first 30 days of life) cerebral autoregulation of blood flow may be impaired and this autoregulation of blood flow is very much dependent on the end tidal carbon dioxide level. In turn, cerebral blood flow and cerebral blood volume directly affect intracranial pressure. In pediatric patients there is often a decreased anesthetic requirement in order to achieve sedation, in addition to an increased risk of hemodynamic instability due to organ immaturity. Induction of anesthesia is more rapid in neonates and emergence from anesthesia more delayed than in an older child or adult. This is particularly true with volatile anesthetic agents. Surgery in the neonatal period is often done emergently and neonatal oxygen demands are higher. There is, therefore, greater morbidity and mortality in surgeries in the neonatal period than in surgeries in the older population.

In older pediatric patients and in patients beyond the neonatal period, autoregulation of cerebral blood flow is not as problematic. In patients with increased intracranial pressure, premedication prior to the surgical case without the use of the sedatives will avoid increased cerebral vasodilatation and worsening intracranial pressure. Inhalational anesthetic effects can decrease cerebral perfusion pressure and mean arterial pressure and increase intracranial pressure at 0.5 MAC or greater and this is more pronounced in the pediatric patient. In Figures 5.3 and 5.4, one sees loss of lower extremity MEPs with decrease in the mean arterial pressure during a scoliosis repair and subsequent return of these potentials after the pressure was raised.

Intravenous anesthetic agents are typically used during induction of and maintenance of anesthesia along with an intravenous opioid and are sometimes used to supplement anesthesia when a halogenated agent is used. Total intravenous anesthesia (TIVA) is increasingly being used instead of anesthesia with halogenated agents when neurophysiologic monitoring is needed (Scheufler and Zentner, 2002). These agents may include propofol, ketamine, midazolam, dexmedetomidine, or, less commonly, a barbiturate, such as thiopental or

Table 5.2 Effect of varying agents on neurophysiologic signals

Agent	SEPs	MEPs	BAER	EMG
Halogenated agent	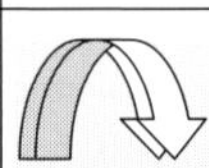	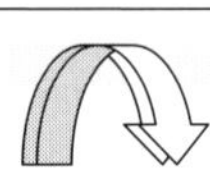	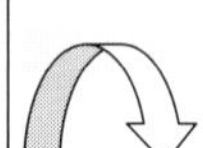	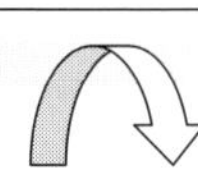
Opioid	—	—	—	—
Paralytic agent	—	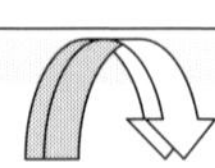	—	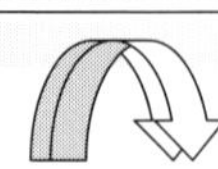
Intravenous anesthetic-high dose	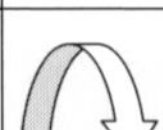	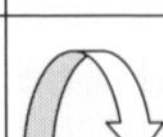	—	—

pentobarbital. Propofol is the most commonly used agent in TIVA, owing to the ease of titration at onset and ability to terminate its use at the end of the procedure (Chen, 2004). The level of propofol infusion needed varies greatly and may be as low as 20 (μg/kg)/min or may be as high as 100 (μg/kg)/min or greater (Sekimoto *et al.*, 2006). Table 5.2 illustrates the effect of varying anesthetic agents on monitoring modalities, including SEPs, MEPs, BAER, and EMG potentials.

In contrast, barbiturates are less often used in surgical cases involving neurophysiologic monitoring, owing to their profound negative effect on the ability to record MEPs adequately. However during intravenous induction of anesthesia, the use of thiopental can assist in allowing loss of consciousness and cardiovascular stability at doses of 3–5 mg/kg (Kawaguchi and Furaya, 2004). Ketamine is an anesthetic agent that has been found to enhance the amplitude of EP potentials but it also has as a side effect, the production of hallucinations and delirium in some patients as well as increased intracranial hypertension, and therefore it is often not used during NIOM (White *et al.*, 1982). Etomidate at 0.1–0.3 mg/kg may increase the amplitude of EP potentials and has other significant side effects, including the production of interictal spikes on EEG and the theoretical risk of increased clinical seizure activity, thus limiting its use in NIOM.

The benzodiazepines, such as midazolam, are used as preinduction agents to relieve anxiety and produce mild sedation in preparation for endotracheal intubation. These agents produce rapid onset anxiolysis and sedation as a preinduction agent. If continued during the operative procedure they can have mild negative effects on prolongation of SEP latency and decrease in SEP and MEP amplitudes.

Intravenous opioids are also used perioperatively to control pain and assist in sedation without interfering in the ability to perform NIOM. These include fentanyl, alfentanil, sufentanil, remifentanil, hydromorphine, and morphine. They have minimal effect on EEG monitoring, producing mild slowing of the EEG only. They produce little effect on SEP amplitude and continuous infusions are preferred over boluses as having the least effect. Compared with fentanyl, remifentanil reduces the cortical SEP amplitude less although it is reduced to a small degree with both agents (Samra *et al.*, 2001). Remifentanil has the benefit of very rapid onset as part of the total TIVA regimen; however, its use has been associated with opioid intolerance. Effective and reliable intraoperative recordings are well described with remifentanil and remifentanil–propofol anesthesia without complication. Hermanns and colleagues (2007) described propofol–remifentanil anesthesia use in spine surgery in patients with neuromuscular as well as idiopathic scoliosis with good results.

Nitrous oxide may be used in operative cases; however, its use in posterior fossa surgeries in which the sitting position of the patient is used may warrant more consideration due to the potential risk of venous air embolism. Nitrous oxide may cause venous air embolism due to its ability to expand air bubbles. Whether or not this is truly a risk – and the degree of such a risk – has been a subject of debate in the anesthesiology literature for some time. Cranial nerve monitoring may be utilized during posterior fossa surgeries due to the risk of cranial nerve injury. The cranial nerve is stimulated at the operative site and recordings are made from muscles with cranial nerves having a motor component. Cranial nerves V, VII, VIII, XI, and XII are most frequently monitored; however, other cranial nerves, including VI and III, may be monitored depending on the clinical situation and operative procedure. In addition SEP, BAEP, and EMG may be utilized and therefore the judicious use of anesthetic agents and muscle relaxants is necessary.

The goal of emergence from anesthesia is quick awakening of the patient to assess neurological function and confirm pulmonary and hemodynamic stability. The choice of anesthetic regimen is crucial to this goal and should involve communication between the neurophysiologic monitoring team, surgical staff, and anesthesiology team for any given patient, to achieve satisfactory neurophysiologic monitoring. Changes seen in NIOM can predict a delay in emergence from anesthesia or complications in motor and sensory pathways that may correlate with clinical deficits upon awakening, or may suggest more generalized complications of cerebral hypoxia or ischemia. The ability to monitor adequately (NIOM) is therefore necessary to ensure a smooth transition to normal emergence from anesthesia without complication.

Questions

1. Which of the following anesthetic agents would have little to no effect on NIOM?
   (a) Isoflurane
   (b) Sevoflurane
   (c) Nitrous oxide
   (d) Sufenta

2. Which of the following agents should be limited in use or not used at all, to monitor transcranial electric motor stimulation adequately?
   (a) Opioids
   (b) Nitrous oxide
   (c) Halogenated anesthesia
   (d) Neuromuscular blocking agents
3. Halogenated anesthetics have effects on which of the following monitoring modalities?
   (a) BAER
   (b) SEPS
   (c) TCES
   (d) Pedicle screw stimulation evoked CMAPs
   (e) All of the above
4. Which of the following anesthetic agents has been found to enhance the amplitude of evoked potentials?
   (a) Opioids
   (b) Halogenated anesthetics
   (c) Ketamine
   (d) Nitrous oxide
5. Which of the following agents must be used cautiously during posterior fossa surgeries because of the risk of venous air embolism?
   (a) Nitrous oxide
   (b) Opioids
   (c) Halogenated anesthetics
   (d) Barbiturates
6. The goal of the preinduction stage is which of the following?
   (a) Emergence from anesthesia
   (b) Deep sedation and paralysis
   (c) Anxiolysis and sedation
   (d) Coma
7. The anesthetic protocol most compatible with NIOM is which of the following?
   (a) Combination of halogenated agent and nitrous oxide
   (b) Total intravenous anesthesia (TIVA)
   (c) 80% halogenated agent in combination with a barbiturate
8. Inhalational anesthetics have which of the following effects on cerebral perfusion pressure and intracranial pressure at 0.5 MAC or greater?
   (a) Increase cerebral perfusion pressure and mean arterial pressure and decrease intracranial pressure
   (b) Decrease cerebral perfusion pressure and mean arterial pressure and increase intracranial pressure
   (c) Do not affect cerebral perfusion pressure or mean arterial pressure but increase intracranial pressure
   (d) Do not affect cerebral perfusion pressure, mean arterial pressure, or intracranial pressure
9. Which statement about induction of anesthesia in the neonate compared with the older pediatric patient is true?
   (a) Induction is delayed
   (b) Induction is not possible in a neonate
   (c) Induction is more rapid
   (d) There is no difference in the neonate compared with the older pediatric patient
10. Which of the following statements is false?
   (a) Halogenated anesthetics impair or decrease evoked potential amplitudes
   (b) Paralytic agents should be minimized during electromyographic recording
   (c) Nitrous oxide improves the ability to record evoked potentials when used in combination with halogenated anesthetics
   (d) Opioids have little to no effect on evoked potential recordings

References

Anschel, D. J., Aherne, A., Soto, R. G., *et al.* (2008). Successful intraoperative spinal cord monitoring during scoliosis surgery using a total intravenous anesthetic regimen including dexmedetomidine. *Journal of Clinical Neurophysiology*, **25**(1), 56–61.

Ben-David, B., Taylor, P. D., and Haller, G. S. (1987). Posterior spinal fusion complicated by posterior column injury. A case report of a false-negative wake-up test. *Spine*, **12**(6), 540–543.

Boisseau, N., Madany, M., Staccini, P., *et al.* (2002). Comparison of the effects of sevoflurane and propofol on cortical somatosensory evoked potentials. *British Journal of Anaesthesia*, **88**(6), 785–789.

Chen, Z. (2004). The effects of isoflurane and propofol on intraoperative neurophysiological monitoring during spinal surgery. *Journal of Clinical Monitoring and Computing*, **18**, 303–308.

Frigon, C., Sedeek, K., Poulin, C., Brown, K., and Farmer, J. P. (2008). Does ketamine affect intraoperative electrophysiological monitoring in children undergoing selective posterior rhizotomy? *Pediatric Anesthesia*, **18**, 831–837.

Gibson, P. R. (2004). Anesthesia for correction of scoliosis in children. *Anaesthesia and Intensive Care*, **32**(4), 548–559.

Hermanns, H., Lipfert, P., Meier, S., *et al.* (2007). Cortical somatosensory-evoked potentials during spine surgery in patients with neuromuscular and idiopathic scoliosis under propofol–remifentanil anesthesia. *British Journal of Anaesthesia*, **98**(3), 362–365.

Hu, Y., Luk, K. D., Wong, Y. W., Lu, W. W., and Leong, J. C. (2001). Effect of stimulation parameters on intraoperative spinal cord evoked potential monitoring. *Journal of Spinal Disorders*, **14**(5), 449–452.

Iwasaki, H., Tamaki, T., Yoshida, M., *et al.* (2003). Efficacy and limitations of current methods of intraoperative spinal cord monitoring. *Journal of Orthopaedic Science*, **8**(5), 635–642.

Kawaguchi, M. and Furuya, H. (2004). Intraoperative spinal cord monitoring of motor function with myogenic motor-evoked potentials: a consideration in anesthesia. *Journal of Anesthesia*, **18**(1), 18–28.

Konya, D., Gercek, A., Dagcinar, A., Baykan, N., and Ozek, M. M. (2009). Prevention of brisk hyperactive response during selective dorsal rhizotomy in children with spasticity: isoflurane versus sevoflurane maintenance anesthesia. *Journal of Clinical Neuroscience*, **16**, 241–245.

Nathan, N., Tabaraud, F., Lacroix, F., *et al.* (2003). Influence of propofol concentrations on multipulse transcranial motor-evoked potentials. *British Journal of Anaesthesia*, **91**, 493–497.

Samra, S., Dy, E., Welch, K., Lovely, L. K., and Graziano, G. P. (2001). Remifentanil and fentanyl based anesthesia or intraoperative monitoring of somatosensory evoked potentials. *Anesthesia and Analgesia*, **92**, 1510–1515.

Scheufler, K.-M. and Zentner, J. (2002). Total intravenous anesthesia for intraoperative monitoring of the motor pathways: an integral view combining clinical and experimental data. *Journal of Neurosurgery*, **96**, 571–579.

Sekimoto, K., Nishikawa, K., Ishizeki, J., *et al.* (2006). The effects of volatile anesthetics on intraoperative monitoring of myogenic motor-evoked potentials to transcranial electrical stimulation and on partial neuromuscular blockade during propofol/fentanyl/nitrous oxide anesthesia in humans. *Journal of Neurosurgery and Anesthesiology*, **18**, 106–111.

White, P. F., Way, W. L., and Trevor, A. J. (1982). Ketamine: its pharmacology and therapeutic uses. *Anesthesiology*, **56**, 119–136.

Zaarour, C., Engelhardt, T., Strantzas, S., *et al.* (2007). Effect of low dose ketamine on voltage requirement for transcranial electrical motor-evoked potentials in children. *Spine*, **32**(22), E627–E630.

6 Somatosensory evoked potential monitoring with scalp and cervical recording

Marc R. Nuwer

Somatosensory evoked potential (SEP) intraoperative monitoring (IOM) of the lower extremities uses stimulation at the ankles over the posterior tibial nerve with recording over the neck and scalp. Less commonly, the common peroneal nerve may be used. Similarly, SEP IOM of the upper extremities usually uses stimulation of the median nerve at the wrists with recording over the neck and scalp. Sometimes the ulnar nerve is used instead. These types of monitoring provide an assessment of the peripheral conduction, spinal cord posterior column, brainstem medial lemniscus, and hemisphere lemniscal sensory pathway.

The goal is to provide an early warning of complications, thereby allowing the surgeon the opportunity to intervene and correct the problem if possible. A second goal is to provide reassurance that no notable complication has occurred so far, thereby allowing the surgeon to provide additional or greater degrees of a procedure.

Techniques

Ankle or wrist to scalp and neck SEP monitoring techniques are similar to SEPs commonly found in the outpatient testing. Table 6.1 shows common simple techniques for SEP monitoring.

Stimulation

Repetitions

About 300 stimulations produce a well-defined EP that can be measured and compared with the baseline. Ideally, this is accomplished once per minute. In the more realistic setting a larger number of stimuli is used, e.g., 500–1000. Several minutes are needed to produce a new EP because of electrocautery, movement, muscle, and other artifacts that interrupt or add noise to the EP recordings.

Sites

Stimulation is delivered to the lower extremity at the posterior tibial nerve at the ankle where it lies superficially just posterior to the medial malleolus. For patients with a peripheral neuropathy, the alternate site is the peroneal nerve near the knee just below the fibular head. Stimulation to the upper extremity is delivered to the median or ulnar nerve at the wrist.

Stimulation is performed unilaterally with testing for both the left and right side separately for each nerve. This monitors each side of the spinal cord or brain. Measuring both sides is helpful because impairment occasionally affects only one side (Lesser *et al.*, 1986; Molaie, 1986). Left and right sides can be tested in an alternating overlapping manner during the same period of time. Most monitoring equipment allows for more or less simultaneous averaging from each of several stimulation sites. This can be accomplished using a protocol in which stimulation of one side is delayed. For example, the left ankle is stimulated at 0 ms, the right ankle is stimulated at 90 ms, both use an 80 ms duration of recording, and the whole cycle is repeated 5.11 times per second.

Median nerve stimulation is included in thoracic and lumbar cases, because it serves as a control for systemic or anesthetic causes of changes. It can also help detect an incidental brachial plexus impairment, e.g., from arm positioning.

For cervical procedures, median or ulnar nerve pathways are the primary pathways monitored. Because ulnar nerves enter the spinal cord at a lower level, they provide a better way to assess the mid- to lower-cervical cord during cervical surgery. Posterior tibial nerve channels are often used in cervical cases to give good coverage to the lowest spinal levels in the field. The use of lower extremity SEPs in cervical cases allows for detection of a high thoracic or low cervical spinal cord injury.

Table 6.1 Summary of techniques for somatosensory evoked potential monitoring in surgery

Stimulation	Lower extremity: posterior tibial nerve at the ankle (or peroneal nerve at the knee) Upper extremity: median (or ulnar) nerve at the wrist Intensity to cause a 1–2 cm movement Stimulate at 5.1 per second per nerve, adjusting the rate as needed
Recording	For lower extremity stimulations: Two cortical channels, adjusting the sites as needed to find optimal peaks for an individual patient Active electrodes at Cz′, CPz, C1′, C2′, C3′, C4′ Reference electrodes at Fz, Fpz, Erb's point, or scalp contralateral to the active electrodes For example, channels C3′–C4′ and Cz′–Fpz A cervical channel: CSp5–Fpz For upper extremity stimulations: Three cortical channel: C3′–Fpz, C4′–Fpz, C3′–C4′ One cervical channel: CSp5–Fpz One peripheral channel: Erb's ipsilateral–contralateral Filters 30 and 3000 Hz, notch filter off, adjusting the filters as needed 300 trials per EP, more if needed
Criteria for alarm	50% drop in amplitude 5–10% increase in latency

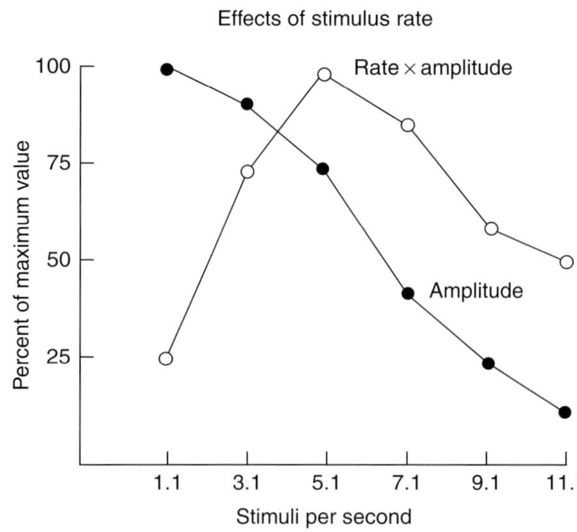

Figure 6.1 Effects of increasing stimulus rate for lower extremity SEPs. As the rate increases, the EP amplitude decreases. The product of rate and amplitude shows the trade-off between advantageous increase in speed of testing and disadvantageous loss of amplitude. In this case, 5.1 per second appeared to be the best compromise between speed and attenuation (from Nuwer and Dawson, 1984a, with permission).

Rate and intensity

Stimuli should be above the motor threshold, sufficient to cause a 1–2 cm movement in the appropriate muscle groups. Median and ulnar nerve stimuli produce thumb and fifth finger twitches. Posterior tibial and peroneal nerve stimuli produce foot twitches. The intensity can be set before neuromuscular blockade or else a default stimulus intensity is chosen, such as 20 mA. During neuromuscular blockade, a stimulus artifact should be seen in the recording channels.

Needle electrodes, disks, or paste-on electrodes are all commonly used for nerve stimulation. A pair of electrodes is secured over the nerve. Needles avoid skin resistance changes during long cases. For disks or paste-on electrodes, the electrogel paste should be free of calcium to avoid chemical burns from iontophoresis into the skin.

The optimal stimulus repetition rate is often about five stimuli per second, as illustrated in Figure 6.1 (Nuwer and Dawson, 1984a). Faster stimulation produces EPs more quickly but the EP peaks have lower amplitude. Slower stimulation can produce larger amplitude EPs but at the trade-off of slower production of new EPs. When baseline EPs are large, the stimulation rate can be faster. When baseline EPs are small, the stimulation rate can be slowed. Stimulation rates should avoid exact multiples of 60 Hz (or 50 Hz), such as 5.00 per second, to average out any residual line noise artifact. Rates of 4.11 or 5.11 per second are recommended.

Recording

Scouting and flexibility

Recording from several different sites is helpful. It is difficult to predict which sites will be most useful for monitoring an individual patient. Once monitoring begins, the best potentials should be sought and chosen for that particular case. This searching process involves "scouting" for the best recording channels, rates, and filters for that patient.

Near-field scalp recording channels are those recording from near the generator source. Near-field SEP channels have active electrodes on the scalp, and usually use a reference on the scalp. Channels with a nearby scalp reference are less affected by background muscle and movement artifacts. Reference electrodes placed far away, e.g., on the neck, ear, or shoulder, can record far-field potentials from the brainstem or other deep generator sources but are more susceptible to muscle and other artifacts. Far-field potentials recorded from the cervical channel are less affected by changes in anesthetic concentrations.

The surgical site may mandate choosing alternate recording sites. This is most often encountered in neurosurgical cases in which the craniotomy itself lies where the optimal recording electrodes would have been. Flexibility is needed in that case. Sometimes several different sites need to be tried to see which will work best.

Preoperative studies are useful in the similar ways. They can determine the location of optimal peak amplitudes. They can identify patients in whom pre-existing impairment has eliminated EPs, so that IOM may not be a viable option during surgery. When done on the day prior to surgery, they can also allow for placement of electrodes in advance of surgery so as to save OR time during set-up.

Site nomenclature

The International Federation of Clinical Neurophysiology's 10–20 System names the scalp sites used for recording. The sites used for IOM include the 10% extension of the 10–20 System (Nuwer, 1987). For example, in the 10% system, the location half way between Cz and C3 is known as C1. In additional to the well-known F, C, P, T, and O rows, the rows half way between them are referred to by letter combinations. CP is the row half way between C and P and FC is half way between F and C. Site letters followed by a prime are located 2 cm posterior the named site, e.g., C3′ is 2 cm posterior to C3. C3′ is pronounced C3-prime. The name Ci′ refers to whichever of C3′ or C4′ is ipsilateral to the limb stimulated, and Cc′ refers to whichever is contralateral. Channel Cc′–Ci′ is the contralateral site referenced to the ipsilateral site for the sites C3′ and C4′.

Traditional SEP sites over spines, at Erb's point, and other locations are also used. Erb's point is located above the clavicle, 2 cm lateral to the insertion of the sternocleidomastoid muscle. Sites over vertebrae are referred to with their spinal level, sometimes including the term Sp for *spine*. Using that convention, CSp5 is located on the back of the neck over the fifth cervical spine's posterior spinous process.

Some recommended recording sites are listed in Table 6.1.

Lower extremity SEP recording channels

For lower extremity stimulation, recordings are made from skin posterior to the cervical spine and from the scalp. These monitor the far-field cervical and brainstem activity, and the near-field cortical SEP peaks.

The best scalp active recording site varies among patients. Principal sites used include Cz′, C1′ and C2′, C3′ and C4′, and CPz. The active somatosensory cortical generator orientation varies among patients (Figure 6.2). The dipole crosses the midline, resulting in a "paradoxical localization" of the primary cortical peak. That primary cortical peak is seen best from the scalp *ipsilateral* to the leg stimulated.

Several reference electrode sites are commonly used. Scouting among channels early in the case is the best way to find good channels for use in an individual patient. References include the forehead, ear, or mastoid or scalp contralateral to the active electrodes. The latter can produce channels of active-reference pairs, such as C1′–C2′ and C3′–C4′. Short

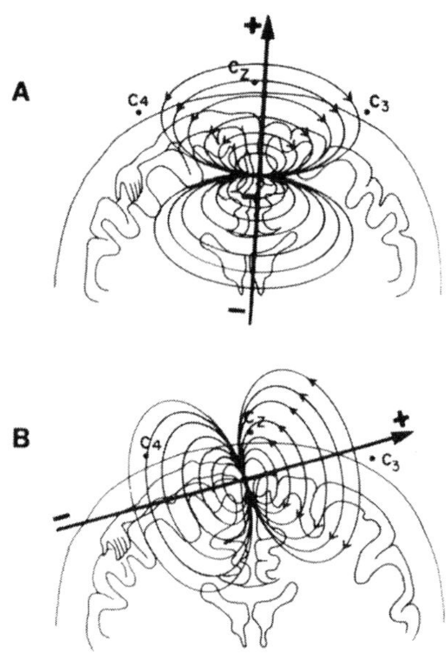

Figure 6.2 Hypothesized variations in the electrical field distribution of the early cortical potentials on stimulation of the posterior tibial nerve. These dipole orientation variations are based on the known variability of the location of the leg area in the cortex in the interhemispheric fissure. Notice how the positive peak generated in one hemisphere can show up best at the scalp vertex (as in **A**), or on the scalp overlying the other hemisphere (as in **B**). The latter phenomenon has been called "paradoxical localization" (from Seyal *et al.*, 1983, with permission).

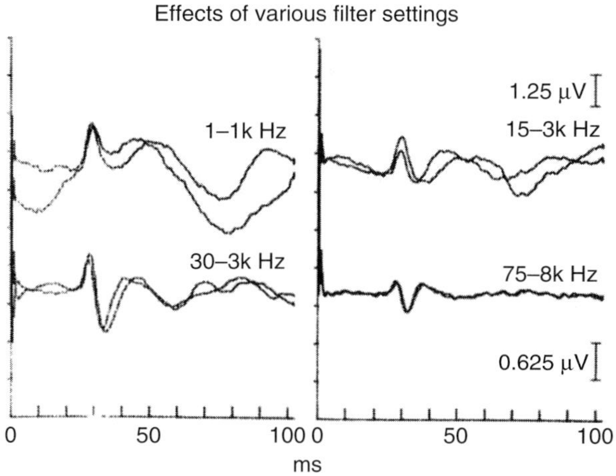

Figure 6.3 Effects of four different filter settings during intraoperative recordings in one patient, taken from a single scalp channel (Cz channel). The four recordings were made simultaneously from the same data. Variability is greatest in the 1 Hz channel and is reduced by higher filter settings. Each pair of EPs is a typical set of two consecutive recordings. The amplitude scale is doubled for the lower two EPs (from Nuwer and Dawson, 1984b, with permission).

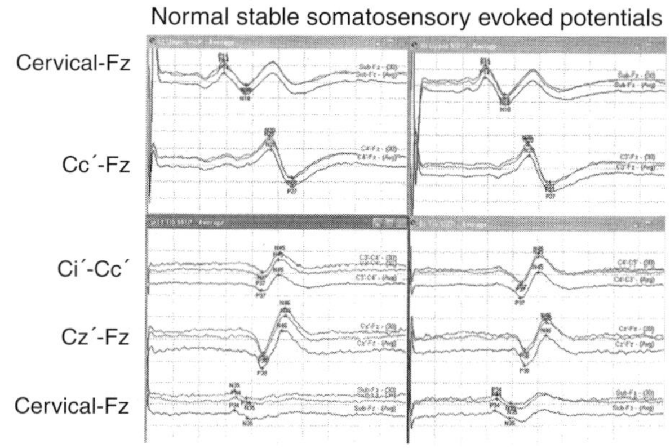

Figure 6.4 Normal stable SEPs (from UCLA Dept. Clinical Neurophysiology, with permission).

distances to the reference reduce noise, but might reduce peak amplitudes too. The optimal recording technique varies among patients.

The subcortical peaks are often recorded from the second cervical spinal level (CSp2). The forehead, ear, or shoulder can serve as the reference. These channels are used because subcortical peaks are less affected by anesthesia. A lumbar or popliteal channel may be used to ascertain that peripheral transmission is intact. For older or obese patients, lumbar potentials may be absent even without any pathology. The commonly used peripheral recording sites are popliteal fossa (PF) over the posterior tibial nerve, a reference site nearby on the knee (K), and sites over the lumbar and thoracic vertebrae, such as T10, T12, L1, or L3.

Upper extremity SEP recording channels

For upper extremity stimulation, recordings are taken from the shoulder, cervical spine, and scalp. The most common active scalp recording sites are C3′ and C4′. Reference electrodes may be at the forehead, ear, or mastoid reference sites. A common reference choice is the contralateral electrode, i.e., a C3′–C4′ channel. For some intracranial cases, modified scalp sites may be used, or subdural or cortical electrodes may be placed in the surgical field.

The subcortical brainstem and cervical peaks can be recorded from the neck, ear, or mastoid. The fifth cervical spinal level (CSp5) is an excellent site. References are placed at the forehead or contralateral shoulder. Far-field peaks recorded with distant references are less affected by anesthesia, and assure monitoring of the mid-cervical, cervicomedullary, or caudal brainstem pathway. An Erb's point channel can ascertain that peripheral conduction reached the brachial plexus.

Filters

The optimal low filter setting usually is 30 Hz. The optimal high filter setting is 1500 to 3000 Hz. This optimizes noise control while retaining SEP peak characteristics. These settings also reduce SEP changes from changes in anesthetic depth (Nuwer and Dawson, 1984a). Figure 6.3 shows the effects of changing the filter settings. The best trade-off was with a 30 Hz low filter. Lower low filter settings, e.g., 1 Hz, had more background variability. Higher low filter settings, e.g., 75 Hz, caused SEP amplitude loss. A good setting will produce reproducible EPs with little variability in the early cortical peak amplitude and latencies (Figure 6.4).

The 60 Hz (or 50 Hz) notch filter should be kept off. When that filter is on, the filter itself can produce a stimulus artifact with a decaying sinusoidal tail. That artifact has a basic frequency of 60 Hz (or 50 Hz), so the artifact shows peaks at 16.6, 33, and 50 ms. Those peaks can be mistaken for stable EPs at 16.6 or 33 ms.

Digital smoothing filters are available on some equipment. They can distort the peak, possibly mixing artifacts with peaks in ways that make interpretation more difficult. The real goal is to find the cause of the noise and eliminate it if at all possible, and to scout to

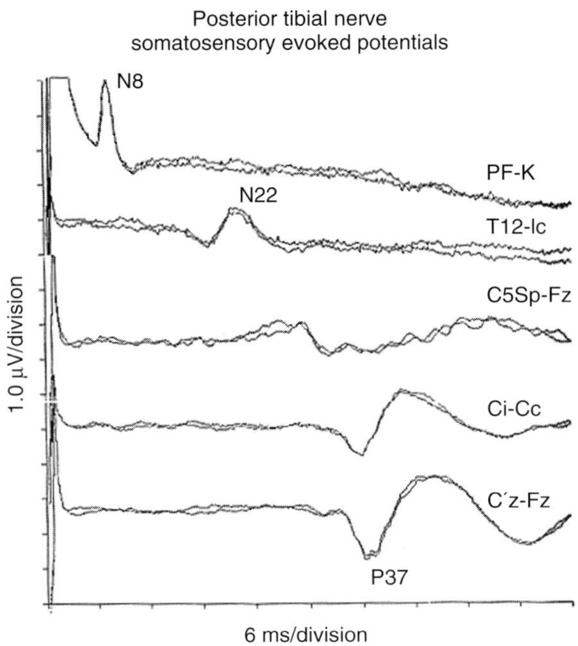

Figure 6.5 Somatosensory evoked potentials from posterior tibial nerve stimulation are shown. Typical peaks are noted at N8, N22, and P37. A cervical peak was also found. These peaks have normal latencies and amplitudes (from Nuwer *et al.*, 1994, with permission).

find channels that are less affected by noise. The monitorist should eliminate the noise or artifacts at their source, not cover them over with filters.

Scouting different recording sites is accompanied by scouting for better filter settings. If the peaks are low amplitude, poorly defined early in the case, then changing the low filter to a lower setting, e.g., 15 Hz, may improve the peak more than the noise.

Peaks, generators, and baseline values

Lower extremity peaks

The P37 peak amplitude and latency are the main measurements for lower extremity SEP IOM. The P37 is the primary cortical peak generated by somatosensory cortex. It is often off-center, paradoxically seen best on the side ipsilateral to the leg stimulated. Latencies should stay within about 5–10% of baseline values, e.g., within 3–4 ms of baseline value. Amplitudes should stay within 50% of baseline values after the anesthetic fade. Figure 6.5 shows typical lower extremity SEP peaks and nomenclature.

A cervical peak is a far-field signal found best from cervical recording electrodes. The cervical peak, seen around 30 ms for lower extremity monitoring, is generated by the gracilis nucleus at the cervicomedullary junction. An opposite polarity trough follows the cervical peak, which may represent conduction up the brainstem's medial lemniscus. Amplitude and latency of the cervical peak is a secondary measurement for lower extremity SEP IOM. The cervical peak is used when the cortical P37 is poorly suited for monitoring. Anesthetic effect is distinguished from clinical impairment partly by observing the cervical peaks, since the cervical peak is more stable than the P37 when anesthesia effects are seen.

The N22 peak is a negative polarity near-field potential seen over the T12 spinal level at about 22 ms after stimulation. It is generated in the lumbar spinal cord, which lies anatomically around the T12 or L1 spine. Other monitorists use the popliteal N8 peak in the same way. These peripheral peaks are monitored to assure that the peripheral pathway is intact. That can help to clarify that decreased cortical potentials are due to a surgical problem rather than a problem with the stimulus.

Upper extremity peaks

The N20 peak amplitude and latency are the primary measurements for upper extremity SEP IOM. The N20 is generated at somatosensory cortex on the posterior side of the central fissure. The N20 is a large well-defined near-field potential seen over the C3′ or C4′ scalp site. Latencies should stay within about 5–10% of baseline values, e.g., within 1–2 ms of baseline values. Amplitudes should stay within 50% of baseline values.

Among the subcortical peaks, the cervical N13 peak is seen using far-field recording from the neck region to a distant reference. The cervical spinal cord gray matter at the mid to lower cervical spinal cord generates an N13 peak. A subsequent P14 peak is generated around the cervicomedullary junction. It is followed by an N18 peak from the brainstem's medial lemniscus. Both the P14 and N18 peaks are far-field potentials seen broadly over the scalp using a noncephalic reference.

A peripheral N9 peak is a near-field potential seen over the proximal brachial plexus. The peak is seen best at Erb's point. Figure 6.6 illustrates the typical routine upper extremity SEP peaks and their nomenclature.

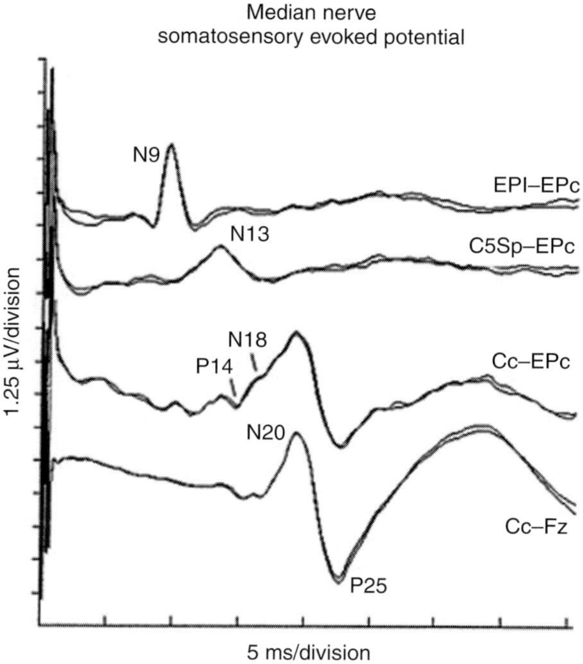

Figure 6.6 Somatosensory evoked potentials from median nerve stimulation are shown. Typical peaks are shown in each of four recording channels. The result is normal (from Nuwer *et al.*, 1994, with permission).

Interpretation of changes, alarms, and considerations

Monitoring teams use amplitude and latency criteria to identify significant changes. A 50% amplitude drop is the threshold for raising an alarm (Nuwer, 1986; Nuwer *et al.*, 1995). A latency increase of 5–10% is also a threshold for alarm. When changes happen, one must assess quickly if these have an anesthetic, technical, systemic, or surgical cause.

Anesthetic effects are common. Amplitudes can decrease due to anesthetic effects, especially for the cortical potentials. Anesthetic fade is the term used for the gradual reduction in cortical peak amplitudes as the nervous system adapts to anesthesia. Fade is especially seen during the first 30 minutes after induction, and, to a small extent, over subsequent hours of a long case. Boluses of some medications may also cause transient drops in SEP amplitude. The same effect can recur if anesthetic depth is increased partway through a case. Anesthetic fade is seen more often with inhalation anesthetics and nitrous oxide. Total intravenous anesthesia produces much less of such an anesthetic fade.

In thoracic or lumbar cases, it is helpful to monitor also the upper extremity SEPs so as to assess anesthetic effects. Anesthetics usually affect all cortical peaks, but cause little change in subcortical peaks. Preexisting impairment may magnify effects of anesthesia. Figure 6.7 shows an asymmetric anesthetic effect, seen in the setting of a partial preexisting unilateral impairment in one pathway.

Technical problems can cause changes. Electrodes can become disconnected. Equipment may malfunction. The monitoring team should check that the stimulus artifact is still present as a check for stimulus integrity. Impedance checks and reviews of the unaveraged raw data recording can also check the recording system's integrity.

Systemic factors that cause changes include hypothermia, hypotension, and hypoxia. Preexisting neurologic problems, e.g., spinal cord compression, may leave a patient sensitive to minor degrees of hypotension. Latency can increase when the limb or core temperature is cool. Cooling the spinal cord with room temperature irrigation may slow latencies.

Surgical events that cause changes include direct cord trauma, excessive traction, blunt trauma, excessive compression, stretching of the cord from spinal distraction, vascular insufficiency from compression, embolus, thrombus, or occasionally other clinical problems. Not all amplitude attenuations herald postoperative neurologic problems. The worse the loss, the worse the risk. A 50–80% decrease for a few minutes poses a small to modest risk of postoperative neurologic deficits if the SEPs return promptly to baseline levels of amplitude and latency. Higher risk occurs with abrupt changes, complete loss, and persistent attenuation. The gravest situation is the abrupt persistent loss of previously easily detected SEPs. Even with an abrupt persistent loss, the risk of a new neurologic postoperative impairment may be about 50–75% (Nuwer *et al.*, 1995). A new neurologic deficit is not a foregone conclusion.

False results

False-positive monitoring is said to occur when the SEP shows changes but the patient suffers no new postoperative neurologic deficit. Despite the name "false", many of these events are true clinical changes that were detected by monitoring, and in which the monitoring change prompts clinical interventions. Those interventions may have been sufficient so that

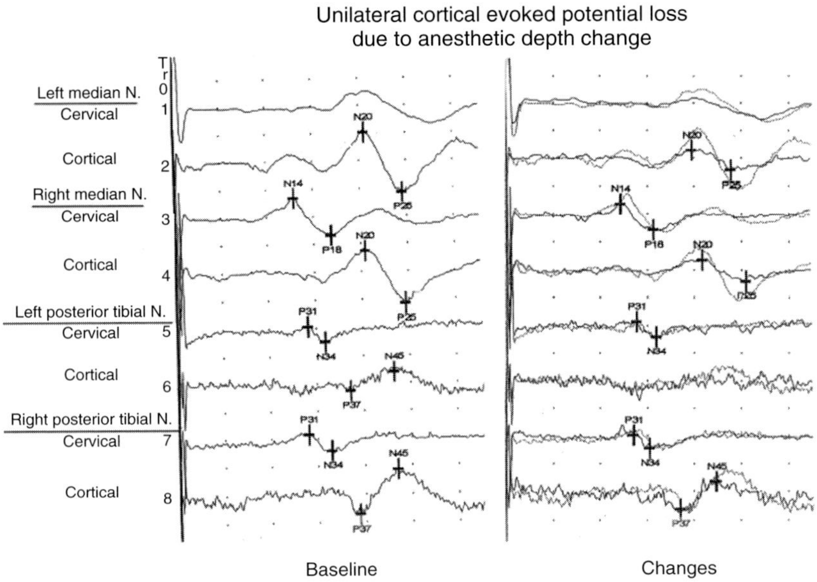

Figure 6.7 Baseline testing shows a relatively attenuated left lower extremity cortical peak (left tracings). After an increase in anesthetic depth (right tracings), that channel no longer shows a reliable EP. (The baseline is superimposed on the newly acquired tracings at the right.) An anesthetic effect is probably the cause of the change, as suggested by both the preserved subcortical peaks for the affected pathway and somewhat attenuated cortical peaks in all other pathways (from UCLA Dept. Clinical Neurophysiology, with permission).

the patient awakens without a deficit because the interventions were successful. Other false-positive events are false alarms due to technical or anesthetic issues.

A false-negative monitoring event is one in which the patient awakens with a new postoperative neurologic deficit despite normal SEP monitoring throughout the case. False-negative events are rare but do occur (Tamaki *et al.*, 1984; Wilber *et al.*, 1984; Ginsburg *et al.*, 1985; Johnston *et al.*, 1986; Ben-David *et al.*, 1987; Harper *et al.*, 1988; More *et al.*, 1988). Rare cases result in unforeseen postoperative paraparesis or paraplegia. Some reports of false-negatives are for patients with sensory dysesthesias lasting for days or weeks postoperatively, possibly due to root injury. Lesser and colleagues (1986) reported six cases from four institutions who had a new postoperative neurologic deficit despite stable intraoperative SEPs. Three reported patients probably developed impairment postoperatively, referred to as "delayed onset" deficits. Delayed onset cases are those in which the patient awoke from surgery initially with good strength but that was lost hours to days later. Monitoring SEPs is not particularly useful to predict what will happen hours or days later. Nor is it sensitive for identifying single spinal root lesions.

A large multicenter study assessed the likelihood of false-negative and false-positive surgical outcome for spinal cord monitoring (Table 6.2). False-negative outcomes were rare, and many of those were delayed onset. False-positive monitoring was more common. Equivocal results in that report and table showed minor or transient changes. Table 6.3 shows the monitoring sensitivity and specificity of that multicenter outcome study.

Outcomes for spinal surgery

The multicenter scoliosis survey is the largest study of SEP IOM outcomes. That study of approximately 100 000 surgical cases compared the outcomes for cases that used monitoring against cases where monitoring was not used. It found significantly fewer neurological deficits, fewer persistent deficits, and fewer severe deficits with monitoring. Paraplegia and paraparesis were reduced by 60% with monitoring (Nuwer *et al.*, 1995). Results are presented in Table 6.2.

Many other studies have reported that SEPs accurately predict postoperative neurological deficits. Many of those reports also discuss how concurrent motor-evoked potential (MEP) monitoring is an additional way of identifying problems and preventing complications, and can be used in the same cases as SEPs. The advantages of SEPs over MEPs are that they are less strict in their anesthetic requirements, cause no movements during surgery, and can easily be done continuously throughout surgery. In some patients, the anesthetic requirements or movements during surgery are enough for the surgeon to opt to have only SEPs. It is best to decide what to monitor on a case-by-case basis.

Table 6.2 Neurologic outcome prediction rates for SEP monitoring in spinal surgery

Total procedures monitored	**51 263**	**(100%)**
False-negative (FN) rate: neurologic postoperative deficits despite stable SEPs		
Definite	34	(0.063%)
Equivocal	13	(0.025%)
Delayed onset	18	(0.035%)
Total:	65	(0.127%)
False-positive rate: no neurologic deficits despite SEP changes		
Definite	504	(0.983%)
Equivocal	270	(0.527%)
Total:	774	(1.510%)
True-positive (TP) rate: neurologic deficits predicted by SEP changes		
Definite	150	(0.293%)
Equivocal	67	(0.131%)
Total:	217	(0.423%)
Neurologic deficits (FN plus TP)		
Definite	184	(0.356%)
Equivocal	80	(0.156%)
Delayed onset	18	(0.035%)
Total:	282	(0.550%)
True-negative rate: no neurologic deficit and stable SEPs		
Total	50 207	(97.94%)

These data are from the multicenter outcome study of SEP spinal-cord monitoring in scoliosis organized through the Scoliosis Research Society (Nuwer *et al.*, 1995). They were obtained from 153 US surgeon respondents. Note the very low rate of definite false-negative cases (0.063%).

Table 6.3 SEP monitoring validity measures

Sensitivity: 417/451	92%
Specificity 50 207/50 781	98.9%
Positive predictive value 417/991	42%
Negative predictive value 50 207/50 241	99.93%

These outcome measures are from the multicenter outcome survey of 153 US surgeons organized through the Scoliosis Research Society (Nuwer *et al.*, 1995). The table shows a breakout of validity measures from that survey. The very high negative predictive value here indicates the high reliability of the monitoring when the SEP remains normal and stable. The outcome survey report (Nuwer *et al.*, 1995) discusses in detail these data and related assumptions.

Cunningham *et al.* (1987) evaluated 33 consecutive patients undergoing surgical procedures for lesions of the descending thoracic or thoracoabdominal aorta. The criteria were persistent loss of the SEPs. Surgical interventions to identify and correct the problem were undertaken unless the SEP loss was transient. Each patient served as their own control. The study compared preoperative to postoperative motor status, looking for the presence of any new postoperative motor deficit. The SEPs showed no change in 17 patients, and 0/17 had new motor deficits. The SEPs did show a change in 16 patients, and 5/16 had a new motor deficit. The difference was significant ($p < 0.02$).

Meyer *et al.* (1988) conducted a retrospective review of 150 patients treated surgically for acute thoracic or lumbar spinal injuries whose SEPs were monitored. Surgical interventions were carried out to identify and correct the impairment when SEPs changed. The study compared preoperative and postoperative motor status, looking for any new postoperative motor deficit. The SEPs showed no change in 144 patients, and 0/144 had new motor deficits. The SEPs did show a change in 6 patients, and 1/6 had a new motor deficit. The difference was significant ($p < 0.05$).

May *et al.* (1996) assessed 191 patients undergoing cervical spine procedures. Median nerve SEPs were monitored in all cases. In addition, ulnar or posterior tibial nerve SEPs were monitored in 12 cases. The SEP criterion for change was a greater than 50% amplitude reduction in N20, occurring suddenly and persisting without an apparent anesthetic or systemic cause. When SEPs changed, surgical interventions were carried out to identify and correct the impairment. The study compared preoperative and postoperative motor status, looking for the presence of any new postoperative motor deficit. The SEPs showed no change in 144 patients, and 1/144 had new motor deficits. The SEPs did show a change in 33 patients, and 9/33 had a new motor deficit. The difference was significant ($p < 0.0001$).

Khan *et al.* (2006) conducted a retrospective review of 508 patients who underwent anterior cervical fusion with single-level or multilevel cervical corpectomies. The criteria for SEP change were a greater than 50% amplitude decrease or more than 10% latency increase. Motor outcome was determined without knowledge of SEP changes. When SEPs changed, surgical interventions were carried out to identify and correct the impairment. The surgical interventions included increasing blood pressure, repositioning a retractor, reducing the distraction, removing graft, removing or adjusting instrumentation, and stopping the surgical operative maneuver preceding the change.

The study compared preoperative and postoperative motor status, looking for the presence of any new postoperative motor deficit. The SEPs showed no change in 481 patients, and 8/481 had new motor deficits. The SEPs did show a change in 27 patients, and 4/27 had a new motor deficit. The difference was significant ($p < 0.003$).

Leung *et al.* (2005) carried out a retrospective cohort survey of 871 patients who had had anterior spinal deformity surgery. Patients who had had previous spinal surgery were excluded. The criteria for change were greater than 50% amplitude decrease or more than 10% latency increase.

The study compared preoperative and postoperative motor status, looking for the presence of any new postoperative motor deficit. The SEPs showed no change in 845 patients, and 0/845 had new motor deficits. The SEPs did show a change in 26 patients, and 5/26 had a new motor deficit. The difference was significant ($p < 0.00001$).

Kelleher *et al.* (2008) published a prospective analysis of 1055 consecutive patients during cervical spine surgery. When SEPs changed, surgical interventions were carried out to identify and correct the impairment. The surgical interventions included releasing the distraction, changing the decompression, ceasing tumor debulking until EPs improved, raising the blood pressure, and undoing instrumentation immediately preceding the deterioration. The study compared preoperative and postoperative motor status, looking for the presence of any new postoperative motor deficit. The SEPs showed no change in 1012 patients, and 6/1012 had new motor deficits. The SEPs did show a change in 26 patients, and 10/26 had a new motor deficit. The difference was significant ($p < 0.00001$). Among the 26 (2.5%) patients who had significant SEP changes, 7 had a complete loss of the SEP whereas 19 showed a greater than 50–90% amplitude loss. All were considered true-positive monitoring events. Among those 26 events, 16 corresponded to an indentified intraoperative event.

Achouh *et al.* (2007) conducted a prospective cohort survey of SEP monitoring in 444 consecutive descending thoracic and thoracoabdominal aortic repair procedures. These included 270 patients with thoracoabdominal aorta procedure and 174 with a descending thoracic aorta procedure. The criteria for change were greater than 50% decrease in amplitude or more than 10% latency increase. When SEPs changed, surgical interventions were carried out to identify and correct the impairment. The surgical interventions included reimplantation of intercostal arteries, lowering of cerebrospinal fluid pressure by free drainage, increasing of hemoglobin level by transfusion, increasing of distal aortic perfusion pressure by either adding volume to the pump to increase flow, or using vasoconstrictors, or both. The study compared preoperative and postoperative motor status, looking for the presence of any disabling degree of new postoperative motor deficit. The SEPs showed no change in 357 patients, and 3/357 had new motor deficits. The SEPs did show a change in 87 patients, and 15/87 had a new disabling motor deficit. The difference was significant ($p < 0.01$). An additional 14 patients had spinal cord deficits with onset days after aorta repair surgery.

Other studies have looked at the combined effects of SEPs and MEPs (Pelosi *et al.*, 2002; Hilibrand *et al.*, 2004; Etz *et al.*, 2006; Lee *et al.*, 2006; Costa *et al.*, 2007; Eggspuehler *et al.*, 2007a, 2007b; Schwartz *et al.*, 2007; Sutter *et al.*, 2007a, 2007b; Weinzierl *et al.*, 2007). The general results are similar to those for SEP alone.

Sala *et al.* (2006) carried out a different kind of study. They published a retrospective historical control study review of 50 patients operated on for intramedullary spinal cord tumor. The criteria for change were an MEP loss, a greater than 50% SEP amplitude decrease, or more than 10% SEP latency increase. The MEPs monitored were both spinal epidural recorded D-waves and transcranial electrical MEPs recorded from muscle. A historical control group of 50 was selected from a database of 301 patients previously operated on without intraoperative neurophysiological monitoring. Control patients were carefully matched to patients in the IOM group by preoperative neurological status, histological results, tumor location, and extent of removal. If IOM showed changes, myelotomy was either temporarily stopped or was moved to a different location along the tumor. The study compared the outcomes for patients with EP monitoring versus patients without EP monitoring. The investigators measured variations in the McCormick grade of weakness. For the patients with intraoperative monitoring, the preoperative to postoperative change in McCormick grade of weakness was +0.28. For the patients without intraoperative monitoring, the preoperative to postoperative change in McCormick grade of weakness was

−0.16. The difference between groups was significant ($p < 0.002$).

Overall, the literature shows that SEP IOM provides a useful alarm function for surgery. Interventions are possible when the IOM raises an alarm. Those interventions are successful at reducing postoperative neurological deficits. False-negative monitoring cases are rare.

Intracranial monitoring

The same SEP monitoring protocols can apply to intracranial neurosurgical procedures. When applied in posterior fossa cases, SEPs are used along with brainstem auditory evoked potentials and cranial nerve EMG monitoring. These provide a good survey of the various pathways in and around the brainstem. When applied for cerebral hemisphere surgery, SEPs are used along with EEG to monitor the various regions of cortex and its vascular supply.

Typical applications include cerebropontine angle tumor resection in which retractor stretching of the brainstem can cause brainstem injury or ischemia. For aneurysm clipping, IOM helps to demonstrate that the arterial lumen remains unimpeded and the vascular territory is not ischemic. The SEPs are also monitored for many other types of intracranial procedure (Nuwer, 2008).

Monitoring tactics

For spine surgery, both upper and lower extremity SEPs are usually monitored. For brainstem or cerebral hemisphere procedures, sometimes only upper extremity SEPs are needed. For procedures around the anterior cerebral artery, lower extremity SEPs are needed because that artery supplies only the lower extremity portion of somatosensory cortex.

At the beginning of the case, place electrodes at many recording sites. Scout among the SEPs from the various recording sites to find the best ones for that patient. Once the best recording sites are found, try adjusting the filters or stimulation rates to get the best SEPs that can be produced as quickly as possible during surgery. Don't just set up a "standard case" protocol used for every patient.

Reduce artifacts as much as possible by eliminating them at their sources, not just filtering the artifacts. Braid wires to reduce 60 Hz and other noise. Identify equipment that is producing electrical noise by unplugging – not just turning it off.

Document the interactions between the in-room technologist, the clinical neurophysiologist supervisor, and the surgeon or anesthesiologist.

Conclusions

Monitoring SEPs intraoperatively uses ankle or wrist stimulation and recording from scalp and neck with techniques familiar from outpatient SEP testing. False-negative cases are rare. False-positive results occur occasionally. Recording sites and filter settings should be tailored to the individual patient. A monitoring team can apply these techniques to orthopedic, neurosurgical, and cardiovascular procedures to provide early warnings of possible deficits and improve patient outcomes.

Questions

1. Recording from an active electrode near the generator source is referred to as:
   (a) Close electrode recording
   (b) Far-field recording
   (c) Closed-field recording
   (d) Near-field recording
2. The optimal SEP stimulation rate is often around
   (a) 3 per second
   (b) 5 per second
   (c) 7 per second
   (d) 9 per second
3. When cortical SEPs are low amplitude at the beginning of the case, the monitorist can try
   (a) Faster stimulation rates
   (b) Lowering the low filter setting
   (c) A smaller sample size to produce EPs more quickly
   (d) Turning on the notch filter
4. In the 10% system, electrode site CP1 is located
   (a) Halfway between Cz and P3
   (b) Halfway between Cz and C3
   (c) Halfway between Pz and P3
   (d) Halfway between C3 and P3
5. Erb's point is located at
   (a) 5 cm above the mid-clavicle just lateral to the sternocleidomastoid
   (b) 2 cm above the mid-clavicle just lateral to the sternocleidomastoid
   (c) Above the clavicle 2 cm lateral to the insertion of the sternocleidomastoid
   (d) Above the clavicle 5 cm lateral to the insertion of the sternocleidomastoid

6. The most likely location to find the P37 peak for left posterior tibial SEP testing is
   (a) C1′
   (b) C2′
   (c) Cz′
   (d) CPz
7. Recording site PF is at
   (a) Posterior frontal
   (b) Parieto-frontal
   (c) Para-frontal
   (d) Popliteal fossa
8. The 60 Hz notch filter produces a sinusoidal artifact with peaks at
   (a) 10, 20, and 30 ms
   (b) 15, 30, and 45 ms
   (c) 16.6, 33.3, and 50 ms
   (d) 25, 50, and 75 ms
9. Criteria for change in posterior tibial SEPs are commonly
   (a) 10% amplitude loss or 4 ms latency increase
   (b) 30% amplitude loss or 4 ms latency increase
   (c) 50% amplitude loss or 4 ms latency increase
   (d) 70% amplitude loss or 4 ms latency increase
10. The greatest amplitude decreases in cortical SEPs are commonly associated with
   (a) Cooling to 32 °Celsius
   (b) 1.0 MAC use of inhalation anesthetics
   (c) Too high a setting of the stimulus intensity
   (d) Too low a low filter setting

References

Achouh, P. E., Estrera, A. L., Miller, C. C., *et al.* (2007). Role of somatosensory evoked potentials in predicting outcome during thoracoabdominal aortic repair. *Annals of Thoracic Surgery*, **84**, 782–788.

Ben-David, B., Haller, G., and Taylor, P. (1987). Anterior spinal fusion complicated by paraplegia. A case report of a false-negative somatosensory evoked potential. *Spine*, **12**, 536–539.

Costa, P., Bruno, A., Bonzanino, M., *et al.* (2007). Somatosensory- and motor-evoked potential monitoring during spine and spinal cord surgery. *Spinal Cord*, **45**(1), 86–91.

Cunningham, J. N. Jr., Laschinger, J. C., and Spencer, F. C. (1987). Monitoring of somatosensory evoked potentials during surgical procedures on the thoracoabdominal aorta. IV. Clinical observations and results. *Journal of Thoracic and Cardiovascular Surgery*, **94**, 275–285.

Eggspuehler, A., Sutter, M. A., Grob, D., Jeszenszky, D., and Dvorak, J. (2007a). Multimodal intraoperative monitoring during surgery of spinal deformities in 217 patients. *European Spine Journal*, **16**(suppl. 2), S188–S196.

Eggspuehler, A., Sutter, M. A., Grob, D., *et al.* (2007b). Multimodal intraoperative monitoring (MIOM) during cervical spine surgical procedures in 246 patients. *European Spine Journal*, **16**(2), S209–S215.

Etz, C. D., Halstead, J. C., Spielvogel, D., *et al.* (2006). Thoracic and thoracoabdominal aneurysm repair: is reimplantation of spinal-cord arteries a waste of time? *Annals of Thoracic Surgery*, **82**, 1670–1678.

Ginsburg, H. H., Shetter, A. G., and Raudzens, P. A. (1985). Postoperative paraplegia with preserved intraoperative somatosensory evoked potentials. *Journal of Neurosurgery*, **6**, 296–300.

Harper, C. M., Daube, J. R., Lichy, W. J., and Klassen, R. A. (1988). Lumbar radiculopathy after spinal fusion for scoliosis. *Muscle and Nerve*, **11**, 386–391.

Hilibrand, A. S., Schwartz, D. M., Sethuraman, V., Vaccaro, A. R., and Albert, T. J. (2004). Comparison of transcranial electrical motor and somatosensory evoked potential monitoring during cervical spine surgery. *Journal of Bone and Joint Surgery*, **86**A, 1248–1253.

Johnston, C. E., II, Happel, L. T., Jr., Norris, R., *et al.* (1986). Delayed paraplegia complicating sublaminar segmental spinal instrumentation. *Journal of Bone and Joint Surgery (America)*, **68**, 556–563.

Kelleher, M. O., Tan, G., Sarjeant, R., and Fehlings, M. G. (2008). Predictive value of intraoperative neurophysiological monitoring during cervical spine surgery: a prospective analysis of 1055 consecutive patients. *Journal of Neurosurgery: Spine*, **8**, 215–221.

Khan, M. H., Smith, P. N., Balzer, J. B., *et al.* (2006). Intraoperative somatosensory evoked potential monitoring during cervical spine corpectomy surgery: experience with 508 cases. *Spine*, **31**, E105–E113.

Lee, J. Y., Hilibrand, A. S., Lim, M. R., *et al.* (2006). Characterization of neurophysiologic alerts during anterior cervical spine surgery. *Spine*, **31**, 1916–1922.

Lesser, R. P., Raudzens, P., Lueders, H., *et al.* (1986). Postoperative neurological deficits may occur despite unchanged intraoperative somatosensory evoked potentials. *Annals of Neurology*, **19**, 22–25.

Leung, Y. L., Grevitt, M., Henderson, L., and Smith, J., (2005). Cord monitoring changes and segmental vessel ligation in the “at risk” cord during anterior spinal deformity surgery. *Spine*, **30**, 1870–1874.

May, D. M., Jones, S. J., and Crockard, H. A. (1996). Somatosensory evoked potential monitoring in cervical surgery: identification of pre- and postoperative risk factors associated with neurological deterioration. *Journal of Neurosurgery*, **85**, 566–573.

Meyer, P. R. Jr., Colter, H. B., and Gireesan, G. T. (1988). Operative neurological complications resulting from thoracic and lumbar spine internal fixation. *Clinical Orthopedics and Related Research*, **237**, 125–131.

Molaie, M. (1986). False-negative intraoperative somatosensory evoked potentials with simultaneous bilateral stimulation. *Clinical Electroencephalography*, **17**, 6–19.

More, R. C., Nuwer, M. R., and Dawson, E. G. (1988). Cortical evoked potential monitoring during spinal surgery: sensitivity, specificity, reliability, and criteria for alarm. *Journal of Spinal Disorders*, **1**, 75–80.

Nuwer, M. R. (ed.). (1986). *Evoked Potential Monitoring in the Operating Room*. New York: Raven Press.

Nuwer, M. R. (1987). Recording electrode site nomenclature. *Journal of Clinical Neurophysiology*, **4**, 121–133.

Nuwer, M. R. (ed.). (2008). *Intraoperative Monitoring of Neural Function. Handbook of Clinical Neurophysiology*. Vol. **8**. Amsterdam: Elsevier.

Nuwer, M. R. and Dawson, E. (1984a). Intraoperative evoked potential monitoring of the spinal cord: enhanced stability of cortical recordings. *Electroencephalography and Clinical Neurophysiology*, **59**, 318–327.

Nuwer, M. R. and Dawson, E. (1984b). Intraoperative evoked potential monitoring on the spinal cord: a restricted filter, scalp method during Harrington instrumentation for scoliosis. *Clinical Orthopaedics*, **183**, 42–50.

Nuwer, M. R., Aminoff, M., Desmedt, J., *et al.* (1994). IFCN recommended standards for short-latency somatosensory evoked potentials. Report of an IFCN committee. *Electroencephalography and Clinical Neurophysiology*, **91**, 6–11.

Nuwer, M. R., Dawson, E. G., Carlson, L. G., Kanim, L. E. A., and Sherman, J. E. (1995). Somatosensory evoked potential spinal-cord monitoring reduces neurologic deficits after scoliosis surgery: results of a large multicenter survey. *Electroencephalography and Clinical Neurophysiology*, **96**, 6–11.

Pelosi, L., Lamb, J., Grevitt, M., *et al.* (2002). Combined monitoring of motor and somatosensory evoked potentials in orthopaedic spinal surgery. *Clinical Neurophysiology*, **13**, 1082–1091.

Sala, F., Palandri, G., Basso, E., *et al.* (2006). Motor-evoked potential monitoring improves outcome after surgery for intramedullary spinal-cord tumors: a historical control study. *Neurosurgery*, **58**, 1129–1143.

Schwartz, D. M., Auerbach, J. D., Dormans, J. P., *et al.* (2007). Neurophysiological detection of impending spinal-cord injury during scoliosis surgery. *Journal of Bone and Joint Surgery (America)*, **89**, 2440–2449.

Seyal M., Emerson, R. G., and Pedley, T. A. (1983). Spinal and early scalp-recorded components of the somatosensory evoked potential following stimulation of the posterior tibial nerve. *Electroencephalography and Clinical Neurophysiology*, **55**, 320–330.

Sutter, M., Eggspuehler, A., Grob, D., *et al.* (2007a). The validity of multimodal intraoperative monitoring (MIOM) in surgery of 109 spine and spinal-cord tumors. *European Spine Journal*, **16**(suppl. 2), S197–S208.

Sutter, M., Eggspuehler, A., Grob, D., *et al.* (2007b). The diagnostic value of multimodal intraoperative monitoring (MIOM) during spine surgery: a prospective study of 1,017 patients. *European Spine Journal*, **16**(suppl. 2), S162–S170.

Tamaki, T., Noguchi, T., Takano, H., *et al.* (1984). Spinal cord monitoring as a clinical utilization of the spinal evoked potentials. *Clinical Orthoptics*, **184**, 58–64.

Wilber, R. G., Thompson, G. H., Shaffer, J. W., Brown, R. H., and Nash, C. L., Jr. (1984). Postoperative neurological deficits in segmental spinal instrumentation. *Journal of Bone and Joint Surgery (America)*, **66**A, 1178–1187.

Weinzierl, M. R., Reinacher, P., Gilsbach, J. M., and Rohde, V. (2007). Combined motor and somatosensory evoked potentials for intraoperative monitoring: intra- and postoperative data in a series of 69 operations. *Neurosurgical Review*, **30**, 109–116.

7 Transcranial electric motor stimulation: technique and exclusionary criteria

Gloria M. Galloway

Methods and background

Transcranial electric stimulation (TES) is employed in the operating room on an anesthetized patient to assess the motor pathways that may be at risk of injury. These motor pathways originate in the pyramidal neurons and axons of the motor cortex and descend through the corticospinal pathways to the alpha motor neuron in the periphery. Transcranial electric stimulation serves not only as a mechanism to monitor these motor pathways but as a secondary safeguard for cortical and spinal monitoring if somatosensory tract monitoring is either not obtainable or the somatosensory signals are lost or changed during the surgical procedure (López, 2004). In this way, as with somatosensory monitoring, transcranial electric stimulation for motor tract monitoring is invaluable for certain surgical situations. Unlike motor stimulation in the epidural or subdural region, transcranial electric stimulation allows for recording of pure motor signals without the concern that antidromic sensory volleys are intermixed with the signals being reviewed (Minahan *et al.*, 2001).

Spinal cord stimulation involves placing a stimulating electrode in the epidural space and recording from an electrode in the subarachnoid space by using a Tuohy needle for insertion into the lower lumbar area. Using this approach, inhalational anesthetic agents, as well as neuromuscular blocking agents, can be used without the limitations imposed by the use of TES (Erb *et al.*, 2005). These epidural and subarachnoid needles may become dislodged during surgery and are difficult to reapply during the procedure; the recording will then not be as reliable in comparison with the baseline.

With this type of recording from the spinal cord, a direct or D wave is obtained along with I waves (indirect waves), which follow the D wave. The direct wave is so called because it results from direct stimulation of corticospinal tract neurons, in contrast to the indirect waves, which result from transsynaptic activation of these corticospinal tract neurons. This spinal-cord evoked response or potential (SCEP) is the result of both ascending and descending impulses and therefore does not reliably detect a pure motor signal. The resulting summated response is compromised by the large amplitude sensory response (Dawson *et al.*, 1991). In this way, impairment in responses derived from the motor tract pathway would be masked by the intervening sensory signals and so not be detected. This summated response is a mixed sensory and motor response and probably only monitors the integrity of the dorsal columns (Minahan *et al.*, 2001). The recording may be made from muscle groups (the most common technique employed), the spinal cord, or peripheral nerves. This technique could then result in potentially false-negative monitoring, which has been reported (López, 2004).

Because of numerous reports (Ben-David *et al.*, 1987) of motor deficits in the setting of unremarkable SEP monitoring in the literature, the development and more widespread use of transcranial electrical motor stimulation (TES) occurred. Transcranial electrical stimulation involves the use of either a separate electrical stimulator (Digitimer, UK) or electrical stimulation through standard EP equipment modules so that constant-current, varying voltage stimuli can be applied to the motor strip. The most widely used independent TES stimulator is manufactured in the UK.

The voltage can typically be increased as needed up to a maximum of 1000–1200 volts with varying durations of 0.5–3 ms and inter-stimulus interval of 1–5 ms. Anodal stimulation may allow for better responses, although anodal block can occur with a blocking of stimuli. Recording EEG disk or cup electrodes over the scalp are placed over the typical location for maximizing the motor strip. A typical placement for upper limb stimulation involves placing the anode 1 cm

Table 7.1 Montages used to record MEPs

C3′–Cz / C1–Cz upper and lower extremity MEPs
C4′–Cz / C2–Cz upper and lower extremity MEPs
C3′–C4′ or C1–C2 lower extremity MEPs

anterior to the C3 and C4 electrode sites and the cathode at Cz or Cpz. Collodium is used in the same way as for securing standard EEG electrodes for EEG recording. Alternatively, needle electrodes can be used but this increases the risk of scalp burns. The use of needle electrodes decreases preparation time compared with the use of surface electrodes, which require abrasion of the skin and application of collodium.

In the pediatric setting and increasingly even in the adult setting, electrodes are placed after induction in the operation room, for patient comfort. In this setting, needle electrodes allow for a more rapid application than surface electrode placement. The standard 10–20 EEG electrode placement is used. Electrode impedance should be below 5 kΩ. For lower limb stimulation these sites are reversed, so a typical placement of the cathode is 1 cm anterior to the C3 or C4 electrode site, referred to as M3 or M4, and the anode at Cz or Cpz, or 1 cm anterior at Mz. Alternatively, electrodes placed 1 cm medial to M3 at M1 or to M4 at M2 can be used and this may be the optimum site depending on the particular patient (see Table 7.1).

Scouting of the best electrode placement may be necessary in the operation room if this cannot be done preoperatively. Magnetic stimulation done preoperatively can help localize the optimum site that may be utilized in the operation room and save time. Although this is not routinely done in most centers at this time, it can be a very helpful approach. This preoperative scouting may also determine that MEP responses are not obtainable and therefore multiple attempts in the operating room can be averted, in this regard, if after some brief attempts with TES in the operation room, no MEP responses are seen after not obtaining these with magnetic stimulation in the unanesthetized patient preoperatively.

When responses cannot be obtained outside of the operating room it may in fact be more difficult to obtain these responses in the operating room, where other factors come into play. However, a brief attempt should be made with TES in the operating room, since it may be effective even in the absence of reliable responses with magnetic stimulation.

Recording responses depends on good muscle selection bilaterally and obtaining at least two muscle groups in each limb allows for redundancy of recording. Surface or subdermal needle electrodes can be used. Needle electrodes can be preferred, because they are less likely to be dislodged and typically have less impedance than surface electrodes. MEP recording will alternate with SEP recording, beginning once opening is complete and continuing during the period of decortication, and placement of instrumentation. The last MEP response may typically be obtained after all instrumentation is done and closing begins.

Clinical indications

Transcranial electric motor stimulation is indicated in surgeries in which the motor pathways are at risk of compromise (Chung *et al.*, 2009; Deletis and Kothbauer, 1998). In fact, the use of TES and multimodality monitoring in general allows a more aggressive and expansive surgical approach to occur when needed and a warning system for potential neurological injury in any complex surgery involving these pathways. These surgeries include scoliosis repair, intramedullary and extramedullary spinal tumor resection, disectomies and laminectomies, tethered cord repair, abdominal aorta repair, vascular surgeries of the spine, and any other surgery in which the motor pathways may be inadvertently compromised (Koyanagi *et al.*, 1993). In these cases, after induction of anesthesia, and once an anesthetic regimen is stabilized, intraoperative baseline motor-evoked potentials can be obtained and these ongoing potentials can be compared with the baseline potentials throughout the procedure.

Factors, other than the surgical technique, that may interfere with motor monitoring need to be taken into consideration, including the neuromuscular and neurological status of the patient, whether or not any anesthetic boluses were given, temperature of the patient, degree of neuromuscular blockade, blood pressure, or blood loss. All of these factors can effectively reduce the desired potential amplitudes. Figure 7.1 illustrates a loss of motor-evoked potentials from the lower extremities when the mean arterial pressure was reduced below a critical level, while Figure 7.2 shows an improvement, with the

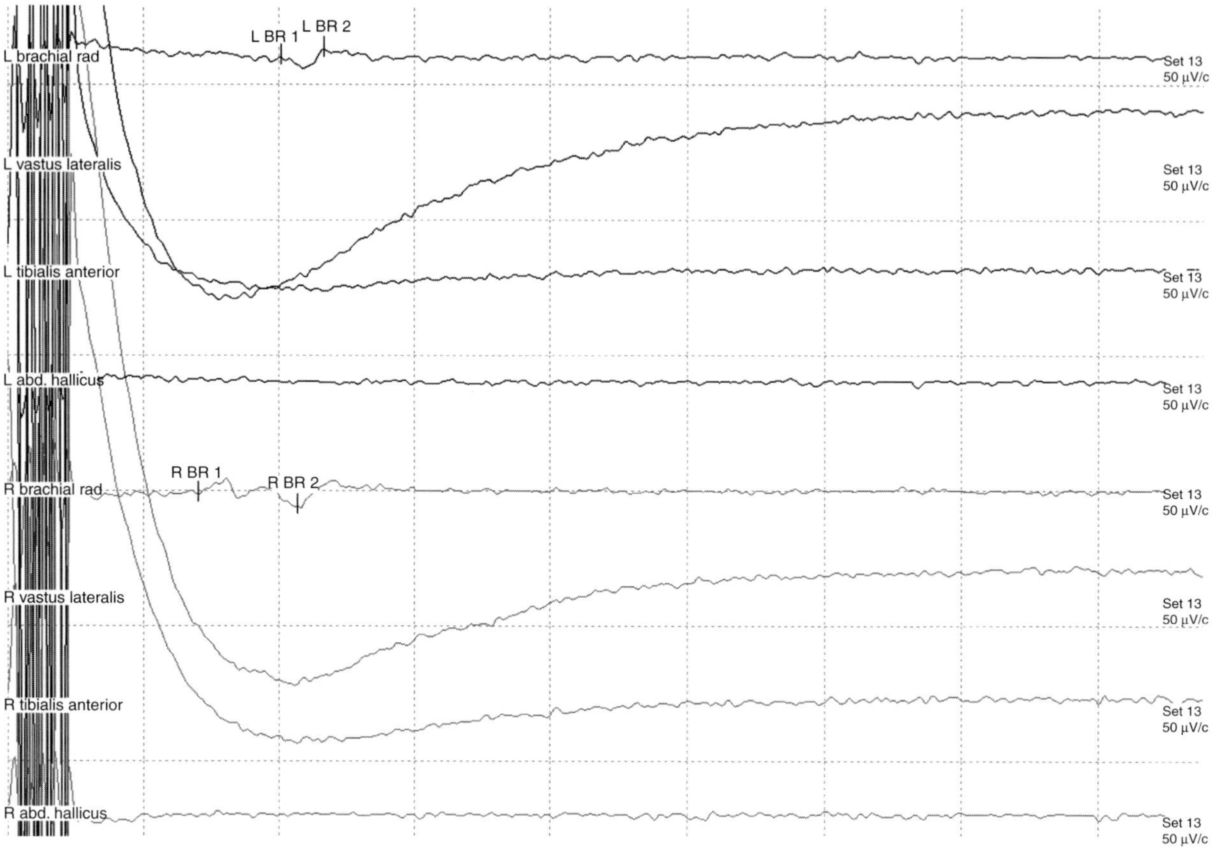

Figure 7.1 Loss of MEPs from the lower extremities following reduction in the mean arterial pressure. For color version, see plate section.

reappearance of motor potentials when the mean arterial pressure was raised.

Limitations in the use of TEP NIOM relate to the fact that at the present time and with the current techniques and equipment involved, significant variability of MEP amplitude can be expected within the same patient during a surgical case. This variability of MEP amplitude is also seen with the use of transcranial magnetic stimulation. This limits somewhat the ability to use amplitude of the MEP response as a component of the warning criteria. In this way, an "all-or-none" response will serve as the basis for warning of abnormality. A warning based on an all-or-none response can theoretically delay the warning that could otherwise be achieved if the MEP amplitudes were more stable throughout a case on a particular patient.

Although in general an "all-or-none" response serves as the basis for abnormality, in some cases where the MEP amplitude is relatively constant during the course of a case, so that appreciable amplitude variability is not seen, then a clear and persistent change in amplitude or distinctly progressive decrement in the amplitude could serve as an early warning prior to the impending loss of response altogether. This is important because delay in discovering a true change in MEPs can increase patient morbidity and the risk of paralysis or paraplegia.

Risks of TES

The side effects of TES are few and the most common side effects are lip or tongue lacerations. These lacerations are typically prevented by using a soft bite block and this is in common usage during these surgeries. More serious side effects, including seizures, cardiac arrhythmias, scalp burns, and muscle hematomas, are very uncommon. In fact, in patients without a history of prior seizure, both the theoretical risk and the apparent risk of developing a seizure after single-pulse transcranial electrical stimulation has

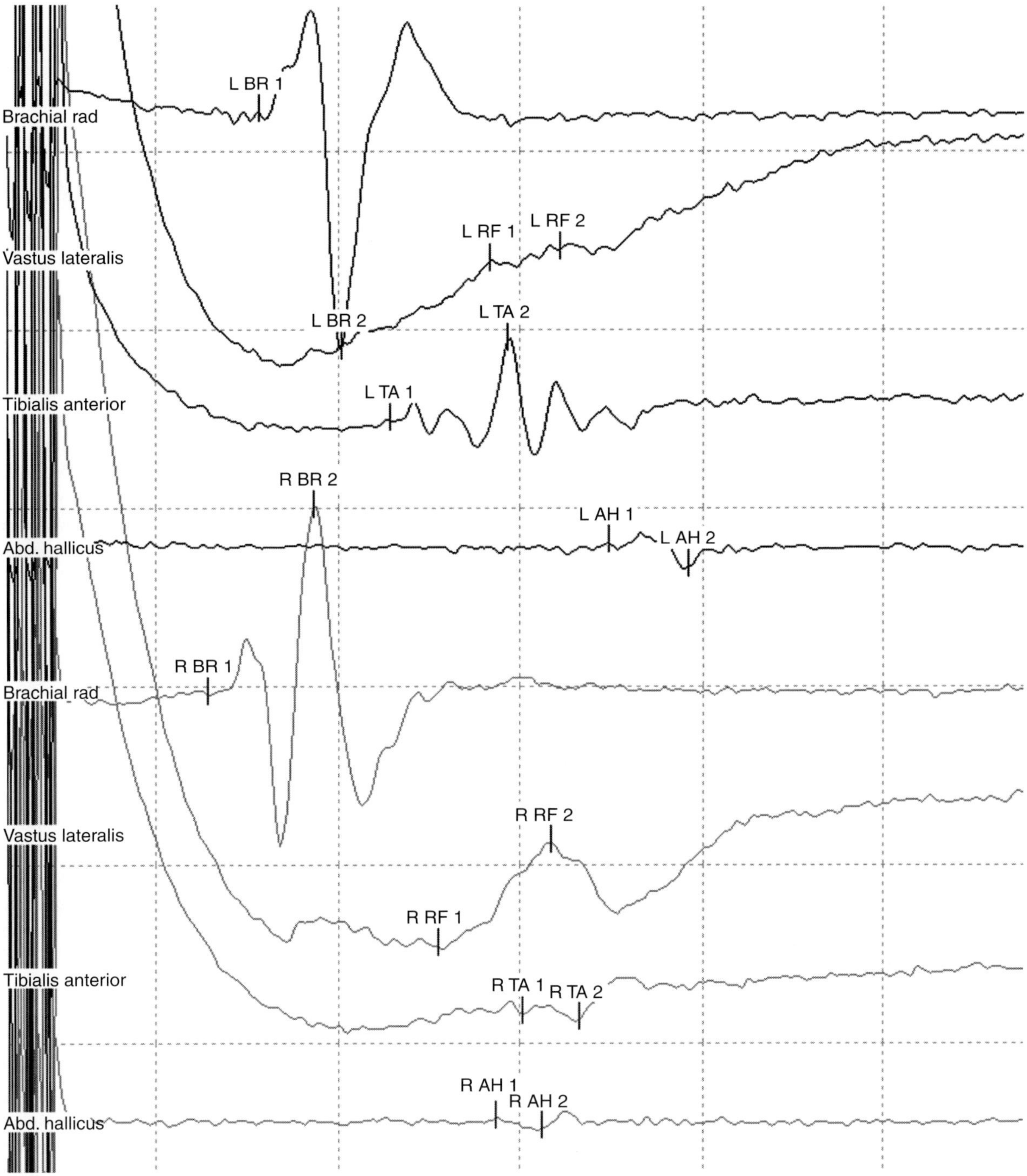

Figure 7.2 Return of MEPs in the lower extremities following an increase in the mean arterial pressure. For color version, see plate section.

no support in the current literature. Because of "epileptic kindling" and because of the lack of data on the use of TES in children at our institution we exclude patients from undergoing TcMEP monitoring if they have a history that suggests the potential risk for recurrent or uncontrolled seizures, as will be described. Candidates who may not be able to undergo TES should be selected prior to the procedure, if possible during preoperative testing. At our institution, and based on potential risks with the use of this procedure, patients with uncontrolled seizures or significant seizures are excluded, as are patients with any implantable devices, including a baclofen pump or cardiac pacemaker; patients with a prior history of intracranial surgery or vascular clips. Valvular cardiac disease is not necessarily a contraindication as long as a pacemaker is not present. The benefits of MEP monitoring should be weighed against the potential risk to the patient (Warner *et al.*, 1989) of the MEP procedure. For example, in a patient with well controlled seizures undergoing complicated spine surgery, the added potential benefit in performing TES during the surgery can be thought to outweigh the small risk of inducing a seizure during the surgical procedure. In this way, during the preoperative assessment, and discussions with the patient or family, a medical and neurological history is obtained, so that potential exclusions to the use of TES can be determined prior to the surgical procedure.

Effects of anesthetic and neuromuscular blocking agents on MEP responses

Muscle MEP responses are sensitive to the effects of anesthesia with effects at the neuromuscular junction and anterior horn cell. The use of continuous steady infusions of intravenous anesthesia with propofol or ketamine are best since decreases in the muscle MEP responses are seen with boluses >100 mg/kg. Intravenous benzodiazepines do not have a significant effect on MEP responses, although if large doses are used this can cause difficulty in awakening the patient easily either after the procedure is over or if a problem arises during the procedure and a wake-up test is done. Generally a train of MEP responses will help abate the effects of anesthesia and allow reliable effective MEP monitoring not achieved with single-pulse stimulation. After induction of anesthesia, neuromuscular blocking agents should be avoided so that they do not hamper the ability to obtain MEP responses. However, boluses of anesthetic agents and continuous high dose infusions of halogenated anesthetics greater than 0.5 MAC may have a detrimental effect on the recording of MEPs (Pelosi *et al.*, 2001). This may occur as the surgery continues, and an accumulation effect will occur that can be seen over time as a slowly progressive loss of MEP amplitude in association with prolongation of SEP latency and amplitude decrement (Iwaskaki *et al.*, 2003). The traces in Figure 7.3 were recorded when an increase in the sevoflurane infusion level to >2% occurred during NIOM and this was associated with a loss of all MEPs. Subsequent recovery of MEPs occurred when the anesthetic level was decreased, as seen in the traces shown in Figure 7.4.

In most cases, the effects seen on MEPs will be seen in the SEP recording as well but this is not always the case. Changes in the SEP recordings may be seen later than those shown in the MEP recording. When a progressive change in NIOM is seen and thought attributable to anesthetic effect, communication with the anesthesiology team is necessary so that the anesthetic level can be lowered if possible in order to adequately monitor neurophysiological signals. The surgical team must be made aware of any changes in neurophysiological monitoring, including when the changes are thought to be due to anesthesia, temperature, or arterial pressure, so that necessary interventions can occur.

Ketamine is typically avoided in the operating room with NIOM due to its potential to increase the amplitude of cortical SEP potentials and maybe also MEPs, which would make interpretation more complex and possibly confusing. However, it might be useful in patients in which attempting to obtain adequate MEPs for monitoring is problematic. Unfortunately, this information may not be known ahead of the surgical procedure. The suggestion of this possibility in the operation room may be obtained in a preoperative assessment if transcranial magnetic stimulation results in small MEPs and technical factors are excluded. This information will be useful because an electric field is induced by transcranial magnetic stimulation. However, transcranial electrical stimulation and transcranial magnetic stimulation do not necessarily yield exactly the same findings and therefore attempts to extrapolate the findings on one study to the other should be done cautiously. Using ketamine to facilitate NIOM with MEPs including in children has been studied by several researchers (Erb *et al.*, 2005).

Baseline MEPs are obtained before or after baseline SEPs are obtained. Once opening begins, evoked potential collections will typically stop and not restart until after opening is completed, owing to the use

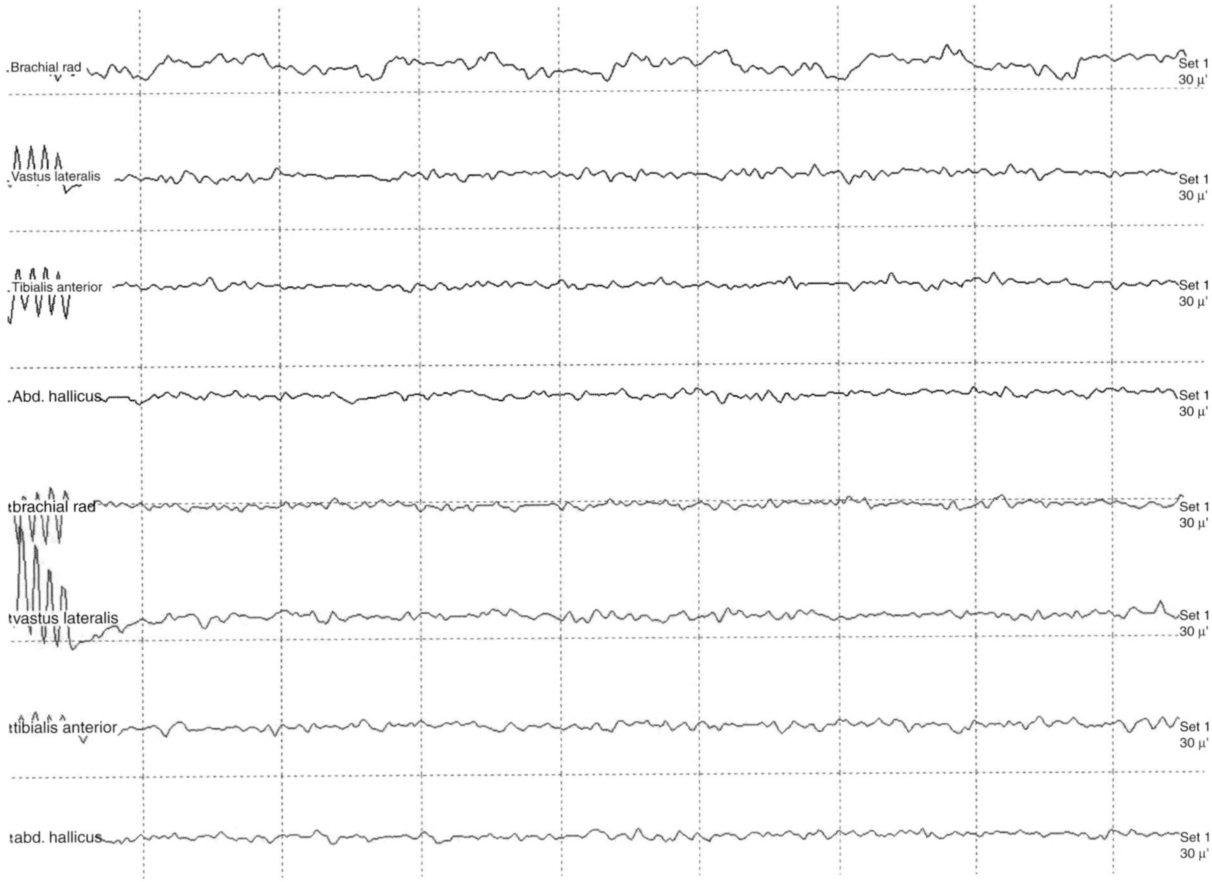

Figure 7.3 Increasing the level of sevoflurane anesthesia resulted in a loss of MEP potentials from legs and arms. For color version, see plate section.

of bovie or electrocautery devices, which do not allow collecting and averaging of EP responses reliably. Motor evoked potential responses may vary depending on operating room variables, and therefore it is best to monitor at least two muscle groups in each limb. At our institution, in our protocol, a distal muscle in the lower extremity, such as the abductor hallicus, is easily recognized for needle insertion. Other muscles will typically include the vastus lateralis in the lower extremities and in the upper extremities a hand muscle, such as the abductor pollicus brevis or thenar muscle group, as well as the brachioradialis or biceps muscle. These muscles are usually easily recognizable in patients even with poorly developed muscles. Table 7.2 illustrates a typical recording muscle protocol used at our institution. Variations will be made based on the particular patient and surgery involved. Using this protocol allows for two muscle groups in each limb, in case one signal is lost, and a more distal as well as a more proximal muscle group to be used. Depending on the specific levels of the surgical procedure, muscles will be selected that are appropriate for the surgical procedure and levels involved.

Warning criteria and patterns of abnormality

Because of the variability of MEP response latencies and amplitudes and their dependence on the alpha motor neuron pool excited, typically an all-or-none response is monitored. If a response that was previously present is no longer present, this would indicate a significant change (Bose *et al.*, 2004). This result should be reliably reproduced after more than one stimulation train. If bilateral upper and lower

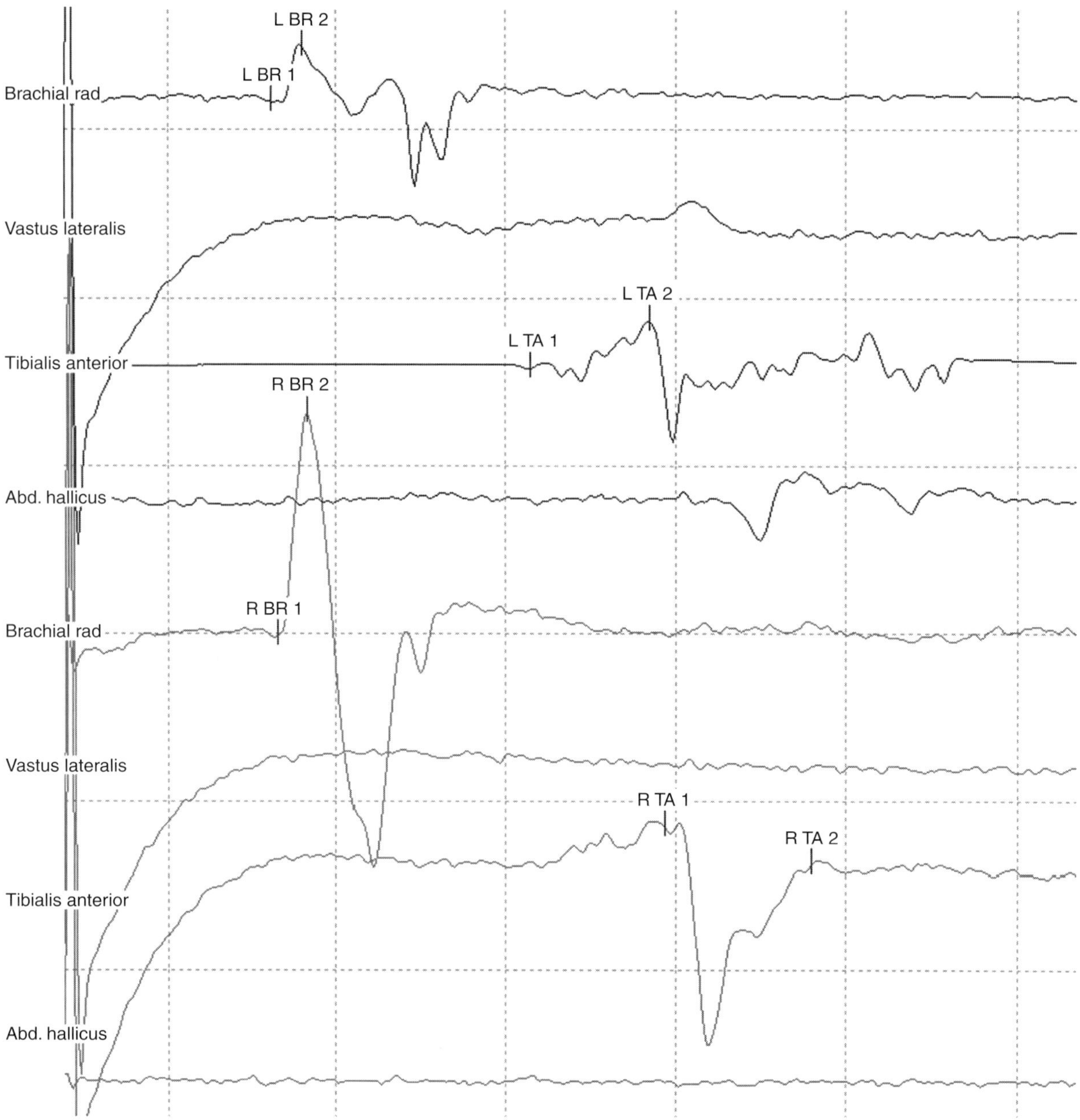

Figure 7.4 With a reduction in sevoflurane level, the MEP responses returned in all limbs. For color version, see plate section.

extremity responses are changed, one should consider a systemic etiology, such as blood pressure changes or anesthetic factors (Othman *et al.*, 2004). Considerations include boluses or a summated effect referred to as "anesthetic fade", in which over a prolonged surgery there is a small but measurable decrease in evoked response amplitudes due to the summated effects of anesthesia even in the absence of boluses. In contrast to this would be a change that only affects the lower extremities or affects an upper and lower extremity on only one side; in these scenarios a surgical etiology is more likely involved (Jellinek *et al.*, 1991).

Technical factors have to be eliminated, including checking electrode impedances and equipment failures. (Change batteries in the stimulation devices!)

Table 7.2 Muscle groups that can be used in recording MEPs

Thenar muscles	C8, T1
Biceps or brachioradialis	C5, 6
Abductor hallucis	S1, S2
Tibialis anterior	L4–5
Other lower extremity muscles can be used given a particular patient and include:	
Gastrocnemius	S1, S2
Vastus lateralis	L2–4
Rectus femoris	L2–4

Given the fact that an all-or-none response is truly what is being monitored, still a verifiable amplitude decrease when the amplitude had been consistent may also be considered a warning and this is where the experience and expertise of the interpreting neurophysiologist physician is important in deciding whether a warning is warranted or not (Osburn, 2006). In cases in which there is not a significant variability of MEP amplitude and then a significant (possibly 30% or greater although an even smaller decrease if persistent may be warranted) decrease in amplitude occurs and is persistent, this may be a significant enough change to warrant communication to the surgeon and possible intervention.

Sometimes with a thoracic or upper lumbar surgical insult one will see a decrease in the amplitude of the MEP responses from one or both lower extremities over several trials. Other times the loss occurs abruptly and may correlate with tightening of a rod or placement of a hook. The traces in Figure 7.5 show that there is loss of MEPs from one lower extremity during rod placement in a scoliosis surgery. In Figure 7.6, one sees that loosening of the rod is associated with recovery of the MEP waveforms from that extremity. In these cases, rapid identification and communication to the surgical staff will allow for an intervention such as a loosening of that rod or removal of a hook with improvement and reappearance of the MEP potentials.

Motor evoked potentials are typically, although not always, more sensitive to spinal-cord ischemia and manipulation than SEP potentials (Smith *et al.*, 2007). In most cases, soon after the MEP potentials show a change there will be a change in the SEPS correlating to that region. In some cases, SEPs remain unchanged despite MEP disappearance (Deletis and Sala, 2008). In some cases, particularly when the MEP change was only identified once the signal disappeared, the SEPs may change simultaneously. With resolution, the MEPs typically will show improvement earlier than the SEPs. Restoration of MEP potentials with or without SEP involvement is associated with a good postoperative outcome (Jones *et al.*, 1996). In contrast, when MEP potentials do not reappear even in the presence of SEPs there will very likely be motor deficit postoperatively. When MEP potentials are present but SEP potentials do not return (typically much later) transient sensory deficit can be expected without motor compromise (Lee *et al.*, 2006).

When a significant SEP or MEP change occurs during the course of a surgery the surgical response is to attempt to correct or undo whatever surgical manipulation was being done at the time the change was first noticed (Wiedemayer *et al.*, 2002). In addition, increasing the mean arterial pressure to 90 mmHg will help to provide additional blood flow to a possibly compromised spinal cord or cortical or subcortical region, depending on the surgery involved. Dexamethasone may be added following the spinal-cord trauma protocol, although increasing the MAP and undoing the surgical manipulation is typically most helpful. In some centers, attempting to awaken the patient to see if they can move their feet, called the wake-up test, may be done but this can be associated with unintentional extubation, patient anxiety, and a false sense of security as it only tests distal motor function. Monitoring MEPs during this time is the more useful modality than attempting a wake-up test.

Increasing numbers of studies have shown the beneficial utility of motor potential monitoring of motor tract pathways to aid in preventing neurological deficit, most notably paraplegia or paralysis (Weinzierl *et al.*, 2007). Several studies have now shown that MEPs may and often do signal evidence of change prior to a change noted in SEP recordings (Ben-David *et al.*, 1987) and that MEPs may in fact change in the setting of false-negative SEPs.

In contrast, there have been fewer reports of false-negative MEPs when true SEP changes occurred. In addition, when MEPs showed improvement within the first few minutes, paraplegia was typically avoided (Pelosi *et al.*, 2002). Combined monitoring of motor and sensory tracts lessens the risk of false-negative outcomes when monitoring is taking place (Schwartz *et al.*, 2007; Sutter *et al.*, 2007).

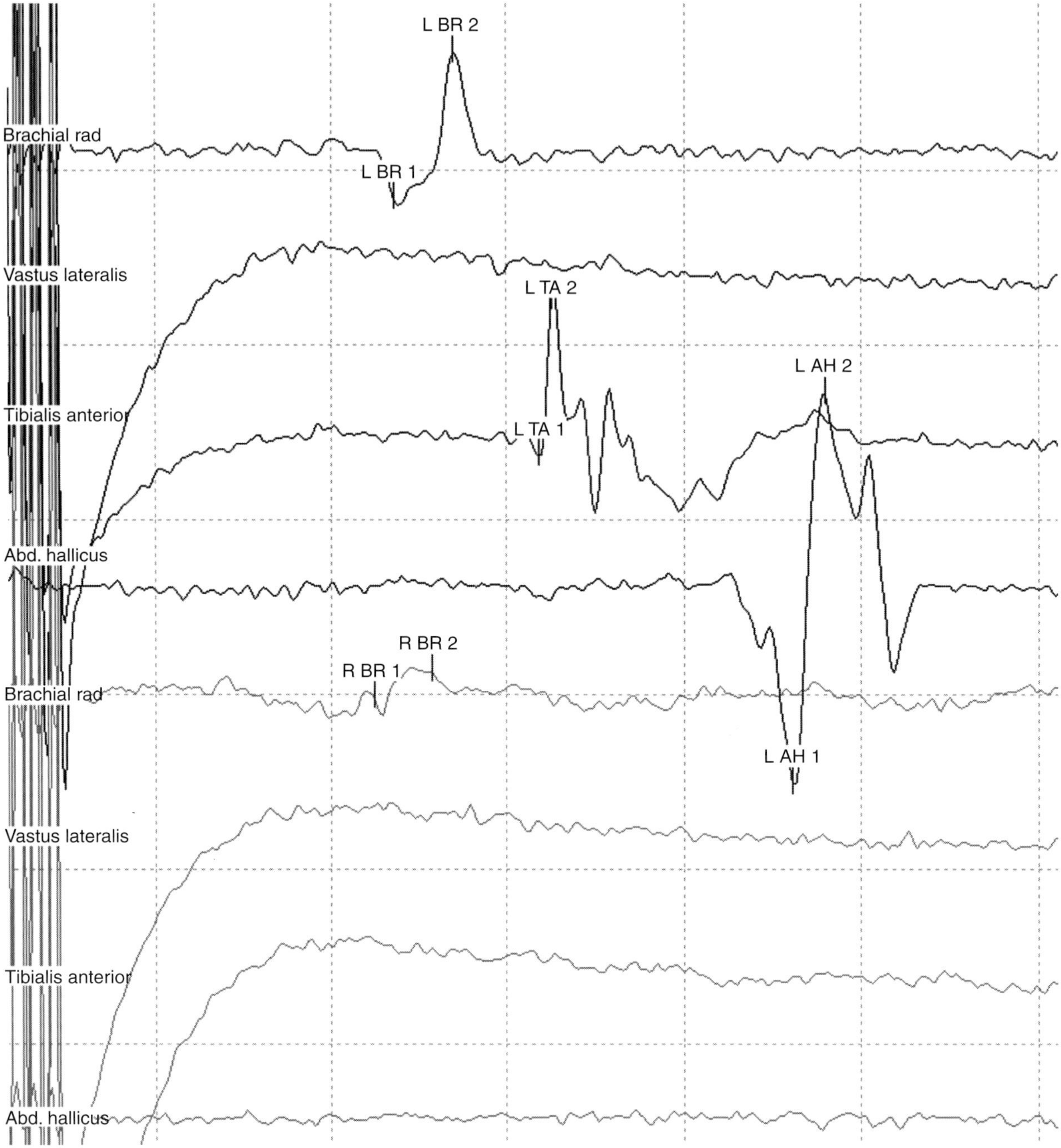

Figure 7.5 Loss of MEPs from the right leg following tightening of a rod during scoliosis surgery. For color version, see plate section.

Conclusion

Transcranial electrical stimulation is a reliable method of motor pathway monitoring and can provide a beneficial impact on the prevention of paraplegia and paralysis during surgeries in which the motor pathways may be compromised. This modality, used in combination with somatosensory evoked potential monitoring, with trained experienced physician supervision and interpretation not only serves to warn of possible ischemic or mechanical injury to

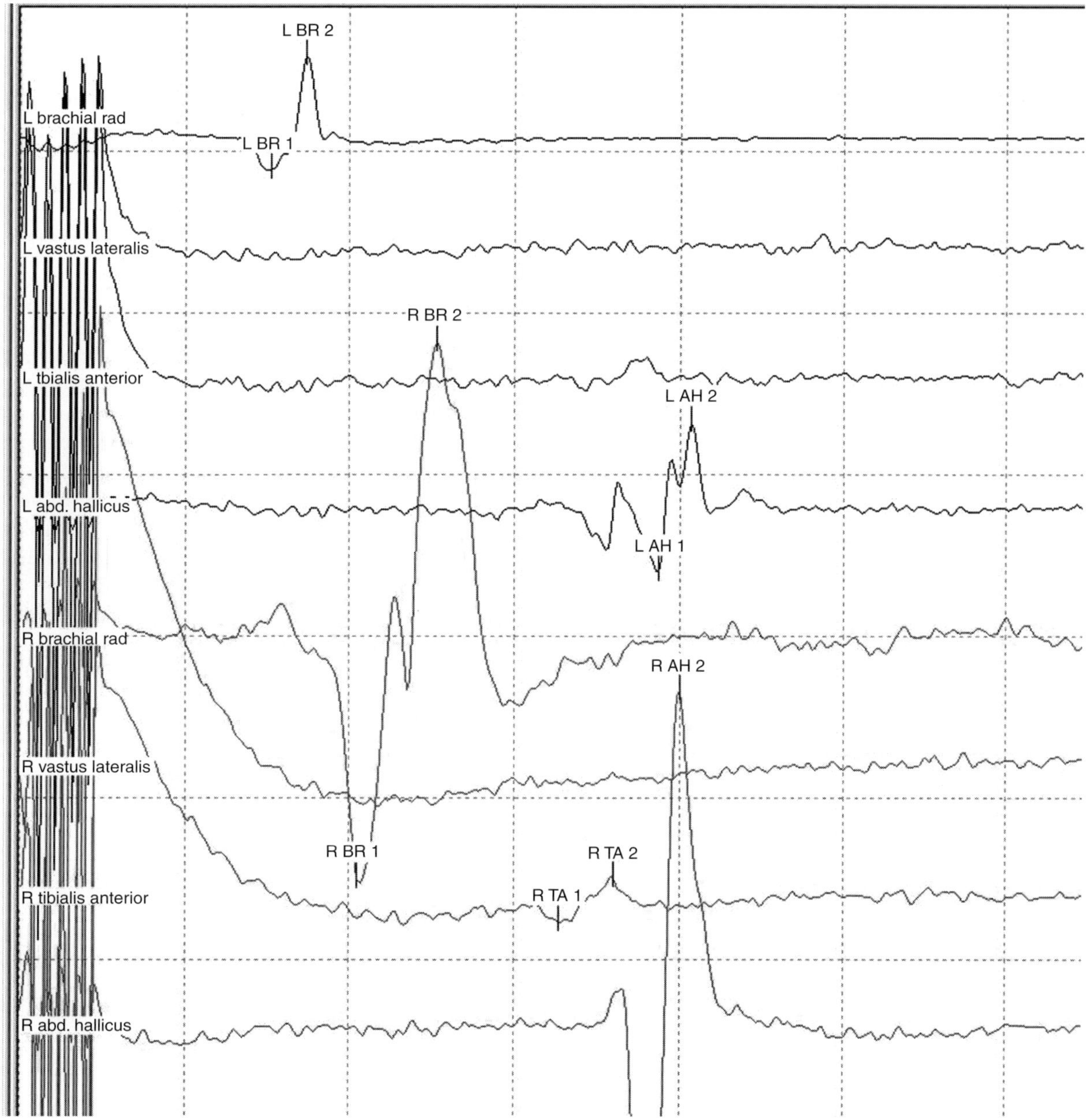

Figure 7.6 Return of MEPs from the right leg with loosening of rod during scoliosis surgery. For color version, see plate section.

the motor pathways involved but allows an expansion of the surgical procedure over what might otherwise be obtained. Continued loss of these potentials for longer than minutes can be associated with paraplegia if systemic and technical factors are eliminated. In contrast, restoration of these potentials within minutes of a change is most often associated with a favorable neurological outcome.

Case examples for Chapter 7

Case 1

Figures 7.5 and 7.6 are taken from the case of an otherwise healthy 14-year-old female patient undergoing a posterior fusion as part of scoliosis repair for idiopathic scoliosis. This condition had been progressive and associated with some mild back discomfort but no weakness. Both SEPs and MEPs from both the upper and lower extremities were monitored throughout the case and compared with baseline values after induction. With placement of the first rod the surgeon proceeded to tighten it, and at that point MEPs were lost from the right leg without loss or delay in SEPs (Figure 7.5). The surgeon was alerted and loosened the rod and the MEPs returned within one minute of intervention and remained unchanged throughout the duration of the case (Figure 7.6). The second rod was placed without incident. The patient awoke without deficit.

Case 2

A 15-year-old female was undergoing an uneventful surgical repair for scoliosis with posterior fusion with SEP and TcMEP monitoring. During this procedure, sevoflurane was increased progressively throughout the case. The neurophysiology team alerted the surgeon and CRNA (certified registered nurse anesthetist) that SEP amplitudes were decreased cortically, a delay in SEP latency was present cortically, and all MEP amplitudes were decreased but still present, probably due to the anesthetic levels. Once the sevoflurane reached 2.5%, MEPs were lost from both legs (Figure 7.3). The sevoflurane was decreased back down to 1.5% and the MEPs returned after 15 minutes (Figure 7.4) and the SEPs after 20 minutes of the change. The patient awoke without deficit.

Case 3

An 11-year-old male was undergoing resection of a spinal angioma that had bled. To assess the expanse of resection possible and ensure integrity of spinal tracts, TcMEPs and SEPs were monitored. Mean arterial pressure decreased to 52 mmHg during the case and subsequently MEPs from the lower extremities were lost but SEPs remained unchanged. As the MAP increased to 70 mmHg the MEPs returned in the lower extremities within several minutes and were robust from all limbs. The surgery proceeded without further event and the patient awoke without deficit.

Questions

1. Unlike motor stimulation in the epidural or subdural region, transcranial electric stimulation allows for which of the following to occur?
   (a) Combined orthodromic sensory and descending pathway motor potentials to be recorded
   (b) Recording of pure motor signals without the concern of involvement of antidromic sensory volleys being mixed in the signals being reviewed
   (c) Pure orthodromic sensory signals to be recorded
2. Which of the following statements about MEPs is true?
   (a) Continued loss of these potentials for more than five minutes can be associated with paraplegia if systemic and technical factors are eliminated
   (b) MEPs are resilient to the effects of hypoxia and ischemia and would not be expected to be lost or affected with these conditions
   (c) The presence or absence of MEPs is very variable and these potentials are not as reliable a monitoring modality as SEPs in most clinical scenarios in which NIOM is utilized
3. The use of MEPs has limited the need for which of the following tests?
   (a) Somatosensory evoked potentials
   (b) QSART
   (c) Wake-up test
   (d) Electromyography
4. An appropriate response to an MEP or SEP change during a spinal surgery may include which of the following?
   (a) Lowering the mean arterial pressure to 40 mmHg
   (b) Hyperventilation
   (c) Increasing the inhalational anesthetic level to increase sedation
   (d) Raising the mean arterial pressure to 90 mmHg
5. "Anesthetic fade" refers to which of the following?
   (a) Gradual decrease in anesthetic levels over the course of a surgery
   (b) Gradual increase in anesthetic levels with summation and use of boluses over the course of a surgery
   (c) Decreasing efficacy of the anesthetic over the course of the surgery
   (d) Shifts in anesthetic personnel during a surgical procedure
6. Which of the following is the correct scalp electrode placement for MEP monitoring of the lower limbs?

(a) The cathode is placed 1 cm anterior to the C3 or C4 electrode site, referred to as M3 or M4, and the anode at Cz or Cpz or 1 cm anterior at Mz
(b) The anode is placed 1 cm posterior to the C or C4 electrode site, referred to as M3 or M4, and the cathode at Cz or Cpz or 1 cm posterior at Mz
(c) The anode is placed 1 cm anterior to the C3 or C4 electrode site, referred to as M3 or M4, and the cathode at Cz or Cpz or 1 cm anterior at Mz

7. Which of the following is true?
(a) Once surgical manipulation occurs, evoked potential collections will typically stop due to the presence of 60 Hz activity, which interferes with accurate interpretation of data
(b) Once opening begins, evoked potential collections will typically stop until after opening, due to the use of bovie or electrocautery devices that do not allow collecting and averaging of EP responses reliably
(c) Once closing begins, evoked potential collections should stop, as no evoked potential changes should be expected at that time

8. The wake-up test is associated with all of the following except which?
(a) Possibility of unintentional extubation
(b) Potential for patient anxiety
(c) Possible false sense of security as it only tests distal motor function
(d) A reliable measure of sensory impairment

9. Which of the following is true?
(a) When MEP potentials are present but SEP potentials do not return (typically much later) transient sensory deficit can be expected without motor compromise
(b) When SEP potentials are present but MEP potentials do not return (typically much later) transient sensory and motor deficits can be expected
(c) When MEP potentials are present but SEP potentials do not return (usually return earlier) then transient motor deficits can be expected without sensory deficits

10. Which of the following is a modality that can be utilized in an awake outpatient setting to evaluate motor pathways?
(a) BAER
(b) Cranial nerve monitoring
(c) Repetitive train transcranial electric stimulation
(d) Transcranial magnetic stimulation

References

Ben-David, B., Haller, G., and Taylor, P. (1987). Anterior spinal fusion complicated by paraplegia: a case report of a false-negative somatosensory-evoked potential. *Spine*, **12**, 536–539.

Bose, B., Sestokas, A. K., and Schwartz, D. M. (2004). Neurophysiological monitoring of spinal-cord function during instrumented anterior cervical fusion. *Spine Journal*, **4**(2), 202–207.

Chung, I., Glow, J. A., Dimopoulos, V., *et al.* (2008). Upper-limb somatosensory evoked potential monitoring in lumbosacral spine surgery: a prognostic marker for position-related ulnar nerve injury. *Spine Journal*, **9**, 287–295.

Dawson, E. G., Sherman, J. E., Kanim, L. A., and Nuwer, M. R. (1991). Spinal cord monitoring: results of the Scoliosis Research Society and the European Spinal Deformity Society survey. *Spine*, **16**(suppl.), S361–364.

Deletis, V. and Kothbauer, K. (1998). Intraoperative neurophysiology of the corticospinal tract. In *Spinal Cord Monitoring*, ed. E. Stalberg, H. S. Scharma, and Y. Olsson. Vienna: Springer, pp. 421–444.

Deletis, V. and Sala, F. (2008). Intraoperative neurophysiological monitoring of the spinal cord during spinal cord and spine surgery: a review focus on the corticospinal tracts. *Clinical Neurophysiology*, **119**(2), 248–264.

Erb, T. O., Ryhult, S. E., Duitmann, E., *et al.* (2005). Improvement of motor evoked potentials by ketamine and spatial facilitation during spinal surgery in a young child. *Anesthesia and Analgesia*, **100**, 1634–1636.

Iwasaki, H., Tamaki, T., Yoshida, M., *et al.* (2003). Efficacy and limitations of current methods of intraoperative spinal-cord monitoring. *Journal of Orthopaedic Science*, **8**, 635–642.

Jellinek, D., Jewkes, D., and Symon, L. (1991). Non-invasive intraoperative monitoring of motor evoked potentials under propofol anesthesia: effects of spinal surgery on the amplitude and latency of motor evoked potentials. *Neurosurgery*, **29**, 551–557.

Jones, S. J., Harrison, R., Koh, K. F., Mendoza, N., Krockard, H. A. (1996). Motor evoked potential monitoring during spinal surgery: responses of distal limb muscles to transcranial cortical stimulation with pulse trains. *Electroencephalography and Clinical Neurophysiology*, **100**, 375–383.

Koyanagi, I., Iwasaki, Y., Isu, T., *et al.* (1993). Spinal cord evoked potential monitoring after spinal cord stimulation during surgery of spinal-cord tumors. *Neurosurgery*, **33**(3), 451–460.

Lee, J. Y., Hilibrand, A. S., Lim, M. R., *et al.* (2006). Characterization of neurophysiologic alerts during anterior cervical spine surgery. *Spine*, **31**(17), 1916–1922.

López, J. R. (2004). The use of evoked potentials in intraoperative neurophysiologic monitoring. *Physical Medicine and Rehabilitation Clinics of North America*, **15**(1), 63–84.

Minahan, R. E, Sepkuty, J. P., Lesser, R. P., Sponseller, P. D., and Kostuik, J. P. (2001). Anterior spinal cord injury with preserved neurogenic "motor evoked potentials". *Clinical Neurophysiology*, **112**, 1442–1450.

Osburn, L. L. (2006). A guide to the performance of transcranial electrical motor-evoked potentials. Part 1. Basic concepts, recording parameters, special considerations, and application. *American Journal of Electroneurodiagnostic Technology*, **46**(2), 98–158.

Othman, Z., Lenke, L. G., Bolon, S. M., and Padberg, A. (2004). Hypotension-induced loss of intraoperative monitoring data during surgical correction of Scheuermann kyphosis: a case report. *Spine*, **29**(12), E258–265.

Pelosi, L., Stevenson, M., Hobbs, G. J., Jardine, A., and Webb, J. K. (2001). Intraoperative motor evoked potentials to transcranial electrical stimulation during two anesthetic regimens. *Clinical Neurophysiology*, **112**, 1076–1087.

Pelosi, L., Lamb, J., Grevitt, M., *et al.* (2002). Combined monitoring of motor and somatosensory evoked potentials in orthopedic spinal surgery. *Clinical Neurophysiology*, **113**, 1082–1091.

Schwartz, D. M., Auerback, J. D., Dormans, J. P., *et al.* (2007). Neurophysiological detection of impending spinal cord injury during scoliosis surgery. *Journal of Bone and Joint Surgery (America)*, **89**, 2440–2449.

Smith, P. N., Blazer, J. R., Khan, M. H., *et al.* (2007). Intraoperative somatosensory evoked potential monitoring during anterior cervical discectomy and fusion in nonmyelopathic patients a review of 1,039 cases. *Spine Journal*, **7**, 83–87.

Sutter, M., Eggspuehler, A., Grob, D., *et al.* (2007). The diagnostic value of multimodal intraoperative monitoring (MIOM) during spine surgery: a prospective study of 1,017 patients. *European Spine Journal*, **16**(2), S162–S170.

Warner, M. A., Offord, K. P., Warner, M. E., *et al.* (1989). Role of preoperative cessation of smoking and other factors in postoperative pulmonary complications: a blinded prospective study of CAB patients. *Mayo Clinic Proceedings*, **64**, 609.

Weinzierl, M. R., Reinacher, P., Gilsbach, J. M., and Rohde, V. (2007). Combined motor and somatosensory evoked potentials for intraoperative monitoring: intra- and postoperative data in a series of 69 operations. *Neurosurgical Review*, **30**, 109–116.

Wiedemayer, H., Fauser, B., Sandalcioglu, I. E., Schafer, H., and Stolke, D. (2002). The impact of neurophysiological intraoperative monitoring on surgical decisions: a critical analysis of 423 cases. *Journal of Neurosurgery*, **96**, 255–262.

8 Electrocorticography and intraoperative electroencephalography

Marc R. Nuwer

Introduction

Electroencephalography (EEG) and electrocorticography (ECoG) are used in surgery to test and monitor the brain.

Intraoperative EEG is used most often to assess for signs of intraoperative ischemia of the cerebral hemispheres. The recording procedures are similar to those used for outpatients. The most common types of surgery using intraoperative EEG are carotid endarterectomy, cerebral aneurysm clipping, and cardiac surgery. Not only can EEG screen for signs of cerebral ischemia but it can also be used to determine if the patient is sufficiently deep in deliberate burst suppression.

Electrocorticography is the procedure in which EEG is recorded directly from the exposed cerebral cortex during a craniotomy. The goal of ECoG is to identify regions suitable for surgical resection and those other areas of cortex that should be left intact. Related procedures are sometimes performed together with ECoG. One related technique is direct cortical stimulation, a procedure for localizing motor, sensory, or language cortex. A related technique is to leave grid or strip electrodes in the subdural space for days or weeks to monitor for epileptic seizures in the inpatient epilepsy monitoring unit.

Electrocorticography was the first technique used in intraoperative neurophysiology (Nuwer, 2008a). Beginning in the 1940s, ECoG was used to localize epileptogenic tissue during neurosurgical treatment of medically refractory partial epilepsy (Walker *et al.*, 1947; Jasper, 1949; Marshall and Walker, 1949; Meyers *et al.*, 1950; Penfield and Jasper, 1954; Ajmone Marsan and Baldwin, 1958). The equipment for ECoG has evolved since then, but the clinical goal remains relatively unchanged. Another common application of ECoG is for the surgical resection of tumors, as it can identify dysfunctional tissue to be resected and physiologically normal regions to be spared. The direct cortical stimulation technique can also identify critical motor and language regions so that they can be spared from resection.

Recording

Scalp EEG

Traditional digital EEG machines are used to record these intraoperative procedures. Routine EEG monitoring in surgery uses the typical 21-electrode International Federation of Clinical Neurophysiology's 10–20 electrode placement system. This includes the 16 electrodes in the parasagittal and temporal chains as well as the three sagittal electrodes and two ear electrodes. All those electrodes can be used in carotid endarterectomy and cardiac surgery. This provides good coverage of the cerebral hemispheres sufficient to identify not only global and hemisphere-wide ischemia but also the smaller, more localized regions of change that can occur with emboli.

In craniotomies, some electrodes might not be placed. The craniotomy site itself includes some scalp sites. Electrodes that would have gone there may be placed at nearby substitute sites. On other occasions, strip or grid electrodes are placed in the surgical site to record directly from cerebral cortex in lieu of the missing scalp sites.

For scalp recordings, disk electrodes are suitable. Collodion or other glue may be used to keep them in place during long cases. When surgery is urgent and insufficient time is available to place disk electrodes or to use collodion, needle electrodes are a suitable quickly placed substitute.

Filters are set to 0.3 Hz and 70 Hz in most cases. The use of a low filter below 1.0 Hz helps to identify the slowing present with ischemic injury. The notch 60 Hz (or 50 Hz) filter can be used for EEG, unlike the usual method of avoiding the notch filters for SEPs. The high filter at 70 Hz or higher is helpful to record completely the 20–30 Hz cerebral activity due to anesthesia. Clinical criteria for change from baseline require accurate recording both of the ischemia-related slow activity and the anesthesia-related fast activity.

Page speed, previously known as paper speed, is slowed so that 30–60 seconds fit onto one digital screen sweep for many routine intraoperative EEGs. This horizontally compresses the EEG signal. The advantage of compression is to help detect changes over a one-minute time frame. A slow sweep speed also makes changes in slow or fast activity or amplitude attenuation more noticeable.

Modern digital equipment allows for simultaneous screen display of two EEG segments. The left window can be set to display a baseline EEG segment. This assists the reader's visual memory of baseline frequency content. The right window can be set to display the most recent 30–60 seconds of EEG. The reader can continually refer between the baseline and the current activity, looking for relevant changes. Such a visual comparison with the baseline is helpful for determining whether changes have developed slowly. The reader can observe quick changes on the current activity screen and slow changes by comparing the current and baseline windows.

Montages are similar to those for outpatient EEG testing. Bipolar or referential montages can be used. Usually a single montage is used throughout a case. Readers have their own preferences for setting the montage. One option is to use 16 channels of temporal and parasagittal bipolar electrode chains along with two channels between pairs of the three sagittal electrodes. This produces 18 channels for monitoring. In this option, the ear electrodes are not used.

For a more extended review of EEG monitoring during surgery, see Nuwer *et al.* 1993; 2008b.

Electrocorticography

Recording ECoGs from the exposed cerebral cortex uses strip and grid electrodes. These have three-millimeter-wide metal disks embedded in a flexible silicon rubber matrix. The disks are one centimeter apart, as measured from disk center to center. Disks are made of stainless steel or a platinum–iridium alloy. A bundle of fine wires exits the silicon rubber strip or grid, and the bundle ends in a multi-pin connector. This can be mated to a multi-pin adaptor with thicker, sturdier wires or ribbon cable. At the other end of this cable is a set of female pins that can be plugged into an ordinary EEG machine's electrode junction box. The strips and grids come in different sizes. Commonly used sizes in surgery are the 1×4 and 1×8 strips and the 4×5 grids. For patients in whom these are left in place for days or weeks to catch epileptic seizures, sizes are available up to 8×8.

Electrocorticography grids and strips tend to curl up or pull slightly away from the cortex at their edges. This is especially a problem when the electrode lies over a sulcus. The curling causes poor contact between the electrode and the underlying cortex. To prevent this curling up, the surgeon should put wet cottonoid over the strips or grid. This weighs down the silicon rubber electrode matrix so that each metal electrode makes good electrical connection with the underlying cortex.

Filters are set at 0.3 Hz and 70–100 Hz. The 60 (or 50) Hz notch filter can be switched on if needed. These filter settings allow for identification of epileptic activity in the higher frequency range as well as identification of slow waves associated with regions of pathology.

For ECoG, the page sweep speed is typical of routine EEG, not slowed as in EEG intraoperative monitoring during carotid endarterectomy. This more familiar page speed allows the reader to indentify epileptic activity in the ECoG more accurately.

Montages may be made bipolar between adjacent contacts on the grid or strip. Alternatively, a referential montage can be used, often one in which each grid or strip electrode is measured compared with a distant reference electrode. A clip electrode placed on muscle or a disk electrode on the scalp can serve as a reference electrode. Referential recordings are often used without excessive contamination of the recordings by an active reference.

The surgeon places the grids and strips over regions needing monitoring or testing. Often they are moved to alternate locations as needed. For epilepsy resections, the strips may be slid under the temporal lobe so as to measure from mesial structures. In tumor resections, grids and strips can be slid under other lobes, under the skull in the subdural space, or into the interhemispheric fissure. During direct cortical stimulation, one or several strips are placed near the stimulation sites.

Stimulation

Direct cortical stimulation requires special instrumentation and precautions. Generally the equipment used is specifically designed for this purpose. The surgeon may hold a bipolar stimulation wand against the cortex. This is used most often in the operating room where the cortex is directly exposed. Alternatively, the stimulation may be delivered though the electrode contacts of the grids or strips. The latter is most often used in the subacute recordings on an epilepsy monitoring unit. The stimulating wand is about 30 cm long, with two metal ball tips separated by a few millimeters. The wand can be moved quickly among cortical locations so as to test many different places in a relatively short time.

Stimuli are bipolar and biphasic. Bipolar means that the current runs between the two adjacent active contacts of the wand. Biphasic means that the current switches polarity half way through each stimulus pulse. The biphasic stimulation method is preferred so as not to leave a net polarization on either metal contact. Polarization could move metal ions onto the brain, and alter the electrical sensitivity of the underlying cortex to the stimulus pulses. If the stimulating wand electrodes are steel, metal ions have the added risk that cortex can be injured by the iron ions. For safety reasons, the noble metal platinum or a platinum alloy is the preferred metal for the stimulating electrode tips, because of concern about transfer of iron ions into the cortex with steel electrodes. Using biphasic stimulation avoids that risk because of the cancellation of charges within each stimulation pulse.

Safety studies have been published. They suggest that monophasic pulses are more likely to cause tissue injury (Mortimer *et al.*, 1970; Pudenz *et al.*, 1975). Studies also suggested that the upper limit of the safe charge densities is around 40 ($\mu C/cm^2$)/phase (Agnew and McCreery, 1987).

A 10 mA stimulating current produces a peak current density of 0.05 amperes per square centimeter (A/cm^2) in the region immediately beneath typical electrodes. That charge density declines rapidly to 0.02 A/cm^2 at 5 mm away from the stimulating electrodes. The current density is inversely proportional to the *square* of the distance from the electrodes, so that current density continues to drop off very rapidly with distance from the stimulating electrodes. The cortical stimulation is within the safe range for patients as long as the stimulating electrodes make good contact with cortex, i.e., have sufficient contact area.

Typical ECoG direct cortical stimuli are composed of stimulus pulses that are 100–300 milliseconds (0.1–0.3 ms) in duration. Pulse durations less than 0.3 ms tend to excite cortical cell bodies, whereas those of 1 ms or more can directly excite axons at the lower layers of cortex. Usually these stimulation pulses are delivered in trains of 50–60 per second. Each train lasts for several seconds. A 5–7 second train is used for language testing, whereas shorter trains are used for motor testing. Durations much longer are discouraged because they are more likely to provoke seizures. Stimulus intensity up to 14 mA (milliamperes) is commonly used for the typical wand designs. This is sufficient to disrupt function at the location of the wand and in the nearby 1 cm of cortex. Often the technique starts with a lower intensity, e.g., 6 mA, and changes gradually to higher intensities if no desired or adverse effects are seen at the lower intensity at a particular site. The settings mentioned above are described in constant-current units, but some users' methods apply constant-voltage stimulation. Current is more directly related to the safety limits, whereas voltages are more directly related to the desired neuronal discharges.

During stimulation, ECoG is recorded from nearby sites. This is to ascertain whether epileptic-like discharges occur with the stimulations. Those discharges can last longer than the stimulation itself, in which case they are referred to as after-discharges. After-discharges can be rhythmic repetitive high-voltage spikes or polyspike–wave complexes. Figures 8.1 and 8.2 show examples of stimulus artifacts followed by after-discharges. They may last for seconds or occasionally minutes, as seen in Figure 8.3. After-discharges can be subclinical or associated with subtle clinical behavioral manifestations when the patient is awake (Stefan *et al.*, 1991). They are an important marker that forewarns that stimulation may have spread not only temporally beyond the end of stimulation but also spatially to other cortical regions. Therefore, stimulation trains with after-discharges should not be used to localize cortical functions. Furthermore, after-discharges herald epileptic seizures occurring with further stimulation at that site with that intensity. This is a reason to move the wand to a different site or to reduce the stimulus intensity.

Seizures can be initiated by cortical stimulation. Cold Ringer's lactate solution can be used to treat seizures during cortical stimulation. The irrigation of the stimulated region rapidly and reliably terminates partial seizures (Sartorius and Berger, 1998).

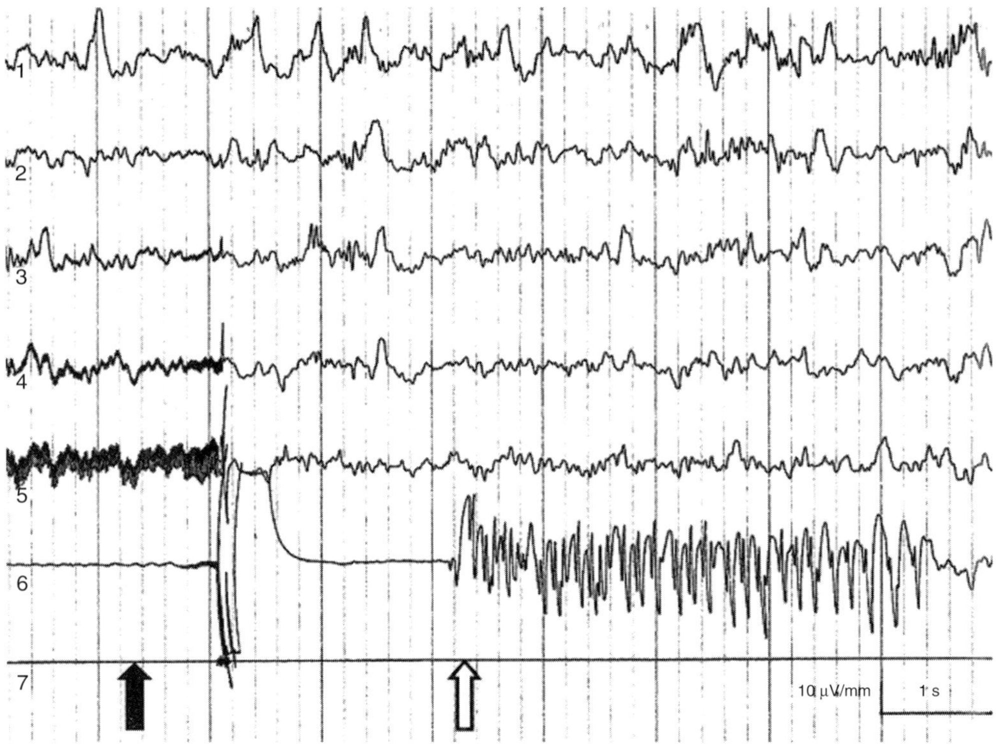

Figure 8.1 After-discharge occurring after direct cortical stimulation. The stimulus artifact is seen in the first two seconds (dark arrow). After that, one channel's amplifier is saturated for two seconds, and then that channel shows an after-discharge lasting at least four seconds, beginning at the open arrow (from Gallantine and Mikati, 2009).

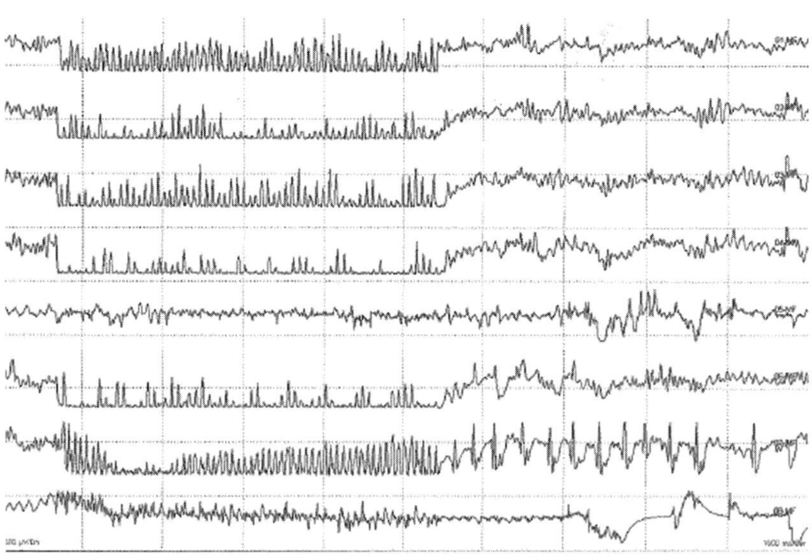

Figure 8.2 After-discharge occurring after direct cortical stimulation. The five-second long stimulus artifact is seen followed by an after-discharge. The after-discharge is mainly seen in the seventh channel.

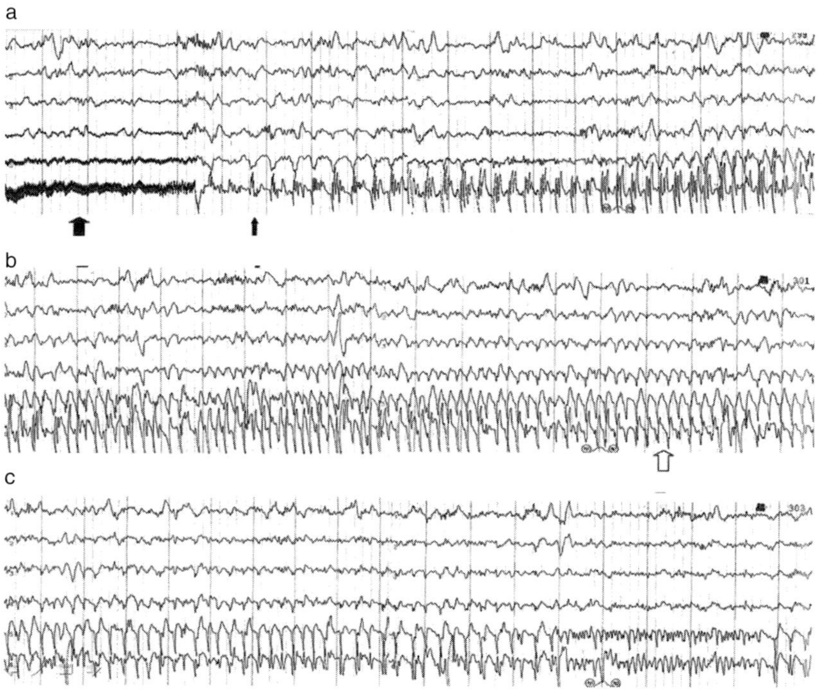

Figure 8.3 An after-discharge develops into a seizure. (a) The stimulus artifact lasts over four seconds (heavy arrow). It is followed by an after-discharge (thin arrow). (b) The after-discharge persists, evolves, and spreads to more channels. (c) The seizure continues, lasting more than a minute overall (from Gallantine and Mikati, 2009).

Interpretation and testing protocols

Electroencephalography

The typical EEG under anesthesia has frontocentral fast activity, often in the 8–12 Hz range, along with frontopolar triangular slow waves and low amplitude generalized slowing. Deeper levels of anesthesia are accompanied by other changes reviewed elsewhere (Jantti and Sloan, 2008).

The earliest EEG sign of cortical ischemia is a loss of EEG fast activity. This occurs when the cerebral blood flow drops to (18 ml/100 g)/min (McKay *et al.*, 1976; Sundt *et al.*, 1981). The criterion for change is a 50% reduction in the background fast activity. The drop in fast activity may be unilateral or bilateral. Loss of fast activity is considered a mild change.

The second level of ischemia is an increase in slow activity. The increase in slow activity may be unilateral or bilateral. The criterion for change is a 50% increase in background slow activity. Increased slowing, especially for very slow delta activity, is considered a moderate change.

The third and worst degree of ischemia is EEG amplitude decrease. The amplitude decrease may progress all the way to an isoelectric (flat) tracing, which occurs when cerebral blood flow drops to (15 ml/100 g)/min. The criterion for change is a 50% reduction in EEG amplitude. This attenuation of the EEG is considered a severe change. These changes may be bilateral or unilateral depending on whether the cerebral arterial circulation shares the ischemia on both sides. Several authors have further defined criteria for the interpretation of ischemic EEG changes during monitoring (Blume *et al.* 1986; Nuwer, 1993; Burke *et al.*, 1999; Pinkerton, 2002).

Some techniques use automated measures of EEG to show changes from baseline more accurately (Minicucci *et al.*, 2000). These can be helpful in identifying change, especially when the change occurs slowly (van Huffelen, 2008). However, care must be taken in the use of automated devices. They can measure activity other than EEG. Therefore, a well trained electroencephalographer is still needed to make sure that the recording measurements are meaningful, not artifactual. Also, some automated techniques compare the EEG from one hemisphere with that from the other hemisphere, and report the difference. The goal of comparing the two hemispheres is to detect ischemia from one hemisphere when a carotid artery or branch is clamped or incidentally occluded. This monitoring theory is flawed, however, because the circle of

Willis often shares ischemia between the two hemispheres, so that finding an interhemispheric asymmetry is a very imperfect way of identifying ischemia. Automated techniques should monitor each of many channels, and report the changes for each channel. Automated techniques can accurately identify even small changes in fast or slow activity, some of which are subclinical and not necessarily of clinical interest during carotid or cardiac surgery (Ahn *et al.*, 1988)

A different use of intraoperative EEG is to assess for burst suppression. This can be achieved under deep anesthesia, such as that obtained with intravenous barbiturates. This deliberate burst suppression reduces the cerebral metabolic rate and helps the brain tolerate the prolonged ischemia that may occur with certain cardiovascular procedures. Similarly, an isoelectric EEG is sought when the patient is cooled in the deliberate hypothermia used in some cardiac procedures. The temperature at which this EEG state is reached differs among patients. To determine whether a sufficiently deep temperature has been reached and maintained, the best way is to assess whether the EEG is in an isoelectric state.

Electrocorticography

The goal of ECoG is to determine locations of epileptic or otherwise impaired cortex. The recording normally shows the same types of EEG features as seen with scalp EEG during anesthesia: frontocentral fast activity, frontopolar triangular slow waves, and low amplitude generalized slowing. Recordings made from sites above the Sylvian fissure show more fast activity than activity from the temporal cortex.

During ECoG, the normal background activity is more sharply contoured than is typical from scalp recordings. This is because the skull attenuates very fast activity more than slow activity, making scalp-recorded EEG look less sharp. The shape of ECoG is similar to scalp recordings made over a skull breach. This more sharply contoured normal background seen in an ECoG needs to be clearly distinguished from epileptic activity. A common mistake of an inexperienced reader is to overinterpret sharply contoured cortical ECoG activity as if it were epileptic and pathological. Examples of epileptic spikes in ECoG are shown in Figures 8.4 and 8.5.

Areas of impaired cortex usually show a relative lack of fast activity, increased slow activity, or epileptic spikes. These can be identified from the recordings by a knowledgeable expert neurophysiologist. Not all areas with epileptic spikes are considered candidates for resection. The areas most suitable for resections are those with the greatest amount of epileptic activity that is most persistent over time and most pathological in its shape.

Direct cortical stimulation

In the unconscious patient, direct cortical stimulation can identify motor cortex. In the awake patient, direct cortical stimulation can identify motor, sensory, and language cortex. An awake patient can be questioned about sensations experienced with stimulations. The primary sensory cortex can be mapped in this way. In the awake patient, the technique can sometimes identify regions that produce the patient's typical epileptic aura. For some patients, stimulation near a seizure focus will produce the patient's typical aura feeling.

When searching for the motor cortex in the unconscious patient, neuromuscular blockade must be limited and anesthesia should usually be light. Finding stimulation-induced movement at some areas demonstrates that the anesthetic depth is adequate and the technique is working properly. Stimulations can then identify face, arm, and leg regions. Observations of responses can be made visually. Alternatively, the patient can be set up with recording electrodes in various muscles to record movements under the surgical drapes.

In the conscious patient during a craniotomy, the patient is extubated while restrained on the table so as to test language (Huncke *et al.*, 1998). Typically, a neuropsychologist or other trained specialist will administer a variety of language tasks that measure different aspects of visual and auditory language processing and short-term memory. Most adult patients tolerate this adequately. Stimulation delivered at a variety of cortical locations can identify the regions of language function so that they can be spared during a resection. Regions without noticeable disruption up to 12–14 mA are considered for resection. The neurophysiologist delivers the stimulations and observes for after-discharges.

Language can be mapped with several tasks. In *visual object naming*, the examiner presents line drawings of Boston Naming Test pictures (Kaplan *et al.*, 1976). The drawings selected are those that the patient could name quickly when tested before surgery (Ojemann, 1979). Naming is a commonly

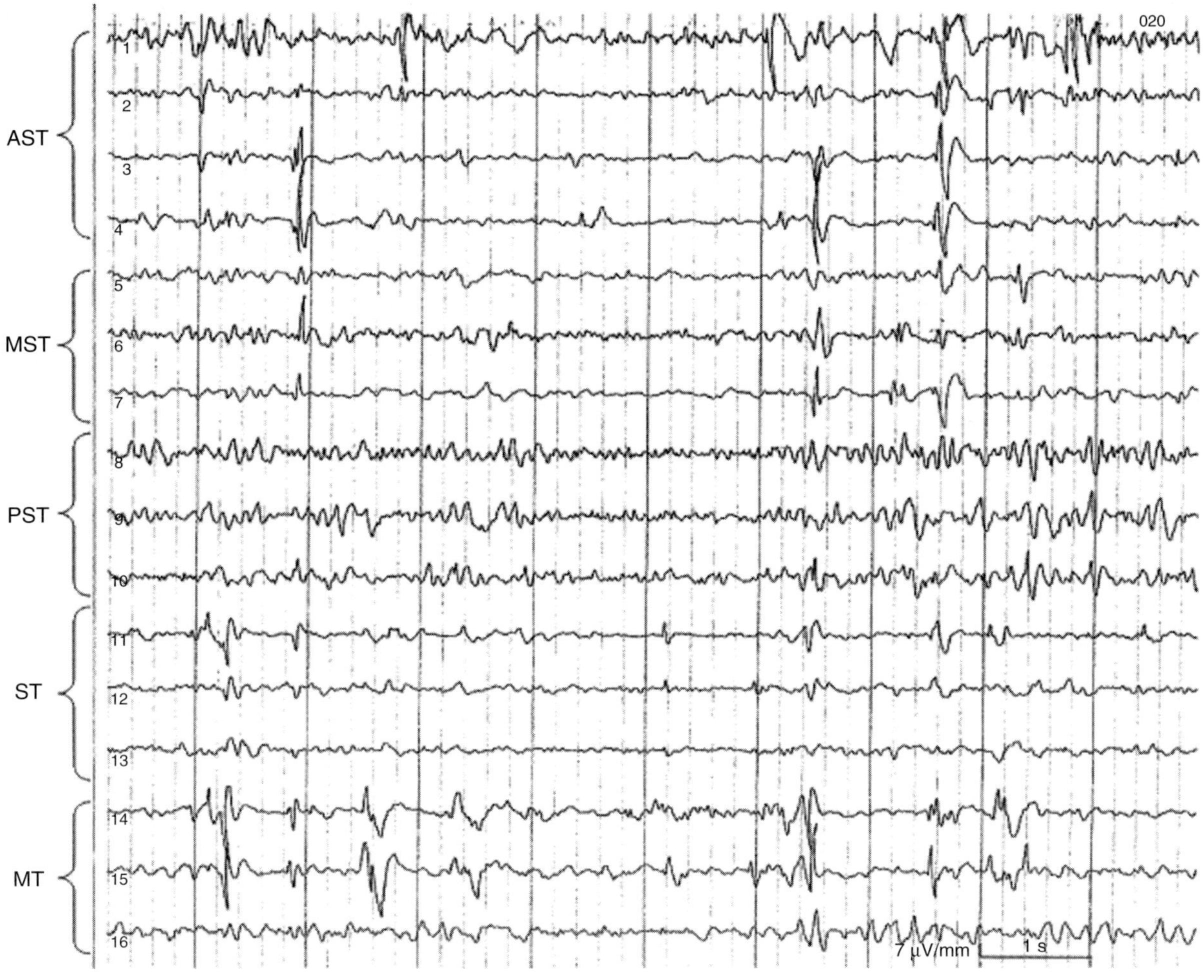

Figure 8.4 Example of epileptic spikes seen in an ECoG recording. A variety of spikes is seen in this recording (from Gallantine and Mikati, 2009).

chosen cortical stimulation mapping language task because naming deficits are seen in many aphasia syndromes. Stimulation can induce a speech arrest or anomia when testing many different cortical regions important for language (Ojemann and Mateer, 1979; Ojemann, 1993). Yet not all cortical language sites cause naming deficits. Stimulation at some cortical language sites causes *sentence-reading* deficits (Ojemann, 1983). In *word-generation* tasks, the examiner asks the patient for a list of words that begin with a certain letter or are of a certain category (e.g., animals). When testing *auditory responsive naming* the examiner asks the patient to name objects or their characteristics, e.g., the answer for "tall pink bird" would be "flamingo", and the answer for "the color of grass" would be "green". Auditory responsive naming and word-generation tasks evaluate frontal language areas (Cuenod *et al.*, 1995; Bookheimer *et al.*, 1998). In contrast to auditory responsive naming, in *visual responsive naming* the question is written on a flash card. While naming is a good screen test for language eloquent cortex, these other language tests should also be performed.

Cortical stimulation produces localized effects that allow function mapping. Changing the stimulating electrodes' position by as little as 0.5–1.0 cm, even along the same gyrus, produces different functional results (Ojemann and Whitaker, 1978; Ojemann, 1983). A site with consistent language function can be located just 1–2 cm from a different site on the same gyrus that does not have a language function (Ojemann, 1979). Language function often exists in

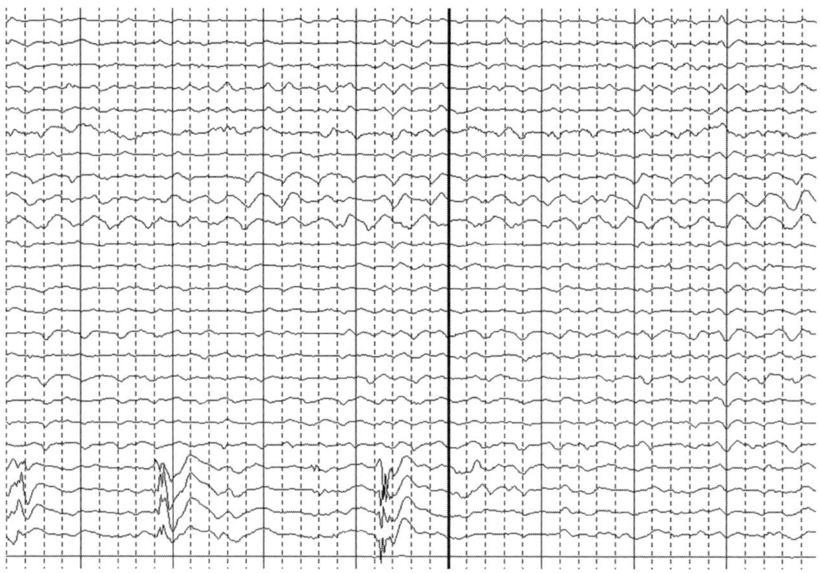

Figure 8.5 Example of epileptic spikes seen in an ECoG recording. Spikes are restricted to the lower four channels in this ECoG (figure courtesy of John Stern, UCLA).

small regions that extend across just 1–2 cm^2 (Ojemann, 1983; 1991). Postoperative language function impairment may occur if the margins of resection come within 1.5–2.0 cm of defined functional language regions. Persistent language deficits are avoided when surgical margins remain >2.0 cm from the sites identified as having language function (Ojemann, 1983; 1991; Haglund *et al.*, 1994).

Anesthetic effects

Anesthesia either excites or depresses the EEG, a pattern described by Winters (1976). The anesthetic agents *propofol* and *etomidate* are activating or exciting. Propofol is popular for ECoG recording and for use before direct cortical stimulation. These agents interfere with brainstem arousal mechanisms by affecting $GABA_A$ receptors (Nelson *et al.*, 2002; Rudolph and Antkowiak, 2004). The EEG resembles slow-wave sleep with vertex sharp waves and spindles. In one theory, propofol-induced EEG excitation consists of an interneuron antisynchrony, one that also produces an EEG beta rhythm at low drug concentrations (McCarthy *et al.*, 2008). Propofol does not produce EEG epileptiform activity (Ebrahim *et al.*, 1994; Samra *et al.*, 1995; Cheng *et al.*, 1996). It is eliminated quickly enough for use during ECoG and awake testing (Soriano *et al.*, 2000; Skucas and Artru, 2006).

Many volatile inhalation anesthetic agents are sedating: *halothane, desflurane, isoflurane*, and *sevoflurane*. They affect $GABA_A$ receptors, glycine receptors, glutamate receptors, and two-pore-domain potassium channels (Grasshoff *et al.*, 2005). They are associated with rhythmic frontocentral alpha activity, but this can be abolished by nitrous oxide (Yli-Hankala, 1990). They have little effect on epileptiform activity at low concentrations (Fiol *et al.*, 1993). These agents can reduce (Asano *et al.*, 2004) or increase (Endo *et al.*, 2002) epileptiform discharges at standard anesthetic levels.

Barbiturates cause fast EEG activity at low doses and burst suppression at higher doses. At low doses, 0.5 mg/kg, the short-acting barbiturate methohexital activates epileptic spike activity. This property has been used deliberately to identify seizure foci during ECoG (Musella *et al.*, 1971; Ajmone Marsan and O'Connor, 1973; Rampil, 1997; Wennberg *et al.*, 1997).

Benzodiazepines cause frontal beta activity, which may be mixed with some epileptiform activity. Higher benzodiazepine doses produce generalized slowing. Benzodiazepines are avoided if ECoG is used to identify a seizure focus.

Nitrous oxide used on its own results in frontocentral fast activity. Combining it with inhalation anesthetic agents produces variable results. It can increase or suppress epileptic spike activity during ECoG, depending on which additional agents it is combined with (Artru *et al.*, 1990; Hosain *et al.*, 1996).

Various other agents are used. *Droperidol* and *dexmedetomidine* are sometimes used during ECoG procedures. Narcotic agents are avoided in general during

ECoG because they can increase epileptiform spiking even from outside the epileptogenic area (Tempelhoff *et al.*, 1992; Cascino *et al.*, 1993; Manninen *et al.*, 1999). *Remifentanyl* may increase epileptiform discharges at higher doses (Wass *et al.*, 2001) but not with lower dose infusions (Herrick *et al.*, 2002).

Overall, keeping the anesthesia light during ECoG increases the chances of finding and localizing epileptiform activity. Propofol can be reduced sufficiently quickly to achieve a light anesthesia, and it does not reduce intraoperative epileptiform abnormalities. Methohexital or similar short-acting drugs can sometimes enhance epileptiform abnormalities during ECoG.

Applications

Carotid endarterectomy (CEA)

The primary purpose of EEG during CEA is to assess whether cerebral blood flow is adequate when the carotid artery is clamped, and whether adequate perfusion is regained when a vascular shunt is placed. A secondary purpose is to assess adequacy of cerebral perfusion during other portions of the procedure when blood pressures may be marginal, given the known carotid stenosis.

Vascular shunts are often indicated and are used when the EEG shows major changes (Blume and Sharbrough, 2005). Placement of the shunt must be associated with EEG recovery. If this is not the case, consider other events such as an embolism or shunt malfunction.

Clear communication between the IOM team and anesthesia is needed so as not to confuse an anesthetic change with a clinical change from ischemia. The monitoring team's responsibility is to assess the fast and slow activity, and identify departures from the baseline. As described already, the commonly used criteria for change are, in order of increasing concern: a 50% decrease in fast activity, a 50% increase in slow activity, a 50% loss of EEG amplitude, and an isoelectric (flat) EEG. Severe changes warrant more definite action to restore perfusion within only a few minutes, whereas moderate changes can be tolerated many minutes longer. Mild changes can be tolerated for many more minutes.

In their reviews of IOM for CEA, Guerit (2008) and Van Huffelen (2008) also discuss the use of SEPs and the optional use of quantified trending graphs to enhance the monitoring of carotid endarterectomy.

Cardiac surgery

The goal of EEG IOM in cardiac surgery is to assure adequate cerebral perfusion and to assess whether hypothermia is adequately deep. Most EEG IOM is performed with open-heart surgery, valve replacement, or repair of congenital malformations performed under cardiopulmonary bypass. Occasionally EEG IOM is carried out during coronary artery bypass graft (CABG) surgery.

The use for detecting perfusion is similar to the use with carotid endarterectomy. Although emboli can occur during cardiac or aortic arch surgery, IOM may be of marginal use for emboli. For perfusion, one should note that brain autoregulation keeps cerebral perfusion adequate over an extended range of systemic blood pressures. Sometimes the cerebral perfusion deteriorates for some patients during cardiac surgery at system blood pressures that would be well tolerated by many others. This individual variability in autoregulation competence is one reason why actual monitoring of cerebral response to perfusion state is clinically valuable. Also, autoregulation may not operate as well with non-pulsatile blood pressures as used with cardiopulmonary bypass in many cardiac surgeries.

Hypothermia is induced to protect brain function during periods of prolonged circulatory arrest. The state of electrocerebral silence (ECS) induced by profound hypothermia is considered a state of cerebral inactivity protecting the brain against the effects of hypoxia–ischemia (Mizrahi *et al.*, 1989). The EEG shows a characteristic evolution during cooling. With cooling, the EEG shows a progressively decreasing amplitude and background slowing. The amplitude attenuation is most pronounced for the frontocentral fast activity. With greater degrees of cooling, the EEG shows a discontinuous suppression–burst type of pattern, and the suppressions become progressively longer. This pattern eventually develops into ECS, the desired state for a patient in prolonged circulatory arrest.

Epilepsy

Electrocorticography is used in some patients undergoing epilepsy surgery. It is especially used when the epileptogenic zone has not been adequately localized using other methods. It is used to further localize the epileptic tissue and define the limits of resection.

For patients with very frequent seizures during either scalp or invasive EEG monitoring, intraoperative ictal ECoG may be useful to demarcate the electrographic ictal onset zone more precisely. An interictal ECoG is much more commonly obtained. Interictal spike discharges usually arise from the epileptogenic zone but also extend to cortex beyond it. Those regions are defined and discussed with the surgeon to tailor the resection to the needs of the individual patient. It is unnecessary to resect the location of each epileptic spike identified during ECoG. Selectivity in resection is better than trying to remove every spike generator.

Sometimes ECoG is performed after resection. Epileptiform discharges observed at the resection margins may simply represent "injury potentials" or "injury spikes", not the location of a probable additional epileptic tissue. Epileptic spikes distant from the resection margin are more concerning, and frontal lobe epilepsy with epileptiform discharges three or more gyri away from the resection site indicates a likelihood of seizure recurrence.

To avoid anesthetic effects, the anesthesia is lightened or minimized during the ECoG. Many anesthetics can alter the presence of epileptiform discharges sought in ECoG. The effects of the anesthetic on the EEG activity are less important than the anesthetic's clearance time in choosing what to use for a patient. Propofol clears adequately to allow good ECoG in most patients. Nitrous oxide and narcotics may be used during the ECoG to avoid pain or consciousness during the procedure.

Tumor resection

There is considerable interindividual variability in the specific anatomic locations of discrete eloquent cortex regions serving language, motor, and sensory functions (Ojemann, 1979; Uematsu *et al.*, 1992; Branco *et al.*, 2003). Therefore, it is important to identify the specific anatomic localization for each eloquent function for each individual patient. Accurate mapping allows the maximal resection of pathological tissue while preserving function (Duffau *et al.*, 1999; 2005). Functional imaging navigation techniques are used in combination with intraoperative direct stimulation mapping, a dual approach found helpful for guiding the resection (Duffau, 2000; Kamada *et al.*, 2005).

Language mapping requires patient participation. An asleep–awake–asleep technique can be used in cooperative patients (Huncke *et al.*, 1998). Initial anesthesia is with propofol or sodium thiopental and endotracheal intubation. Operative sites are infiltrated with local anesthetic. Once the craniotomy has been achieved, the patient is awakened and extubated. After mapping, patients are re-intubated and remain under general anesthesia for the remainder of the operation.

Prior to stimulation testing, a consistent language performance baseline is needed. Otherwise it would be difficult to ascertain whether language disruption is due to the stimulation or simply an inconsistent performance.

Questions

1. In EEG monitoring, the mildest degree of change is:
   (a) 50% increase in slow activity
   (b) 50% decrease in EEG amplitude
   (c) 50% decrease in fast activity
   (d) 50% decrease in the dominant frequency
2. For EEG monitoring, the recommended technique includes:
   (a) Low filter of 5 Hz
   (b) High filter of 35 Hz
   (c) Displaying 30–60 s per page
   (d) Avoiding use of the notch filter
3. ECoG strip and grid electrodes typically have electrodes that are:
   (a) 5 mm across and separated by 1 cm center-to-center
   (b) 5 mm across and separated by 2 cm center-to-center
   (c) 3 mm across and separated by 1 cm center-to-center
   (d) 3 mm across and separated by 2 cm center-to-center
4. Direct cortical stimulation uses pulses that:
   (a) Are bipolar and biphasic
   (b) Are always delivered with platinum electrodes
   (c) Must use a constant-voltage regulator
   (d) Deliver pulses 0.5–1.0 ms long
5. Safe cortical stimulation must be less than:
   (a) 40 ($\mu C/cm^2$)/phase
   (b) 60 (mA/cm^2)/pulse
   (c) 20 (A/cm^2)/phase
   (d) 10 (mm/cm^2)/pulse
6. Constant-current direct cortical stimulator devices are used up to:
   (a) 1 mA
   (b) 6 mA
   (c) 14 mA
   (d) 25 mA

7. A loss of EEG activity can be produced by cerebral blood flow of:
   (a) 18 (ml/100 g)/min
   (b) 28 (ml/100 g)/min
   (c) 38 (ml/100 g)/min
   (d) 48 (ml/100 g)/min
8. An agent commonly used to increase epileptiform activity in an ECoG is:
   (a) Midazolam
   (b) Methohexital
   (c) Halothane
   (d) Propofol
9. An anesthetic commonly used during craniotomies in which ECoG is planned is:
   (a) Ether
   (b) Isoflurane
   (c) Halothane
   (d) Propofol
10. The most common test used to identify language areas is:
   (a) Sentence reading
   (b) Naming tasks
   (c) Word-generation tasks
   (d) Responses to written questions

References

Agnew, W. F. and McCreery, D. B. (1987). Considerations for safety in the use of extracranial stimulation for motor evoked potentials. *Neurosurgery*, **20**, 143–147.

Ahn, S. S., Jordan, S. E., Nuwer, M. R., Marcus, D. R., and Moore, W. S. (1988). Computerized EEG topographic brain mapping: a new and accurate monitor of cerebral circulation and function for patients undergoing carotid endarterectomy. *Journal of Vascular Surgery*, **8**, 247–254.

Ajmone Marsan, C. and Baldwin, M. (1958). Electrocorticography. In *Temporal Lobe Epilepsy*, ed. M. Baldwin and P. Bailey. Springfield, IL: Charles C Thomas, pp. 368–395.

Ajmone Marsan, C. and O'Connor, M. (1973). Electrocorticography. In *Handbook of Electroencephalography and Clinical Neurophysiology*, ed. A. Remond, Vol. 10C. Amsterdam: Elsevier, pp. 3–49.

Artru, A. A., Lettich, E., Colley, P. S., and Ojemann, G. A. (1990). Nitrous oxide: suppression of focal epileptiform activity during inhalation, and spreading of seizure activity following withdrawal. *Journal of Neurosurgery and Anesthesiology*, **2**(3), 189–193.

Asano, E., Benedek, K., Shah, A., *et al.* (2004). Is intraoperative electrocorticography reliable in children with intractable neocortical epilepsy? *Epilepsia*, **45**(9), 1091–1099.

Blume, W. T. and Sharbrough, F. W. (2005). EEG monitoring during carotid endarterectomy and open heart surgery. In *Electroencephalography. Basic Principles, Clinical Applications, and Related Fields*, ed. E. Niedermeijer and F. Lopes da Silva. Philadelphia: Lippincott Williams and Wilkins, pp. 815–827.

Blume, W. T., Ferguson, G. G., and McNeill, D. K. (1986). Significance of EEG changes at carotid endarterectomy. *Stroke*, **17**, 891–897.

Bookheimer, S. Y., Zeffiro, T. A., Blaxton, T. A., *et al.* (1998). Regional cerebral blood flow during auditory responsive naming: evidence for cross-modality neural activation. *Neuroreport*, **9**, 2409–2413.

Branco, D. M., Coelho, T. M., Branco, B. M., *et al.* (2003). Functional variability of the human cortical motor map: electrical stimulation findings in perirolandic epilepsy surgery. *Journal of Clinical Neurophysiology*, **20**, 17–25.

Burke, D., Nuwer, M. R., Daube, J., *et al.* (1999). Intraoperative monitoring. In *Recommendations for the Practice of Clinical Neurophysiology: Guidelines of the International Federation of Clinical Neurophysiology*, 2nd edn., ed. G. Deuschl and A Eisen. *Electroencephalography and Clinical Neurophysiology*. (Supplement no. 52) Amsterdam: Elsevier.

Cascino, G. D., So, E. L., Sharbrough, F. W., *et al.* (1993). Alfentanil induced epileptiform activity in patients with partial epilepsy. *Journal of Clinical Neurophysiology*, **10**(4), 520–525.

Cheng, M. A., Tempelhoff, R., Silbergeld, D. L., *et al.* (1996). Large-dose propofol alone in adult epileptic patients: electrocorticographic results. *Anesthesia and Analgesia*, **83**(1), 169–174.

Cuenod, C. A., Bookheimer, S. Y., Hertz-Pannier, L., *et al.* (1995). Functional MRI during word generation, using conventional equipment: a potential tool for language localization in the clinical environment. *Neurology*, **45**, 1821–1827.

Duffau, H. (2000). Intraoperative direct subcortical stimulation for identification of the internal capsule, combined with an image-guided stereotactic system during surgery for basal ganglia lesions. *Surgical Neurology*, **53**, 250–254.

Duffau, H., Capelle, L., Sichez, J., *et al.* (1999). Intra-operative direct electrical stimulations of the central nervous system: the Salpêtrière experience with 60 patients. *Acta Neurochirurgica*, **141**, 1157–1167.

Duffau, H., Lopes, M., Arthuis, F., *et al.* (2005). Contribution of intraoperative electrical stimulations in surgery of low grade gliomas: a comparative study between two series without (1985–96) and with (1996–2003) functional mapping in the same institution. *Journal of Neurology, Neurosurgery, and Psychiatry*, **76**, 845–851.

Ebrahim, Z. Y., Schubert, A., Van Ness, P., Wolgamuth, B. and Awad, I. (1994). The effect of propofol on the electroencephalogram of patients with epilepsy. *Anesthesia and Analgesia*, **78**, 275–279.

Endo, T., Sato, K., Shamato, H. and Yoshimoto, T. (2002). Effects of sevoflurane on electrocorticography in patients with intractable temporal lobe epilepsy. *Journal of Neurosurgery and Anesthesiology*, **14**, 59–62.

Fiol, M. E., Boening, J. A., Cruz-Rodriguez, R. and Maxwell, R. (1993). Effect of isoflurane (Forane) on intraoperative electrocorticogram. *Epilepsia*, **34**, 897–900.

Gallantine, W. B. and Mikati, M. A. (2009). Intraoperative electrocorticography and cortical stimulation in children. *Journal of Clinical Neurophysiology*, **26**, 95–108.

Grasshoff, C., Rudolph, U. and Antkowiak, B. (2005). Molecular and systemic mechanisms of general anaesthesia: the ‘multi-site and multiple mechanisms’ concept. *Current Opinion in Anaesthesiology*, **18**, 386–391.

Guerit, J. M. (2008). Intraoperative monitoring during carotid endarterectomy. In *Intraoperative Monitoring of Neural Function: Handbook of Clinical Neurophysiology*, ed. M. R. Nuwer. Amsterdam: Elsevier.

Haglund, M. M., Berger, M. S., Shamseldin, M,. Lettich, E., and Ojemann, G. A. (1994). Cortical localization of temporal lobe language sites in patients with gliomas. *Neurosurgery*, **34**, 567–576. Discussion, 576.

Herrick, I. A., Craen, R. A., Blume, W. T., Novick, T., and Gelb, A. W. (2002). Sedative doses of remifentanil have minimal effect on ECoG spike activity during awake epilepsy surgery. *Journal of Neurosurgery and Anesthesiology*, **14**(1), 55–58.

Hosain, S., Nagarajan, L., Fraser, R., Van Poznack, A., and Labar, D. (1997). Effects of nitrous oxide on electrocorticography during epilepsy surgery. *Electroencephalography and Clinical Neurophysiology*, **102**, 340–342.

Huncke, K., Van de Wiele, B., Fried, I., and Rubinstein, E. H. (1998). The asleep-awake-asleep anesthetic technique for intraoperative language mapping. *Neurosurgery*, **42**, 1312–1316. Discussion, 1316–1317.

Jantti, V. and Sloan, T. B. (2008). EEG and anesthetic effects. In *Intraoperative Monitoring of Neural Function*, ed. M. R. Nuwer. Amsterdam: Elsevier, pp. 2–6.

Jasper, H. H. (1949). Electrocorticogram in man. *Electroencephalography and Clinical Neurophysiology* **2**(suppl.), 16–29.

Kamada, K., Todo, T., Masutani, Y., *et al.* (2005). Combined use of tractography-integrated functional neuronavigation and direct fiber stimulation. *Journal of Neurosurgery*, **102**, 664–672.

Kaplan, E., Goodglass, H., and Weintraub, S. (1976) *Boston Naming Test*. Philadelphia: Lea and Febiger.

Manninen, P. H., Burke, S. J., Wennberg, R., Lozano, A. M., and Beheiry, H. (1999). Intraoperative localization of an epileptogenic focus with alfentanil and fentanyl. *Anesthesia and Analgesia*, **88**, 1101–1106.

Marshall, C. and Walker, A. E. (1949). Electrocorticography. *Bulletin of the Johns Hopkins Hospital*, **85**, 344–359.

McCarthy, M. M., Brown, E. N., and Kopell, N. (2008). Potential network mechanisms mediating electroencephalographic beta rhythm changes during propofol-induced paradoxical excitation. *The Journal of Neuroscience*, **28**, 13488–13504.

McKay, R. D., Sundt, T. M., Michenfelder, J. D., *et al.* (1976). Internal carotid artery stump pressure and cerebral blood flow during carotid endarterectomy: modification by halothane, enflurane and Innovar. *Anesthesiology*, **45**, 390–399.

Meyers, H. R., Knott, J. R., Hayne, R. A., and Sweeney, D. B. (1950). The surgery of epilepsy. Limitations of the concept of the corticoelectrographic ‘spike’ as an index of the epileptogenic focus. *Journal of Neurosurgery*, **7**, 337–346.

Minicucci, F., Cursi, M., Fornara, C., *et al.* (2000). Computer-assisted EEG monitoring during carotid endarterectomy. *Journal of Clinical Neurophysiology*, **17**(1), 101–107.

Mizrahi, E. M., Patel, V. M., Crawford, E. S., Coselli, J. S., and Hess, K. R. (1989). Hypothermic-induced electrocerebral silence, prolonged circulatory arrest, and cerebral protection during cardiovascular surgery. *Electroencephalography and Clinical Neurophysiology*, **72**, 81–85.

Mortimer, J. T., Shealy, C. N., and Wheeler, C. (1970). Experimental nondestructive electrical stimulation of the brain and spinal cord. *Journal of Neurosurgery*, **32**, 553–559.

Musella, L., Wilder, B. J., and Schmidt, R. P. (1971). Electroencephalographic activation with intravenous methohexital in psycho-motor epilepsy. *Neurology (Minneapolis)*, **21**, 594–602.

Nelson, L. E., Guo, T. Z., Lu, J., *et al.* (2002). The sedative component of anesthesia is mediated by GABA(A) receptors in an endogenous sleep pathway. *Nature Neuroscience*, **5**, 979–984.

Nuwer, M. R. (1993). Intraoperative electroencephalography. *Journal of Clinical Neurophysiology*, **10**, 437–444.

Nuwer, M. R. (2008a). Overview and history of neural monitoring during surgery. In *Intraoperative Monitoring of Neural Function*, ed. M. R. Nuwer. Amsterdam: Elsevier, pp. 2–6.

Nuwer, M. R. (ed.). (2008b). *Intraoperative Monitoring of Neural Function*. Amsterdam: Elsevier.

Nuwer, M. R., Daube, J., Fischer, C., Schramm, J., and Yingling, C. D. (1993). Neuromonitoring during surgery. Report of an IFCN committee. *Electroencephalography and Clinical Neurophysiology*, **87**, 263–276.

Ojemann, G. A. (1979). Individual variability in cortical localization of language. *Journal of Neurosurgery*, **50**, 164–169.

Ojemann, G. A. (1983). Brain organization for language from the perspective of electrical stimulation mapping. *The Behavioral and Brain Sciences*, **2**, 189–230.

Ojemann, G. A. (1991). Cortical organization of language. *The Journal of Neuroscience*, **11**, 2281–2287.

Ojemann, G. A. (1993). Functional mapping of cortical language areas in adults. Intraoperative approaches. *Advanced Neurology*, **63**, 155–163.

Ojemann, G. and Mateer, C. (1979). Human language cortex: localization of memory, syntax, and sequential motor-phoneme identification systems. *Science*, **205**, 1401–1403.

Ojemann, G. A. and Whitaker, H. A. (1978). Language localization and variability. *Brain and Language*, **6**, 239–260.

Penfield, W. and Jasper, H. H. (1954). *Epilepsy and the Functional Anatomy of the Human Brain*. Boston: Little, Brown.

Pinkerton, J. A., Jr. (2002). EEG as a criterion for shunt need in carotid endarterectomy. *Annals of Vascular Surgery*, **16**, 756–761.

Pudenz, R. H., Bullara, L. A., Dru, D., and Talalla, A. (1975). Electrical stimulation of the brain. II. Effects on the blood-brain barrier. *Surgical Neurology*, **4**, 265–270.

Rampil, I. J. (1997). Electroencephalogram. In *Textbook of Neuroanesthesia with Neurosurgical and Neuroscience Perspectives*, ed. M. A. Albin. New York: McGraw-Hill, pp. 193–220.

Rudolph, U. and Antkowiak, B. (2004). Molecular and neuronal substrates for general anaesthetics. *Nature Reviews Neuroscience*, **5**, 709–720.

Samra, S. K., Sneyd, J. R., Ross, D. A., and Henry, T. R. (1995). Effects of Propofol sedation on seizures and intracranially recorded epileptiform activity in patients with partial epilepsy. *Anesthesiology*, **82**, 843–851.

Sartorius, C. J and Berger, M. S. (1998). Rapid termination of intraoperative stimulation-evoked seizures with application of cold Ringer's lactate to the cortex. Technical note. *Journal of Neurosurgery*, **88**, 349–351.

Skucas, A. P. and Artru, A. A. (2006). Anesthetic complications of awake craniotomies for epilepsy surgery. *Anesthesia and Analgesia*, **102**, 882–887.

Soriano, S. G., Eldredge, E. A., Wang, F. K., *et al.* (2000). The effect of propofol on intraoperative electrocorticography and cortical stimulation during awake craniotomies in children. *Paediatric Anaesthesia*, **10**, 29–34.

Stefan, H., Quesney, L. F., Abou-Khalil, B., and Olivier, A. (1991). Electrocorticography in temporal lobe epilepsy surgery. *Acta Neurologica Scandinavica*, **83**, 65–72.

Sundt, T. M., Jr., Sharbrough, F. W., Piepgras, D. G., *et al.* (1981). Correlation of cerebral blood flow and electroencephalographic changes during carotid endarterectomy. *Mayo Clinic Proceedings*, **56**, 533–543.

Tempelhoff, R., Modica, P. A., Bernardo, K. L., and Edwards, I. (1992). Fentanyl induced electrocorticographic seizures in patients with complex partial epilepsy. *The Journal of Neurosurgery*, **77**, 201–208.

Uematsu, S., Lesser, R., Fisher, R. S., *et al.* (1992). Motor and sensory cortex in humans: topography studied with chronic subdural stimulation. *Neurosurgery*, **31**, 59–71. Discussion, 71–72.

Van Huffelen, A. C. (2008). Electroencephalography (EEG) used in monitoring neural function during surgery. In *Intraoperative Monitoring of Neural Function: Handbook of Clinical Neurophysiology*, ed. M. R. Nuwer, Amsterdam: Elsevier.

Walker, A. E., Marshall, C., and Beresford, E. M. (1947). Electrocorticographic characteristics of the cerebrum posttraumatic epilepsy. *Association for Research in Nervous and Mental Disease*, **26**, 502–515.

Wass, C. T., Grady, R. E., Fessler, A. J., *et al.* (2001). The effects of remifentanil on epileptiform discharges during intraoperative electrocorticography in patients undergoing epilepsy surgery. *Epilepsia*, **42**, 1340–1344.

Wennberg, R., Quesney, F., Olivier, A., and Dubeau, F. (1997). Induction of burst-suppression and activation of epileptiform activity after methohexital and selective amygdalo-hippocampectomy. *Electroencephalography and Clinical Neurophysiology*, **102**, 443–451.

Winters, W. D. (1976). Effects of drugs on the electrical activity of the brain: anesthetics. *Annual Review of Pharmacology and Toxicology*, **16**, 413–426.

Yli-Hankala, A. (1990). The effect of nitrous oxide on EEG spectral power during halothane and isoflurane anaesthesia. *Acta Anaesthesiologica Scandinavica*, **34**, 579–584.

9 Brainstem auditory evoked potential monitoring

Khaled M. Zamel

Introduction

Since its first description in 1967, intraoperative monitoring of brainstem auditory evoked potentials (BAEPs) has been widely used by neurosurgeons as a non-invasive and simple technique for continuous surveillance of the functional integrity of the peripheral and central auditory systems. This is valuable during surgeries involving the brainstem or the cranial nerves, in particular the auditory (vestibulocochlear, eighth) nerve (Harper *et al.*, 1992), the most commonly monitored cranial sensory nerve. The therapeutics and technology subcommittee of the American Academy of Neurology stated that the use of BAEPs in surgeries of the brainstem or the inner ear is useful and non-investigational (Lopez, 2004).

In response to the delivery of an abrupt brief acoustic stimulus, a series of neuroelectrical responses is generated that can be recorded using different techniques. The earliest responses, within the first 10 ms, are termed the short-latency auditory evoked potentials (SAEPs). These include two types of responses, the electrocochleogram (ECochG) and BAEPs (American Clinical Neurophysiology Society, 2006). The ECochG is generated at the cochlea and can be recorded by placing electrodes either directly in the middle ear or at the tympanic membrane (Schlake *et al.*, 2001). The BAEPs are the short-latency auditory evoked responses that occur within the first 10 ms from stimulation, which, in addition to reflecting the conduction in the auditory pathways, also provide valuable information about the general function of the brainstem. Middle- and long-latency auditory evoked potentials are also produced, with latencies between 10 and 50 ms and above 50 ms, respectively. These are mostly generated at the level of the cerebral cortex and are very vulnerable to the effect of anesthesia; therefore, they are not suitable for the purpose of NIOM.

Finally, in certain cases when the appropriate structures are surgically exposed, direct recording from the auditory nerve or nucleus can be very helpful. It is advantageous over BAEPs in that it provides almost instantaneous feedback about changes in conduction in these structures.

This chapter discusses primarily BAEPs, which are most commonly used to monitor the auditory nerve. For many years, monitoring BAEPs remained a preferred method for NIOM of the auditory pathways and the brainstem, mainly due to the high resistance to most anesthetics as well as stability and reproducibility when a constant acoustic stimulus is repetitively delivered to the same subject.

Obtaining an interpretable BAEP record in the operating room is more challenging than in the outpatient setting, therefore, although the technique largely resembles that used in extraoperative BAEPs, some minor but important modifications should be applied for recording intraoperative BAEPs.

During BAEP monitoring, careful assessment of the changes in the pattern of BAEPs and analysis of their relationship to the surgical maneuvers applied should determine whether the changes reflect damage to or potentially reversible dysfunction of the auditory pathways in the ear, auditory nerve, or brainstem and help exclude changes that are due to technical errors or other physiologic mechanisms, such as anesthesia, hypothermia, or acoustic masking from drilling noise (Legatt, 2002).

The auditory system and the components of BAEPs

To correctly obtain and interpret BAEPs, it is essential to fully understand the normal anatomy and physiological function of the auditory generators and pathways, as well as to understand the changes expected with certain surgical procedures or manipulations of the peripheral and central components of the auditory system.

In response to an acoustic stimulus, a series of near and far-field potentials is generated that corresponds

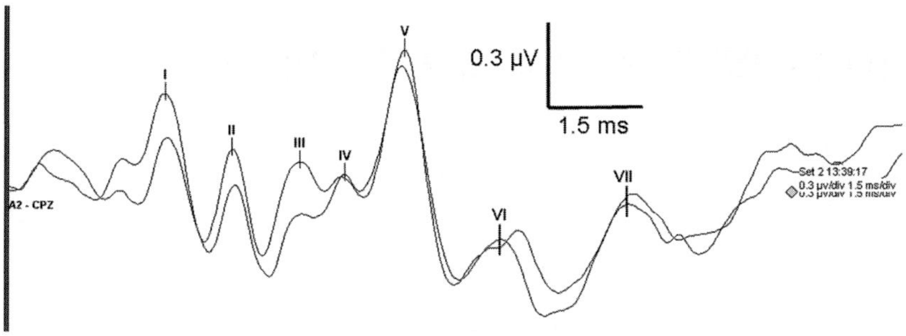

Figure 9.1 Typical recording of brainstem auditory evoked potentials (BAEPs) in a 16-year-old girl with normal hearing following monaural acoustic stimulation, demonstrating the characteristic vertex-positive I–VII waveforms.

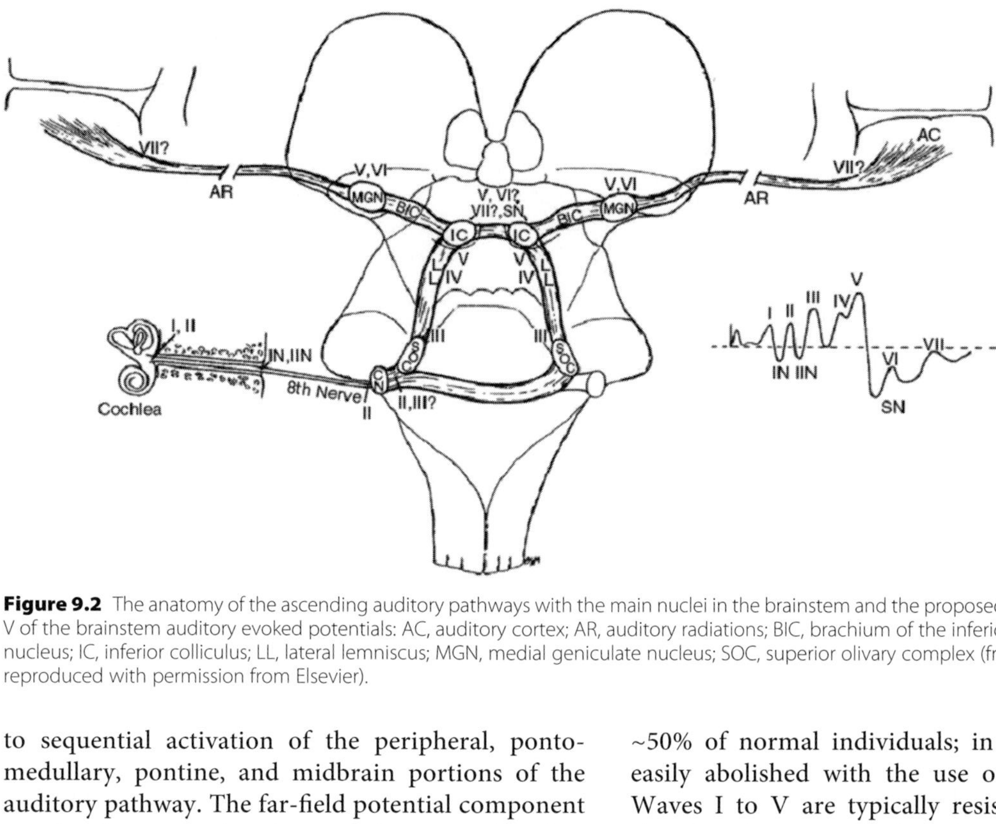

Figure 9.2 The anatomy of the ascending auditory pathways with the main nuclei in the brainstem and the proposed generators of waves I to V of the brainstem auditory evoked potentials: AC, auditory cortex; AR, auditory radiations; BIC, brachium of the inferior colliculus; CN, cochlear nucleus; IC, inferior colliculus; LL, lateral lemniscus; MGN, medial geniculate nucleus; SOC, superior olivary complex (from Legatt *et al.*, 1988, reproduced with permission from Elsevier).

to sequential activation of the peripheral, pontomedullary, pontine, and midbrain portions of the auditory pathway. The far-field potential component can then be captured and recorded by a set of surface recording electrodes on the scalp. These far-field potentials are composed of a series of short-latency, vertex positive and negative waves, also known as BAEPs. The normal BAEPs consists of seven vertex-negative peaks (Figure 9.1), usually numbered using the Roman numerals I through VII. The first identifiable peak, wave I, is an exception from the other components as it is actually a near-field potential that is recorded by a set of electrodes that are in the vicinity of the ears. The first five waves occur in all normal individuals at ~1 millisecond (ms) intervals. Waves VI and VII are less defined and occur in only ~50% of normal individuals; in addition, they are easily abolished with the use of anesthetic drugs. Waves I to V are typically resistant to anesthesia, therefore they are reliable and well suited to be used for NIOM. Wave V has the most prominent peak, always occurs after 5 ms and is usually followed by a prolonged downward deflection. Therefore, wave V is the easiest to identify so that most of the criteria proposed for the use of BAEPs in NIOM depends on either the latency or amplitude of wave V (James and Husain, 2005). The exact generators for waves I to V have not been confirmed and it is now known that several generators probably contribute to each of these, making it difficult to correlate each individual BAEP component with a specific potential source (Figure 9.2). However, for practical reasons it is

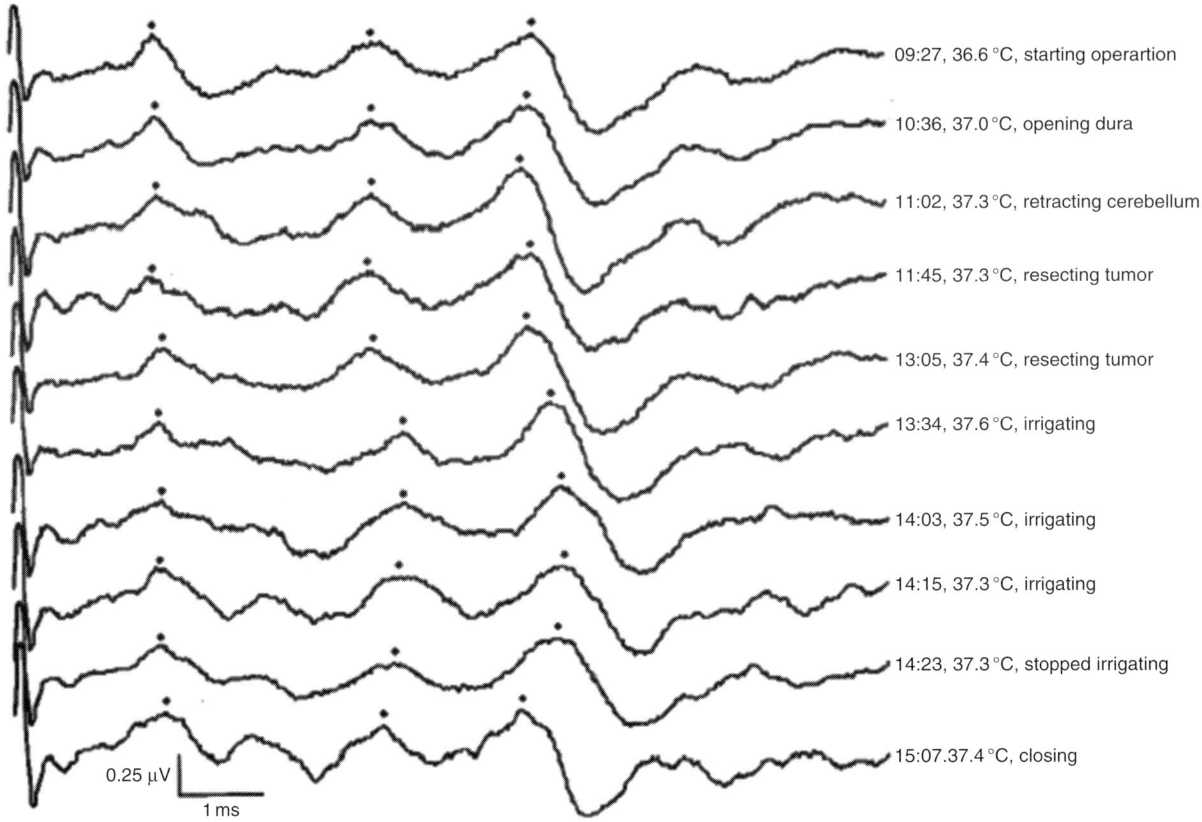

Figure 9.3 The effect of local hypothermia is demonstrated in this recording of brainstem auditory evoked potentials (BAEPs) during surgery for a meningioma in the right cerebellopontine angle. Copious irrigation of the surgical field with cold fluids produced a transient increase of wave III latency with no change in wave I, leading to prolongation of the I-to-III interpeak interval with relatively little change in the III-to-V interpeak interval. This reflects the relatively rapid reduction in neural conduction within the eighth nerve resulting from local cooling. The peak latencies of waves I, III, and V are marked by the small diamonds, and the clock times, esophageal temperatures, and surgical procedures corresponding to each of the BAEP waveforms are noted at the right (reprinted from Legatt, 2002, with permission from Lippincott, Wolters Kluwer Health.)

important to use and rely on specific anatomic structures to correlate with individual sequential waves. It is generally agreed that wave I of the BAEP correlates well with an action potential generated at the distal part of the auditory nerve and appears as a negative potential at the ipsilateral ear electrode. A dysfunction of the eighth nerve toward its proximal end at the cochlea or more rostral in the brainstem auditory pathways will, therefore, not affect wave I and will only cause a prolongation of the later components of the BAEPs (Figures 9.3 and 9.4). Wave II is probably generated by both the cochlear nucleus and the proximal part of the auditory nerve. Wave III is predominantly generated in the lower pons at the ipsilateral superior olivary nucleus with some contribution from the cochlear nucleus. Wave IV is probably generated by the lateral lemniscus. Finally, wave V is probably generated in the mesencephalon at the inferior colliculus.

In summary, even though several generators contribute to each of the BAEP waves, specific anatomic loci have been proposed as the predominant source for the generation of each of these waveforms. These include:

Wave I	distal auditory nerve,
Wave II	cochlear nucleus and proximal auditory nerve,
Wave III	superior olivary nucleus (lower pons),
Wave IV	lateral lemniscus,
Wave V	inferior colliculus (mesencephalon).

Technique

The technique used for recording BAEPs in the operating room largely resembles that used in the

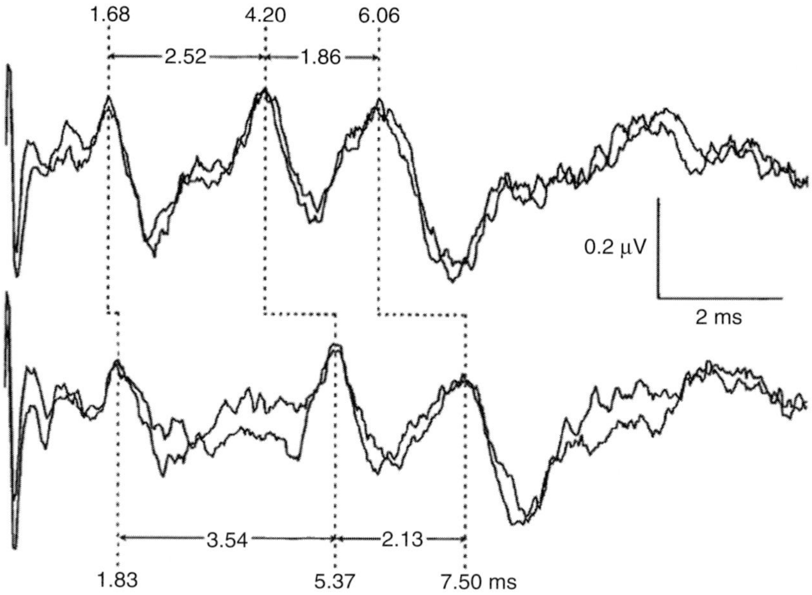

Figure 9.4 Intraoperative brainstem auditory evoked potentials (BAEPs) recorded during surgery for a right acoustic neuroma. The effect of stretching the eighth nerve caused by retraction of the cerebellum is shown in the bottom run. The most prominent change is the marked prolongation of the I-to-III interpeak interval (reprinted from Legatt, 2002, with permission from Lippincott, Wolters Kluwer Health.)

outpatient setting with some important modifications. In contrast with the outpatient lab, the environment in the operating theater is hostile due to electrical noises which make it challenging to obtain reliable signals. Several hundreds of averaged trials are often necessary, which in turn lengthen the duration needed to obtain an interpretable record. Therefore, certain measures are crucial in order to have a successful monitoring session and obtain an interpretable record in the shortest time possible. Serious attempts should be made to provide the appropriate acoustic stimulus, reduce electrical interference, and use optimal electrode placement and filtering to be able to reduce the time needed to acquire an interpretable record.

Stimulus source

Bulky headphones used in extraoperative testing are inappropriate in the operating room. Sound stimulus can be better delivered either through miniature electromagnetic earphones inserted in the ear canals and secured with plastic adhesive tape or, through stable sponge (molded foam) ear inserts connected to a transducer by plastic (usually polyvinyl) tubing (Figure 9.5). Dislodgment or kinking of this tubing can lead to complete absence of all BAEP components or, more commonly, to partial reduction of the stimulus intensity reaching the inner ear, leading to amplitude reduction and delayed latencies. Bone wax is often used to provide a tight seal at the entrance of the external ear canal (Legatt, 2002). It is important to recognize that the use of the plastic tube introduces a fixed acoustic delay in the delivery of the sound to the ear canal, the duration of which is directly related to the tube length, which is approximately 1 ms for each foot of tubing. This delay is beneficial in separating the stimulus artifact from wave I. There is a similar fixed shift in all the subsequent component latencies (BAEP waves) but without a change in the interpeak latencies; thus it doesn't interfere with the interpretation of the averaged trials as they are compared with the patient's own baseline.

Stimulation

Well-defined BAEP waves can usually be obtained by delivering series of 100 µs monophasic square-wave pulses to the transducer, producing brief acoustic clicks. The polarity of the first and most prominent wave of the acoustic waveform of the click determines whether a negative or positive pressure is applied in front of the ear speaker diaphragm. Rarefaction clicks apply negative pressure while condensation clicks apply positive pressure (American Clinical Neurophysiology Society, 2006). The so-called alternating clicks provide continuous trains with sequential alternation between the two polarities. The evoked

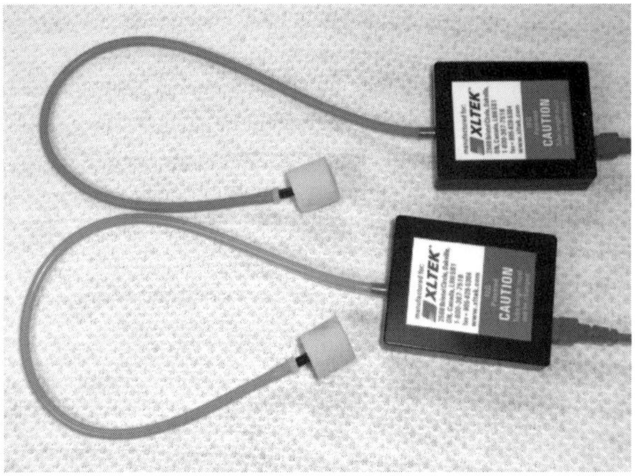

Figure 9.5 Acoustic transducer delivers the sound stimulus through polyvinyl tubing connected to molded foam earplugs. Image courtesy of Natus Medical Incorporated.

responses to either rarefaction or condensation clicks are complex and differ so that both can be tried in the operating room for collecting baseline values, to determine which polarity produces the most reliable waveform. More often, alternating polarity clicks are preferred in NIOM, especially when using electromagnetic earphones as this would help cancel out the large stimulus artifact (Figure 9.6). After deciding on the polarity, the remaining part of the NIOM should always be collected using the same polarity of click stimuli.

A preferred method for delivering the acoustic stimuli is to provide sequential interleaved clicks to both ears. This would allow averaged trials to be obtained in both ears at the same time, therefore avoiding a prolonged gap where one ear is not being tested.

Sound intensity

Like most other sensory evoked potentials, the amplitude of the BAEP is directly proportional to the stimulus intensity. Therefore, in NIOM and since the patient functions as his or her own control, the highest stimulus intensity should be used to produce maximal waveforms, as long as it does not pose a risk of causing hearing loss. An optimal waveform is easier to interpret and would probably allow for shorter averaging trials and faster feedback time. Usually, the stimulus intensity needed is at or above 70 dB nHL but higher intensities are often needed if there is a preexisting hearing loss. Suspending the stimulation is advised when simultaneous loud noises are introduced, for example, from bone drilling or irrigation, as these

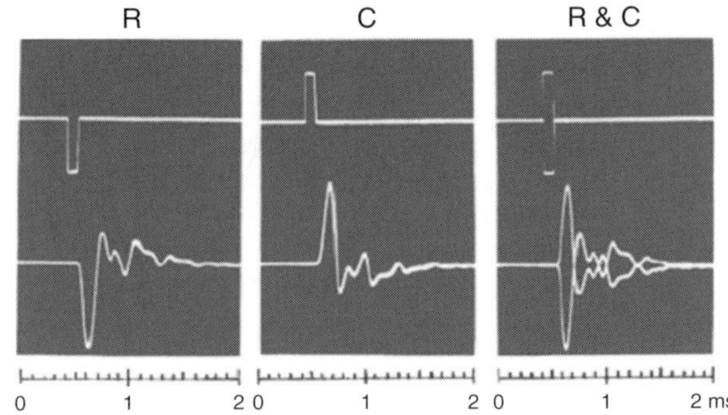

Figure 9.6 Electrical (top) and acoustic (bottom) waveforms of rarefaction (R), condensation (C), and alternating clicks. © American Clinical Neurophysiology Society, reprinted with permission.

can mask the acoustic stimulus causing a decrement in the recorded responses (Legatt, 2002).

Stimulus rate

Faster stimulation rates mean that shorter periods are needed to collect a fixed number of averaged trials, which in turn allows for a faster feedback to the surgeon. On the other hand, amplitudes are inversely proportional to the stimulus rate and the waveform amplitude starts to disintegrate at higher stimulus rates. Therefore, the selected rate should keep a balance between the rate of acquisition and the quality of the signal. A frequently used stimulus rate for the outpatient setting is approximately 10 Hz. This rate can also be used for NIOM but in most cases a faster rate can reliably be used. Exactly 10 Hz or multiples of the used power frequency (60 Hz in the USA) should be avoided to eliminate contamination with electrical artifacts. A faster rate, between 20 and 40 Hz, can be tried while collecting the baseline, in which case the fastest rate that could generate a reliable response can be used, as it would proportionately reduce the time needed to obtain these responses. At our center, we usually use a rate of 33.1 Hz, which is usually adequate; only rarely is a lower rate indicated. Rates higher than 40 Hz are likely to lead to significant decrement in waveform amplitude so that it would require longer time to generate reliable responses. In all cases, the selected rate should be used consistently for the rest of the surgery.

Typically, 1000 averaged trials are sufficient to obtain reliable responses but occasionally 2000 averages are needed if the waveforms are poorly formed.

Recording electrodes and acquisition parameters

Standard electroencephalography (EEG) cup electrodes or subdermal needle electrodes can be used. We found the subdermal needles to be more practical in the operating room as they can be applied quickly after the patient has been anesthetized without the need for skin preparation. They are usually easy to secure and provide stable and steady low electrode–skin impedances. If possible, the addition of more "redundant" electrodes at both recording and ground sites can be useful in case electrodes get dislodged during surgery.

Electrode placement follows the EEG 10–20 International System. The standard set-up should include recording electrode placement as follows: (1) at the vertex (Cz) and (2) over the ipsilateral and contralateral earlobes (Ai and Ac) or mastoid processes (Mi and Mc). An extra electrode placed at a non-cephalic site (e.g., cervical spine or shoulder) is optional but may be helpful under certain circumstances. The ground electrode on the scalp possibly at Fz or FPz. Electrode impedances must be kept below 5 kΩ.

Montage

A standard two-channel montage for BAEP recording as proposed by the ACNS (American Clinical Neurophysiology Society, 2006) is usually sufficient. Two additional channels could be useful and can be easily included with most modern NIOM equipment (Figure 9.7).

Standard

Channel 1: Vertex – ipsilateral earlobe or mastoid (Cz–Ai or Mi),
Channel 2: Vertex – contralateral earlobe or mastoid (Cz–Ac or Mc).

Optional

Channel 3: Contralateral – ipsilateral earlobe or mastoid (Ac–Ai or Mc–Mi),
Channel 4: Vertex – non-cephalic reference (Cz–Nc).

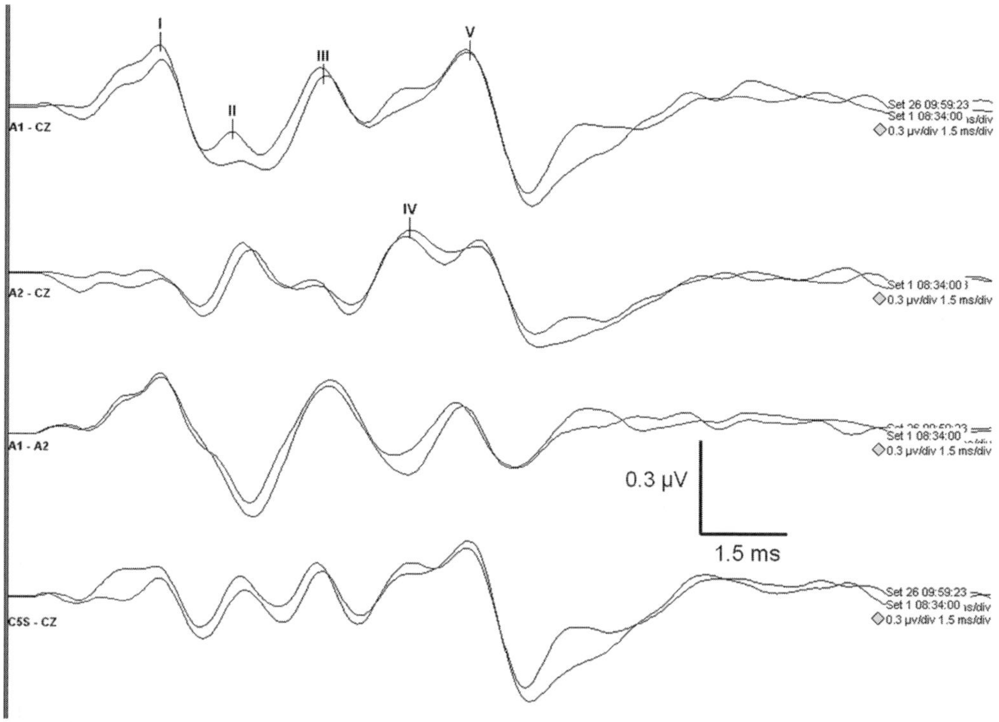

Figure 9.7 Four-channel montage setup for recording brainstem auditory evoked potentials (BAEPs).

Wave I, which represents a near-field potential from the stimulated ear, is usually well displayed on channel 1 of this montage. However, channel 3 (Ac–Ai) can sometimes help with better identification of wave I, especially if there is an abnormal pathology in the auditory nerve. Channel 2 (Cz–Ac) reflects the non-stimulated ear and therefore does not display wave I but provides a good display of the later waves (II–V; far-field potentials). It also helps in better separation of waves IV and V (Figure 9.7) as well as providing a back-up in case the other ear electrode gets dislodged. Channel 4 (Cz–Nc) uses a non-cephalic reference that could further enhance the detection of far-field potentials. This channel can be useful in cases where the wave V morphology and peak are ill-defined for the purpose of monitoring, either because of underlying pathology or because of significant electrical noise. Typical filter settings are a low frequency filter of 100 Hz and a high frequency filter of 3 kHz.

Indications

Surgeries directly involving the auditory pathways

Monitoring surgeries that directly involve the auditory nerve, nucleus, or pathways up to the level of the mesencephalon are the main indications for the use of BAEPs. Tumors arising from the auditory nerve, such as vestibular schwannomas, are a common indication for obtaining BAEPs (Fischer *et al.*, 1987; Legatt, 2002).

The eighth nerve is also at risk during microvascular decompression surgeries in cases of trigeminal neuralgia (Radtke *et al.*, 1989; Wilkins *et al.*, 1991; James and Husain, 2005), hemifacial spasm (Acevedo *et al.*, 1997), or glossopharyngeal neuralgia. The BAEPs are also suitable for monitoring the auditory nerve when it is at risk of ischemia from over stretching when cerebellar retractors are used, for example, during surgeries in the cerebellopontine angle (Wiedemayer *et al.*, 2002). Changes in BAEPs with cerebellar retraction are frequently reversible (Grundy and Jannetta, 1981; Grundy *et al.*, 1981) although they can also be permanent.

Surgeries indirectly involving the auditory pathways

Monitoring BAEPs provides very useful feedback about the general condition of the brainstem (Kalmanchey *et al.*, 1986). BAEPs are routinely monitored during brainstem surgeries for tumors and vascular abnormalities, and manipulation of the brainstem, such as in posterior fossa decompression in Chiari malformation (Zamel *et al.*, 2009). Almost all patients who develop permanent BAEP changes that persist to the end of the surgery are found to have new postoperative neurologic deficits (Manninen *et al.*, 1994). It is important to recognize that BAEPs would not be the testing modality of choice if the brainstem structures at risk are clearly not within or adjacent to the auditory pathways. The BAEPs reflect auditory conduction in the brainstem only between the pontomedullary junction and the superior colliculus. Therefore, monitoring brainstem surgeries that extend above the superior colliculus or below the pons should be complemented by other modalities, such as somatosensory evoked responses (Linden *et al.*, 1997).

Interpretations of brainstem auditory evoked potentials

Clinical neurophysiologists use BAEPs to preserve the brainstem function or to monitor the auditory nerve during surgery in which there is a risk of postoperative hearing deficits. Despite the use of intraoperative BAEPs for over 30 years and the establishment of its benefit in reducing operative morbidity by monitoring the functional integrity of the auditory pathway, no universally accepted criteria are available to determine when a recorded BAEP wave change is significant and thus warrants alerting the surgeon.

Prior to identifying and determining a significant change that reflects injury to neural structures, it is essential to exclude changes that are caused by technical or physiologic factors. Technical factors include instrument failure, dislodgment of the ear insert (Figure 9.8), or breakage of the wires or cables. Numerous sources of electrical interference could exist in the operating room and should be identified. These include but are not limited to electrocautery, surgical drills, a Cavitron ultrasonic surgical aspirator (CUSA), cardiopulmonary bypass machines, the electric operating room table, C-arm fluoroscopes, overhead lighting, surgical microscopes, monitors, intravenous pumps, blood warmers, and patient warmers. Acoustic masking from other sources of noise, such as from surgical drills, also needs to be recognized (Figure 9.9). Significant changes of the

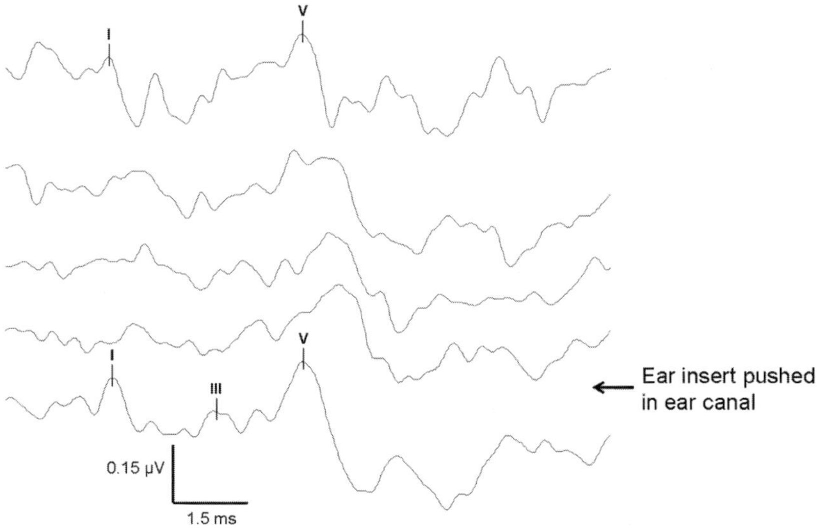

Figure 9.8 Consecutive brainstem auditory evoked potentials (BAEPs) in response to left ear stimulation (earliest waveform at the top), recorded in a patient undergoing posterior fossa decompression surgery. Wave I disappeared with attenuation and prolongation of wave V. The ear plug carrying the acoustic stimulus was found to be dislodged from the patient's ear. The BAEPs reappeared after re-inserting the ear plug.

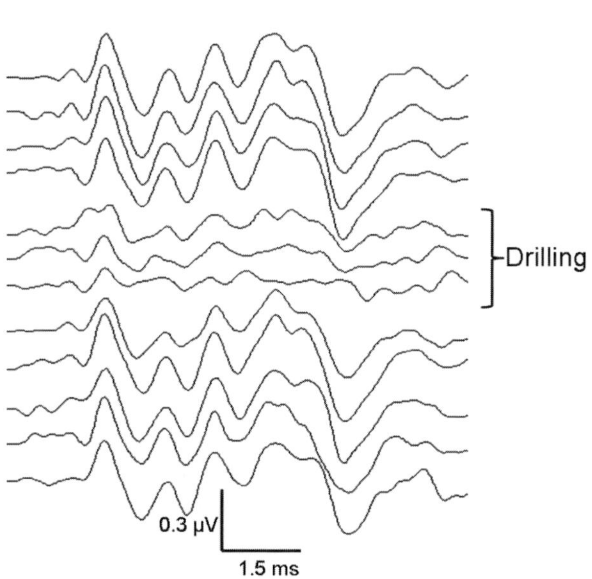

Figure 9.9 Consecutive brainstem auditory evoked potentials (BAEPs) to right ear stimulation recorded in a patient undergoing posterior fossa decompression surgery. Drilling of the occipital bone caused transient attenuation of the BAEPs. To avoid this false-positive BAEP change, BAEP recording should be suspended during bone drilling.

BAEPs could also be due to physiologic factors such as the effect of anesthesia or hypothermia.

As with the 2006 criteria published by the American Clinical Neurophysiology Society for recording standard BAEPs, analysis typically involves monitoring for the presence of waves I, III, and V. The measurements must include the following: (1) wave I peak latency; (2) wave III peak latency; (3) wave V peak latency; (4) I–III interpeak interval; (5) III–V interpeak interval; (6) I–V interpeak interval; (7) wave I amplitude; (8) wave V amplitude; and (9) wave IV–V/I amplitude ratio (American Clinical Neurophysiology Society, 2006).

The two main variables used for monitoring BAEPs are the latencies and the amplitude of the recorded waveforms.

Latencies

Absolute component latencies and interpeak intervals are the standard and most reliable parameters used for monitoring BAEPs. The interpretation of the BAEPs is based on the three most reliable peaks (waves I, III, and V).

The peak of wave V is usually the easiest to identify, as it is the most prominent; therefore changes in the absolute latency of wave V or the interpeak intervals from wave I to V are the most commonly used parameters to predict hearing loss during brainstem surgeries. The values throughout the surgery are compared with the patient's own baseline obtained at the beginning of the surgery using the same acquisition and stimulus parameters. Occasionally, some individuals have an ill-defined wave V peak and a more dominant wave IV peak. In that case it is more practical to rely more on monitoring wave IV latencies (Legatt, 2005). Although arbitrary, the most commonly used latency threshold criteria to warn the surgeon of an impending damage to CNVIII is a

1 ms prolongation of wave V (Watanabe *et al.*, 1989; Manninen *et al.*, 1994). Threshold criteria proposed by others ranged between the recommendation that only complete disappearance of wave V should be used as a criterion for predicting hearing loss (Grundy *et al.*, 1983; Friedman *et al.*, 1985; Schlake *et al.*, 2001) or that the prolongation of wave V by only 0.5 ms or more is clinically significant (Coletti and Fiorino, 1998). To address these differences in practice, James and Husain (2005) reviewed data on 156 patients, who were monitored during surgeries in the cerebellopontine angle (CPA). They found that in 23 patients with CPA tumor, there was no difference in the frequency of hearing loss when four different threshold criteria were compared that ranged from a slight, less than 1 ms prolongation of wave V to complete and persistent loss of wave V. On the other hand, for the 133 patients with non-CPA tumor, hearing loss occurred only when wave V was completely lost. The authors suggested that interpreting changes in wave V should always be in view of the type of surgery and underlying pathology.

The reasons for the lack of universally agreed guidelines for interpreting changes in latencies or waveforms are well understood and it is important to recognize that overrelying on rigid criteria could introduce inaccuracies and unnecessary warnings to the surgeon that could lead to prolongation or even alteration of the surgical procedure. At the same time, in some cases it would be important to notify the surgeon if a gradual but steady deterioration of the waveform morphology or latencies is noted, even though it does not reach the predetermined threshold limits.

Utilizing the BAEP peaks to localize the site of neurologic injury

As discussed above, it is an oversimplification to correlate each peak or trough of the BAEP with an exclusive solitary neural generator. It is more likely that a specific generator is responsible for most of the potential seen for each of the BAEP components in addition to other generators that are contributing to the recorded signal. Therefore, attributing changes in latencies and interpeak intervals to changes in conduction in particular regions of the brainstem should be done with caution. For example, while wave V is usually the most reliable peak, changes in neural conduction of the eighth nerve might cause a larger shift in wave III than in wave V, therefore monitoring interpeak interval I–III would be more sensitive to identify changes than monitoring interval III–V or I–V.

Changes that are seen mainly in interpeak interval III–V usually indicate a conduction change in the ascending auditory pathways rostral to the superior olivary and cochlear nuclei. Additionally, in these cases, Møller suggests that systemic changes, such as changes in cerebral circulation, could be occurring and that the anesthesiologist should be notified (Møller, 2006).

Amplitude

In routine extraoperative testing, unlike latencies, the amplitudes of the BAEP waves vary significantly between human subjects. This restricts the utility of the amplitudes in interpreting the data with the exception of the ratio of the amplitudes of waves V/I, which appears to be more constant (Tusa *et al.*, 1994). This limitation applies more for interpreting routine BAEPs when they are compared with standardized values. For the purpose of NIOM, the amplitudes within the same subject usually remain stable throughout the surgery, allowing for their use in interpreting the findings. While most commonly used criteria depend on changes in waveform peak latencies, some investigators found the amplitudes to be as important or even more useful. In some cases, injury to the auditory pathways can be suggested by changes in the amplitude ratio of waves V/I, which can be seen earlier or even without any change in latencies. A commonly used criterion is a greater than 50% drop in the amplitude of wave V. Hatayama and Møller (1998) have noted wave V amplitude to be a more sensitive predictor of hearing loss than latency. They reported that 61% of 18 cases had statistically significant hearing loss when the amplitude of wave V was decreased by more than 40%. Ramnarayan and Mackenzie (2006) reported hearing loss in 25% of 8 patients who had 50% wave V amplitude loss with increased chances of hearing loss to 45% of 11 patients with 75% amplitude loss and in all patients with 100% amplitude loss ($p < 0.0015$). Others have argued that only a complete disappearance of wave V is associated with hearing loss but not with the 50% amplitude decrement criteria (James and Husain, 2005).

Questions

1. Despite its widespread use in intraoperative monitoring, the utility of brainstem auditory evoked

potentials according to the American Academy of neurology is only investigational.
(a) True
(b) False

2. Unlike short-latency auditory evoked responses, middle and long auditory evoked responses are not suitable for intraoperative use because:
(a) They are highly vulnerable to the effect of anesthesia
(b) They do not reflect the condition of the brainstem well, as they are predominantly generated in the cerebral cortex
(c) They are highly variable across subjects
(d) All of the above
(e) None of the above

3. The brainstem auditory evoked potentials consist of a series of short-latency waves that all represent a far-field potential, except for:
(a) Wave I
(b) Wave II
(c) Wave III
(d) Wave IV
(e) Wave V

4. Wave I of the brainstem auditory evoked potential is best displayed in a channel that combines the following derivations:
(a) Vertex to non-cephalic reference
(b) Ipsilateral ear to vertex
(c) Contralateral ear to vertex
(d) Ipsilateral ear to contralateral ear
(e) None of the above

5. To record BAEPs, it is preferable that the acoustic stimuli be delivered using:
(a) Headphones similar to the one used in the outpatient setting for accurate comparison
(b) Miniature electromagnetic earphones or sponge ear plugs (ear inserts)
(c) The source of the acoustic stimuli can be of any type as long as the sound intensity is consistent
(d) None of the above

6. Wave III of the BAEPs is probably generated at the distal auditory nerve, in the:
(a) Cochlear nucleus
(b) Superior olivary nucleus
(c) Lateral lemniscus
(d) Inferior colliculus

7. When selecting the optimum rate of acoustic stimulation to monitor BAEPs:
(a) A faster rate of stimulation typically improves the quality of waveforms
(b) A slower rate of stimulation requires fewer averaging trials to generate reproducible waves
(c) An exact rate of 10 or 20 Hz is preferable
(d) The wave amplitudes are inversely proportional to the stimulus rate

8. Wave I of BAEP is best displayed in a channel with the following derivations:
(a) Ipsilateral ear and vertex (Cz)
(b) Contralateral ear and vertex (Cz)
(c) Vertex and non-cephalic reference
(d) None of the above

9. All of the following statements regarding the utility of the wave amplitudes in interpreting the BAEPs in NIOM are correct except:
(a) Changes in the absolute amplitudes can occur earlier than changes in the latencies of the waveforms
(b) The BAEPs can typically be monitored using the ratio of the amplitude of waves V/I
(c) Change in wave V amplitude is a sensitive predictor of hearing loss
(d) Changes in the absolute amplitudes are compared with the patient's own baseline values

10. Which of the following is false about the utility of BAEPs in NIOM?
(a) BAEPs can reliably assess the auditory pathways rostral to the mesencephalon
(b) The waveforms in BAEPs can be significantly altered by hypothermia
(c) The standard montage needed to monitor BAEPs includes two channels that link both ears to the vertex
(d) Ear inserts are preferable to headphones as a source of acoustic stimuli
(e) None of the above

References

Acevedo, J. C., Sindou, M., Fischer, C., and Vial, C. (1997). Microvascular decompression for the treatment of hemifacial spasm: retrospective study of a consecutive series of 75 operated patients – electrophysiologic and anatomical surgical analysis. *Stereotactic and Functional Neurosurgery*, **68**, 260–265.

American Clinical Neurophysiology Society (2006). Guideline 9C: guidelines on short-latency auditory evoked potentials. *Journal of Clinical Neurophysiology*, **23**(2), 157–167.

Coletti, V., Fiorino, F. G., Mocella., S., and Policante., Z. (1998). ECochG, CNAP and ABR monitoring during vestibular schwannoma surgery. *Audiology*, **37**, 27–37.

Fischer, C., Bertrand, O., Volman, D., and Larrea, L. G. (1987). Intraoperative BAEP monitoring in posterior-fossa surgery – evaluation of changes in 90 patients. *Electroencephalography and Clinical Neurophysiology*, **67**, 53.

Friedman, W. A., Kaplan, B. J., Gravenstein, D., and Rhoton, A. L. (1985). Intraoperative brainstem auditory evoked-potentials during posterior-fossa microvascular decompression. *Journal of Neurosurgery*, **62**, 552–557.

Grundy, B. L. and Jannetta, P. J. (1981). BAEP monitoring during cerebellopontine angle (CPA) surgery. *Electroencephalography and Clinical Neurophysiology*, **51**, 39.

Grundy, B. L., Lina, A., Procopio, P. T., and Jannetta, P. J. (1981). Reversible evoked-potential changes with retraction of the 8th cranial nerve. *Anesthesia and Analgesia*, **60**, 835–838.

Grundy, B. L., Jannetta, P. J., Procopio, P. T., *et al.* (1983). BAEP monitoring during cerebellopontine angle surgery. *Electroencephalography and Clinical Neurophysiology*, **56**, 30.

Harper, C. M., Harner, S. G., Slavit, D. H., *et al.* (1992). Effect of BAEP monitoring on hearing preservation during acoustic neuroma resection. *Neurology*, **42**, 1551–1553.

Hatayama, T. and Møller, A. R. (1998). Correlation between latency and amplitude of peak V in the brainstem auditory evoked potentials: Intraoperative recordings in microvascular decompression operations. *Acta Neurochirurgica*, **140**, 681–687.

James, M. L. and Husain, A. M. (2005). Brainstem auditory evoked potential monitoring – when is change in wave V significant? *Neurology*, **65**, 1551–1555.

Kalmanchey, R., Avila, A., and Symon, L. (1986). The use of brain-stem auditory evoked-potentials during posterior-fossa surgery as a monitor of brain-stem function. *Acta Neurochirurgica*, **82**, 128–136.

Legatt, A. D., Arezzo, J. C. and Vaughan, H. G., Jr. (1988). The anatomic and physiologic bases of brainstem auditory evoked potentials. *Neurologic Clinics*, **6**, 681–704.

Legatt, A. D. (2002). Mechanisms of intraoperative brainstem auditory evoked potential changes. *Journal of Clinical Neurophysiology*, **19**, 396–408.

Legatt, A. (2005). Brainstem auditory evoked potentials: methodology, interpretation, and clinical application. *Electrodiagnosis in Clinical Neurology*, ed. M. J. Aminoff., New York: Churchill Livingstone, pp. 489–523.

Linden, R., Zappulla, R., and Shields, C. B. (1997). Intraoperative evoked potential monitoring. *Evoked Potentials in Clinical Medicine*, ed. K. H. Chiappa. Philadelphia: Lippincott Williams & Wilkins, pp. 601–638.

Lopez, J. R. (2004). The use of evoked potentials in intraoperative neurophysiologic monitoring. *Physical Medicine and Rehabilitation Clinics of North America*, **15**, 63–84.

Manninen, P. H., Patterson, S., Lam, A. M., Gelb, A. W., and Nantau, W. E. (1994). Evoked-potential monitoring during posterior-fossa aneurysm surgery – a comparison of 2 modalities. *Canadian Journal of Anaesthesia*, **41**, 92–97.

Møller, A. R. (2006). Monitoring auditory evoked potentials. In *Intraoperative Neurophysiological Monitoring*, 2nd edn. New Jersey: Humana Press, pp. 85–124.

Radtke, R. A., Erwin, C. W., and Wilkins, R. H. (1989). Intraoperative brainstem auditory evoked potentials: significant decrease in postoperative morbidity. *Neurology*, **39**, 187–191.

Ramnarayan, R. and Mackenzie, I. (2006). Brainstem auditory evoked responses during microvascular decompression for trigeminal neuralgia: predicting post-operative hearing loss. *Neurology India*, **54**, 250–254.

Schlake, H. P., Milewski, C., Goldbrunner, R. H., *et al.* (2001). Combined intra-operative monitoring of hearing by means of auditory brainstem responses (ABR) and transtympanic electrocochleography (ECochG) during surgery of intra- and extrameatal acoustic neurinomas. *Acta Neurochirurgica*, **143**, 985–995.

Tusa, R. J., Stewart, W. F., Shechter, A. L., Simon, D., and Liberman, J. N. (1994). Longitudinal study of brainstem auditory-evoked responses in 87 normal human subjects. *Neurology*, **44**, 528–532.

Watanabe, E., Schramm, J., Strauss, C., and Fahlbusch, R. (1989). Neurophysiologic monitoring in posterior-fossa surgery. 2. BAEP-wave-I and BAEP-wave-V and preservation of hearing. *Acta Neurochirurgica*, **98**, 118–128.

Wiedemayer, H., Fauser, B., Sandalcioglu, I. E., Schafer, H., and Stolke, D. (2002). The impact of neurophysiological intraoperative monitoring on surgical decisions: a critical analysis of 423 cases. *Journal of Neurosurgery*, **96**, 255–262.

Wilkins, R. H., Radtke, R. A., and Erwin, C. W. (1991). Value of intraoperative brain-stem auditory evoked-potential monitoring in reducing the auditory morbidity associated with microvascular decompression of cranial nerves. *Skull Base Surgery*, **1**, 106–109.

Zamel, K., Galloway, G., Kosnik, E. J., Raslan, M., and Adeli, A. (2009). Intraoperative neurophysiologic monitoring in 80 patients with Chiari I malformation: role of duraplasty. *Journal of Clinical Neurophysiology*, **26**, 70–75.

10 Pedicle screw application and electromyographic recording

Gloria M. Galloway

Background and preparation for the procedure

Pedicle screw placement may be included as part of surgical management during scoliosis repair. Idiopathic scoliosis as defined by the Scoliosis Research Society is a lateral curvature of the spine of greater than ten degrees from the normal vertical line as measured on X-ray (http://www.srs.org). Scoliosis is typically a developmental deformity of the spine and is usually progressive if left untreated. Treatment is often indicated because of the risk of decreasing the chest wall capacity with subsequent pulmonary compromise. For this reason scoliosis screening is often done during physical examinations in adolescents, particularly when the child is being evaluated prior to involvement in competitive sports. Examination of the spine in forward flexion is recommended, as well as frontal and lateral standing radiographs (Kotwicki, 2008).

The surgical treatment of scoliosis is aimed at stopping the progression of the curvature, optimizing balance, and correcting the skeletal deformity so that the top and bottom of the fusion is in alignment at the patient's shoulders and pelvis and assumes the avoidance of complications. There have been reports indicating that there are advantages in the correction achieved in scoliosis fusion surgeries, as well as other corrective surgeries with the use of pedicle screws, over the correction achieved with hook instrumentation at the lumbar level (Dobbs *et al.*, 2008). Surgery involving fusion is described as selective in nature when there is deviation from the midline of the thoracic and thoracolumbar curves and fusion of only the major curve, which is the one with the largest Cobb angle, so that the minor curve is left unfused and mobile. There are numerous potential advantages to pedicle screw placement in the thoracic and lumbar spine. Placement of pedicle screws in the thoracic and lumbar spine can allow a more stable fixation point for some spinal rod instrumentation systems (Shufflebarger *et al.*, 2004). Pedicle screw segmental placement may provide improved major and minor curve correction, and greater balance over that achieved with traditional hooks (Halm *et al.*, 2002). In addition, pedicle screw placement may allow the use of shorter fusion lengths, saving distal motion segments (Barr *et al.*, 1997; Christodoulou *et al.*, 2002).

Procedure

Pedicle screws are placed into the pedicles at the desired vertebral levels and during this placement care is taken not to enter the spinal canal, so that adjacent nerve roots are not irritated outside the medial wall. Studies have shown that pedicle screw placement done without electromyographic monitoring, i.e., blindly, can be erroneous, with entrance into the medial wall of the pedicle 20% of the time (Kim *et al.*, 2009; Laine *et al.*, 2009). This breach can result in the development of radicular symptoms postoperatively (Bjarke *et al.*, 2002). Some surgeons utilize radiography (Figure 10.1) or fluoroscopy after or during the placement as a guide and readjust if necessary (Lenke *et al.*, 2005). However, fluoroscopy has limited sensitivity due to limitations on views available and intraoperative MRI or CT is used instead, in some centers, if available.

Guidance from electromyography during pedicle screw placement is highly sensitive in identifying misplaced screws with resultant nerve root irritation. Preparation for electromyography in these cases can be done utilizing a reasonable and systematic approach. In this approach, the surgeon should tell the NIOM team which spinal levels pedicle screw placement is planned for (Pruijs *et al.*, 2000; Kim *et al.*, 2001). Then the NIOM team can determine which muscle groups are best to utilize. These muscle groups are chosen based on their innervations. Selection should involve those muscles innervated by the nerve roots adjacent to the areas of pedicle screw placement. Needle electrodes are placed in these muscles to record muscle activity (Table 10.1). Surface or recording needle electrodes may be used. Needle electrodes are less likely to get dislodged during patient positioning. For most limb muscles subdermal needle electrodes are sufficient. If a muscle is very deep or a patient has a sizable amount of subcutaneous tissue, a longer EMG recording needle may need to be placed (Figure 10.2).

Table 10.1 Some commonly utilized muscle groups and their innervations in pedicle screw monitoring

Biceps, brachioradialis, deltoid	C5, 6
Triceps	C7, 8
Thenar muscle group (APB primarily)	C8, T1
Hypothenar muscle group	
Intercostals, abdominals, paraspinals	T6–T12
Iliopsoas	L1, 2
Quadriceps, vastus lateralis or medialis	L2, 3 ,4
Tibialis anterior	L4, 5
Abductor hallucis; gastrocnemius	S1, S2
Anal sphincter	S2, 3,4

As with motor-evoked potential testing, muscle relaxants need to be kept completely off or at a minimal level, to identify the absence of a CMAP based on pedicle screw stimulation accurately and reliably. In the presence of too great a degree of muscle relaxation, a CMAP may not be generated due to failure of neuromuscular transmission and a false-negative response will be erroneously relayed to the surgical staff. As with monitoring motor-evoked potentials, monitoring the degree of muscle relaxation with muscle twitches to a train of four stimulations, referred to as TOF (typically at the forearm or wrist), allows for a relatively easy way to know that muscle relaxation is not too great. The NIOM team should be able to visualize these responses directly or utilize the monitoring equipment to set up a repetitive stimulation protocol to access these muscle responses as the surgery progresses and compare them with baseline values for the individual patient.

Interpretation of findings

The responses are best visualized using appropriate filter settings and display gain. If the display gain is set to 50 μV (microvolts) per division it is possible to identify reliably the presence of a CMAP. This may be as low as 50 μV, although it is typically in the order of several millivolts (mV) in amplitude (Figure 10.3). Filter settings include a low-frequency filter set at

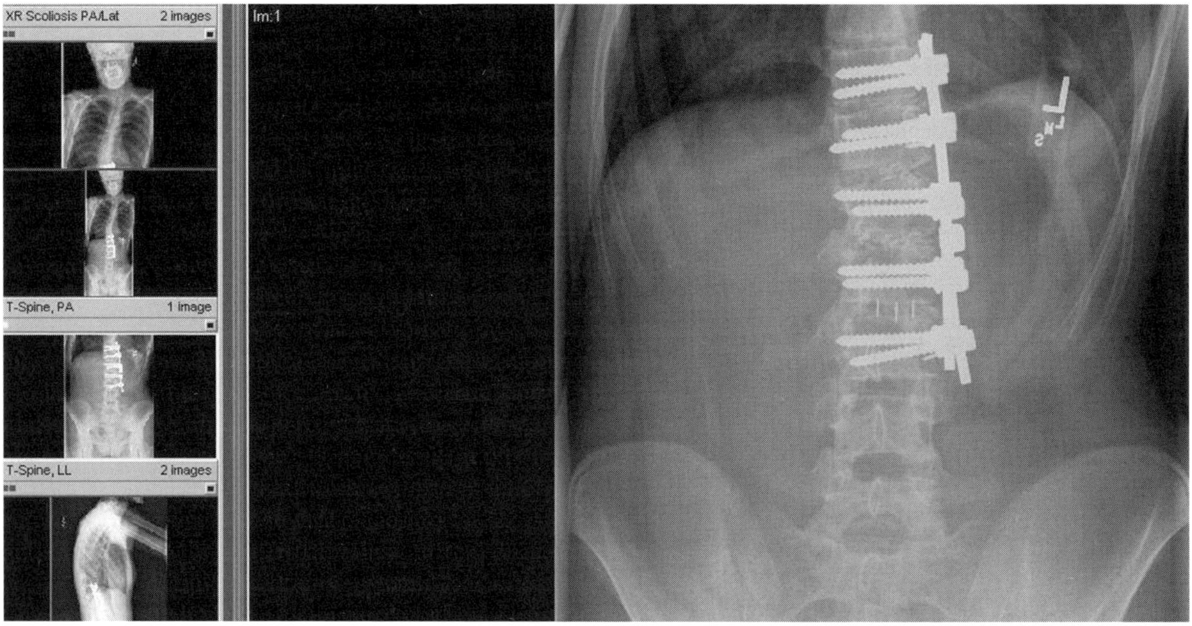

Figure 10.1 Radiograph of a patient after pedicle screw placement.

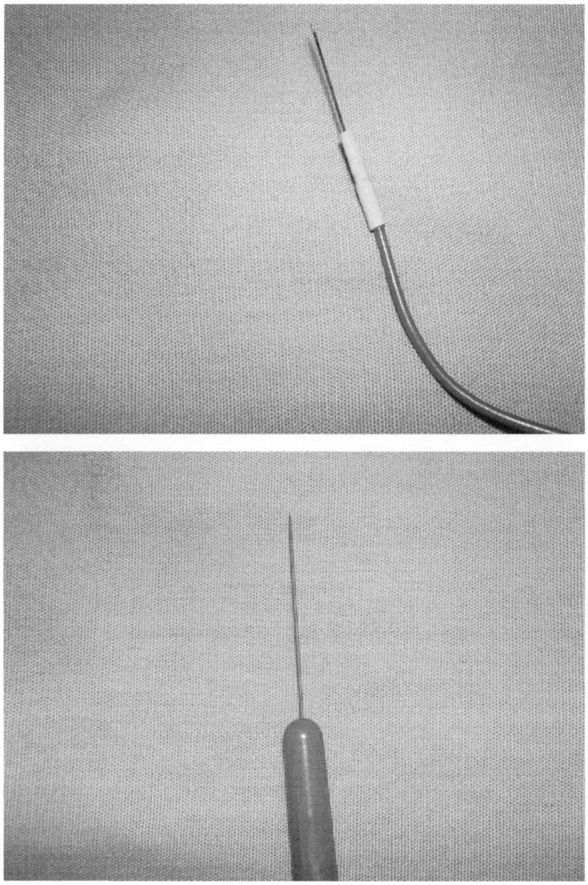

Figure 10.2 Examples of recording and stimulating needle electrodes.

2–3 Hz and a high-frequency filter set at 10 kHz. With recording electrodes in place in the muscles involved in the distribution of pedicle screw placement, a reference electrode is placed 2–3 cm away from the stimulation point. This reference electrode will need to be moved every few levels as the point of stimulation over the vertebral spine will change during the procedure. A ground is placed nearby but away from the stimulation point.

The pedicle hole or track should be stimulated first, using a monopolar stimulating electrode, as this will provide a baseline assessment at increasing stimulation intensity. Then the exposed screw shaft or shank is stimulated at each pedicle screw placement. In the case of a pedicle screw with high impedance, testing the pedicle hole first will prevent false-negative results from occurring with stimulation.

Muscle activity is typically recorded in two ways in these cases and reported to the surgeon or surgeons involved. First, free-run EMG during placement allows a gross assessment of any baseline irritability present. Electromyography is a sensitive and reliable modality for detecting radiculopathy in the operating room during pedicle screw stimulation. In some cases, more disastrous complications on the spinal cord itself can occur with misplaced pedicle screws but this is uncommon (Papin *et al.*, 1999). In this setting, the presence of high-frequency bursts of motor unit potentials referred to as neurotonic discharges are a sensitive indicator of nerve root irritation and therefore of irritation to or injury to the lumbosacral nerve roots involved . This method is quick and reliable and provides continuous or instant feedback during the surgical case (Suk *et al.*, 2003).

Secondly, EMG guidance during pedicle screw placement is very sensitive at evaluating the presence of a medial wall breach, resulting in a compound muscle action potential (CMAP) with stimulation of the screw or pedicle screw hole (Lewis *et al.*, 2002; Shi *et al.*, 2003). With good placement, the bony pedicle, which is normally intact, has high impedance, which resists the flow of current to the adjacent nerve root, and stimulation does not result in the presence of a muscle response or CMAP with malplacement. However, if that resistance is breached, the adjacent nerve root receives current upon electrical stimulation, resulting in a muscle response that can be recorded on the monitoring screen (CMAP); see Figure 10.3.

Typically, one starts to stimulate at a low intensity and adjusts in small increments (such as 1 mA increments) until a predetermined level, based on the baseline testing of that nerve root or on control values, or on the presence of a CMAP, occurs (Raynor *et al.*, 2002). Stimulation often continues to 10 or 14 mA; in some cases, 20 mA is used as the upper intensity limit. There is variability in the literature as to the threshold for CMAP determination. Some centers have used <6 mA others higher at <8 mA as positive for abnormality.

If the stimulation intensity is raised very high, at some point it will be high enough even without a breach to stimulate adjacent nerve roots (Anderson *et al.*, 2002). This is why scouting or searching for the threshold of CMAP generation of the hole is done first if a CMAP is seen with stimulation of the hole, in order to prevent a false-positive or false-negative result. Once found, the stimulus can be reduced incrementally until the threshold for that hole is found.

It is essential to record on the operative log or record which pedicle screws were stimulated, whether a CMAP

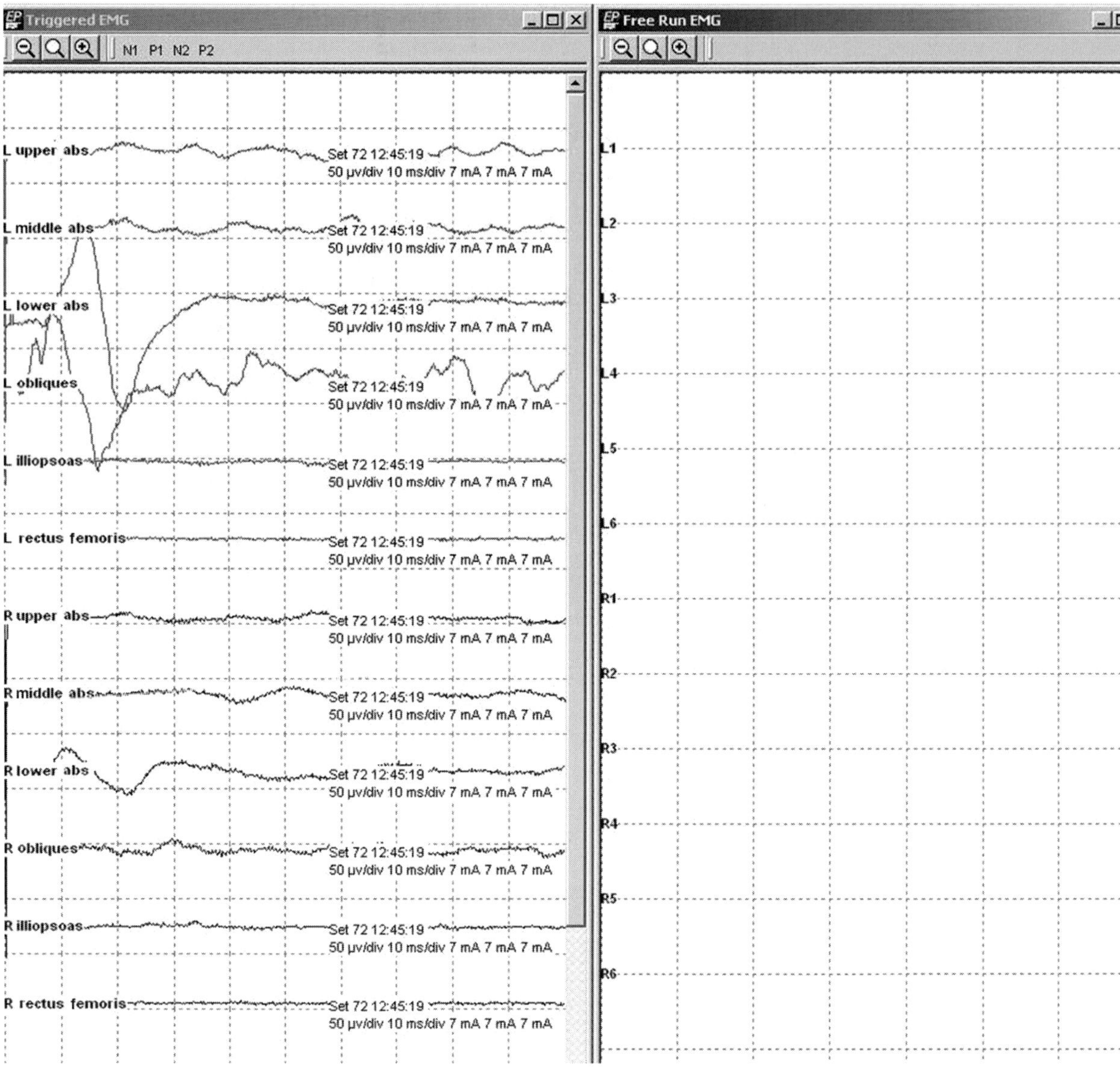

Figure 10.3 A CMAP is generated after thoracic pedicle screw stimulation at 7 mA.

was generated, and at what intensity of stimulation. This is the event log for the procedure. This information can be correlated with the patient's postoperative recovery if needed. It also allows the NIOM team to inform the surgeon accurately at what level of intensity and what anatomical level a CMAP occurred. If testing is repeated after repositioning of the screw, placement of bone wax, or insertion of a different (perhaps smaller) screw, the retesting intensity can be correlated with the intensity of generation of the CMAP prior to these recovery maneuvers (Rosner *et al.*, 2003). Figure 10.4b shows pedicle screw placement in the thoracolumbar spine and Figure 10.4a shows the pedicle screws prior to placement.

The presence of fluid near the pedicle will allow shunting of current and the development of a positive CMAP, even in the absence of a breach. For this reason, it is important that the surgical team frequently suction fluids from the space and keep the pedicle screw area dry during stimulation, to prevent a false-positive result.

Patients with thin cortical bones, such as patients with osteoporosis, may also develop a false-positive CMAP with stimulation as a result of a spread of

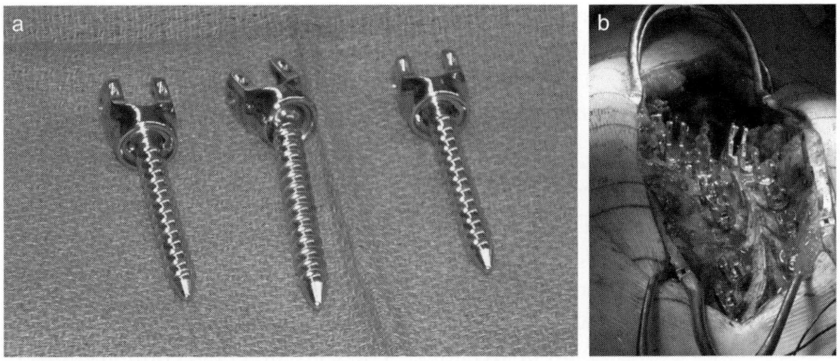

Figure 10.4 (a) Pedicle screws prior to placement; (b) pedicle screw placement in the thoracic and lumbar spine.

electrical current through a thin cortical bone with decreased resistance. If a significant CMAP is seen, the pedicle screw should be removed and the pedicle palpated for evidence of a breach. Bone wax can be placed to assist in plugging the hole (which will later re-close from the bony fragments) and the screw can then be repositioned and retested.

Because scouting for a threshold level of CMAP generation is time consuming, some surgeons prefer only to test the screw itself without testing the hole first, despite the fact that this may lead to a false-positive or -negative result. In nerves that are chronically compressed, the intensity needed to create a CMAP response on stimulation of the pedicle will be less than in a normal nerve. One needs to keep this in mind with pedicle screw stimulation in a patient with chronic nerve root compressions (Jutte and Castelein, 2002).

Pedicle screw monitoring with electromyographic recording needs to be interpreted by trained or skilled physicians experienced in this type of monitoring or errors affecting patient care are possible. This interpretation needs to take into account several factors that influence the results and potential interpretation. The factors that influence pedicle screw interpretation include:

1. The method of monitoring for pedicle screw stimulation depends on utilizing muscles from myotomes innervated by the nerve roots being tested. If the choice of myotomes is inaccurate then, in the absence of a significant CMAP, the surgical staff will believe a pedicle screw to be safely placed when in fact it may not be because the myotome represented was not the one tested.
2. In cases of complete nerve root transection, a CMAP cannot be generated because a neural impulse fails. Since there is discontinuity of innervation, a muscle response cannot be generated.
3. Electromyographic signals can only reliably be obtained with minimal to no neuromuscular blockade. If partial blockade is used, then a sensitive or careful titration of the neuromuscular junction blocking agent should be employed. This is very much dependent on individual patients; in some patients, partial blockade will not be successful and only when no blockade is present can reliable CMAPs be recorded.
4. It may not be possible to know whether a nerve root has been chronically compressed in a given patient. In those patients with chronic compression, the thresholds needed to generate a CMAP with stimulation may be less than in a patient in whom the nerve root is not compressed. Therefore, it is best when possible, that the nerve root be stimulated directly first. In this way, a baseline threshold of the exiting nerve is obtained and can be used as a guide if there is a breach present. Without first performing direct nerve root stimulation, a false-positive result may occur.

Lower cervical and upper thoracic spine levels

Fixation techniques in the upper cervical spine can be problematic given the anatomy in that region and the risk of complications. These complications include vertebral artery penetration, bleeding, radiculopathy, loss of correction, and infection (Yukawa *et al.*, 2009). Rigid postoperative fixation and immobilization is often the result, unless a successful fusion can occur. However, fluoroscopy-guided techniques as well as stereotactic image-guided systems and CT scans have been described in cervical pedicle screw fixation techniques (Abumi *et al.*, 2000). Intraoperative intralaminar screws or wiring techniques may result in a successful outcome in surgeries in the upper cervical spine. The use of C2 intralaminar screws can avoid the necessity of pedicle screw placement and the potential for vertebral artery injury (Dean *et al.*, 2009).

Computer-assisted pedicle screw placement

Stereotaxic computer-aided guidance systems can assist with intraoperative pedicle screw placement and may lower the rate of pedicle violation. These systems are not in widespread use and may be costly, take up space and add time to the surgical procedure. It is also not clear how well they compare with the use of intraoperative evoked electromyography with pedicle screw stimulation (Richter *et al.*, 2004; Kim *et al.*, 2009; Laine *et al.*, 2009).

Complications of pedicle screw placement

Errors in positioning thoracic pedicle screws have been reported in up to 8% of cases (Glaser *et al.*, 2003). Complications from malpositioned screws include nerve root impingement, in rare cases, fracture of the pedicle, and lacerations of the dura with resultant cerebrospinal fluid leaks (Katonis *et al.*, 2003). Cerebrospinal fluid leaks have the potential for introduction of infection (Othman *et al.*, 2006).

Scoliosis in patients with Prader–Willi syndrome

Prader–Willi syndrome was first described in 1956 as a cause of neonatal hypotonia (Prader *et al.*, 1956) but the consensus criteria for diagnosis were not described until 1993 (Holm *et al.*, 1993). It is the most frequent etiology of neonatal syndromic obesity and although it is most often sporadic in nature, cases arising from abnormalities in paternal-derived or maternal-derived chromosome 15q alterations have been well described. Scoliosis is significantly more prevalent in patients with Prader–Willi syndrome than in the general population, with rates of up to 80% in the mature skeletal spine reported and major curves in the thoracic, lumbar, or thoracolumbar regions (Gurd and Thompson, 1981). Kyphosis may be associated and, as in idiopathic scoliosis, the curvature is progressive with advancing age. Growth hormone therapy is often used in these patients to improve lean body mass and osteopenia but may have a contributing effect on the rate of scoliosis progression in these patients (Nagai *et al.*, 2006).

Conclusion

The use of electromyography for operative pedicle screw placement is a reliable, sensitive, and accurate means of determining the presence of a malpositioned screw with prevention of nerve root irritation and root pain or weakness. There are potential complications of pedicle screw placement in any surgery. The risk of complications can be influenced by the patient's presurgical state. In patients with preexisting medical conditions, preparation is needed and a good history should be obtained. An example is the patient with Prader–Willi syndrome, in which scoliosis is more prevalent than in the general population. However, with preparation and communication, pedicle screw stimulation with electromyographic recording can give valuable information to the surgeon, lessening patient morbidity and enhancing the quality of the operative procedure.

Questions

1. Studies have shown that pedicle screw placement done blindly without electromyographic monitoring can be erroneous with entrance into the medial wall what percent of the time?
   (a) 20% of the time
   (b) 50% of the time
   (c) 75% of the time
   (d) 100% of the time
2. Recording of electromyographic signals for pedicle screw stimulation can be done with surface or needle electrodes. Use of needle electrodes has some advantages over surface recording needle electrodes. These advantages include which of the following?
   (a) Needle electrodes are less likely to get dislodged during patient positioning
   (b) Needle electrodes provide a larger recording area than surface electrodes
   (c) Needle electrodes are less expensive than surface electrodes to use
3. Which of the following is true about *neurotonic discharges* seen during electromyographic recording for pedicle screw placement and stimulation?
   (a) They are low frequency potentials, which display in a continuous firing pattern

(b) They are low amplitude, low frequency potentials of brief duration
(c) They are high-frequency bursts of motor unit potentials
(d) They are a sensitive indicator that the adjacent nerve root is unlikely to be compressed or irritated

4. In those patients with chronic compression of nerve roots, what statement about thresholds needed to generate a CMAP with stimulation is false?
   (a) The threshold for generation is probably greater than in a patient in which the nerve root is not compressed
   (b) The threshold will probably be less than in a patient in which the nerve root is not compressed
   (c) Beginning with direct nerve stimulation lessens the risk of false-negative threshold recordings
5. Which of the following statements regarding pedicle screw stimulation and the level of neuromuscular blockade is true?
   (a) Neuromuscular blocking agents should be utilized to ensure that neurotonic discharges do not interfere with stimulation
   (b) The level of neuromuscular blocking agents needs to be minimal and can be measured with trains of four to ensure sensitivity for recording during stimulation
   (c) The level of neuromuscular blocking agents should be gradually increased to prevent movement of the patient at the time of pedicle screw stimulation
6. Which of the following statements regarding muscle selection during pedicle screw stimulation is true?
   (a) The choice of muscle selection is based on muscles used for TES recording
   (b) The choice of muscle selection is based on the myotomes used in pedicle screw placement
   (c) The choice of muscle selection is based on which muscles are most frequently used in patients at your institution
7. Pedicle screw stimulation with electromyographic recording can be done at which of the following levels?
   (a) C1–T5 only
   (b) T5– T10 only
   (c) Wherever pedicle screws are placed and muscles appropriate for those myotomes are available for recording
8. Which of the following statements is true?
   (a) Baseline nerve root stimulation prior to stimulation after pedicle screw placement is time consuming during the surgery and adds little useful information
   (b) Stimulation of the exiting nerve root prior to stimulation of the pedicle screw after placement will help prevent false-positive and false-negative results during electromyographic recording
   (c) In only a limited number of select cases, stimulation of the exiting nerve root prior to pedicle screw stimulation will be helpful and outweigh the risk of adding time to the operative procedure
9. Which of the following recording muscle groups is most useful during thoracic pedicle screw placement?
   (a) Biceps and brachioradialis
   (b) Abdominals and intercostals
   (c) Anal sphincter and detruser muscle
   (d) Thenar and hypothenar
10. Which of the following muscle groups is most useful in lumbar screw placement?
   (a) Thenar and hypothenar
   (b) Abdominals and intercostals
   (c) Tibialis anterior and abductor hallucis
   (d) Biceps and brachioradialis

References

Abumi, K., Shono, Y., Ito, M., *et al.* (2000). Complications of pedicle screw fixation in reconstructive surgery of the cervical spine. *Spine*, **25**(8), 962–969.

Anderson, D. G., Wierzbowski, R., Schwartz, D. M., *et al.* (2002). Pedicle screws with high electrical resistance. *Spine*, **27**, 1577–1581.

Barr, S. J., Schuette, A. M., and Emans, J. B. (1997). Lumbar pedicle screws versus hooks. Results in double major curves in adolescent idiopathic scoliosis. *Spine*, **22**(12), 1369–1379.

Bjarke, C. F., Stender, H. E., Laursen, M., Thomsen, K., and Bunger, C. E. (2002). Long-term functional outcome of pedicle screw instrumentation as a support for posterolateral spinal fusion: randomized clinical study with a 5-year-follow-up. *Spine*, **27**(12), 1269–1277.

Christodoulou, A., Ploumis, A., Zidrou, C., Terzidis, J., and Pournaras, J. (2002). Idiopathic scoliosis. Segmental fusion with transpedicular screw. *Studies in Health Technology and Informatics*, **91**, 433–437.

Dean, C., Lee, M., Robbin, M., and Cassinelli, E. H. (2009). Correlation between computed tomography measurements and direct anatomic measurements of the axis for consideration of C2 laminar screw placement. *Spine*, **9**, 258–262.

Dobbs, M. B., Lenke, L. G., Riley, L. H., *et al.* (2008). Neuromuscular scoliosis. *Emedicine*, www.emedicine.com/orthoped/topic526.htm.

Glaser, J., Stanley, M., Sayre, H., *et al.* (2003). A 10-year follow-up evaluation of lumbar spine fusion with pedicle screw fixation. *Spine*, **28**(13), 1390–1395.

Gurd, A. R. and Thompson, T. R. (1981). Scoliosis in Prader–Willi syndrome. *Journal of Pediatric Orthopedics*, **1**(3), 317–320.

Halm, H., Niemeyer, T., Link, T., and Liljenqvist, U. (2002). Segmental pedicle screw instrumentation in idiopathic thoracolumbar and lumbar scoliosis. *European Spine Journal*, **9**(3), 191–197.

Holm, V. A., Cassidy, S. B., Butler, M. G., *et al.* (1993). Prader–Willi syndrome: consensus diagnostic criteria. *Pediatrics*, **91**(2), 398–402.

Jutte, P. C. and Castelein, R. M. (2002). Complications of pedicle screws in lumbar and lumbosacral fusions in 105 consecutive primary operations. *European Spine Journal*, **11**, 594–598.

Katonis, P., Christoforakis, J., Aligizakis, A. C. *et al.* (2003). Complications and problems related to pedicle screw fixation of the spine. *Clinical Orthopaedics and Related Research*, **411**, 86–94.

Kim, K. D., Johnson, J. P., and Babbitz, J. D. (2001). Image-guided thoracic pedicle screw placement: a technical study in cadavers and preliminary clinical experience. *Neurosurgical Focus*, **10**, E2.

Kim, Y. J., Lenke, L. G., and Cho, Y. S. (2009). Free hand pedicle screw placement in the thoracic spine: is it safe? http://spineuniverse.com/displayarticle.php/article2336.html.

Kotwicki, T. (2008). Evaluation of scoliosis today: examination, X-rays and beyond. *Disability and Rehabilitation*, **30**(10), 742–751.

Laine, T., Lund, T., Ylikoski, M., Lohikoski, J., and Schlenzka, D. (2009). Accuracy of pedicle screw insertion with and without computer assistance – a randomized controlled clinical trial in 100 consecutive patients. www.spineuniverse.com/displayarticle.php/article994.html.

Lenke, L. G., Cheh, G., Riew, K. D., and Kim, Y. J. (2005). *Evaluation of Pedicle Screw Placement in the Deformed Spine Using Intraoperative Plain Radiographs.* American Academy of Orthopedic Surgery 2005 Annual Meeting; Poster Board No: P322.

Lewis, S. J., Lenke, L. G., Raynor, B. L., *et al.* (2002). Triggered electromyographic threshold for accuracy of thoracic pedicle screw placement in a porcine model. www.spineuniverse.com/displayarticle.php/article997.html.

Nagai, T., Obata, K., Ogata, T., *et al.* (2006). Growth hormone therapy and scoliosis in patients with Prader–Willi syndrome. *American Journal of Medical Genetics A*, **140**(15), 1623–1627.

Othman, Z., Lenke, L. G., Bolon, S. M., and Padberg, A. (2004). Hypotension-induced loss of intraoperative monitoring data during surgical corrections of Scheuermann kyphosis: a case report. *Spine*, **29**(12), E258–E265.

Papin, P., Arlet, V., Marchesi, D., Rosenblatt, B., Aebi, M. (1999). Unusual presentation of spinal cord compression related to misplaced pedicle screws in thoracic scoliosis. *European Spine Journal*, **8**, 156–160.

Prader, A., Labhart, A., and Willi, H. (1956). Ein Syndrom von Adipositas, Kleinwuchs, Kryptorchismus und Oligophrenie nach myatonieartigem Zustand im Neugeborenenalter. *Schweizerische Medizinische Wochenschrift*, **86**, 1260–1261.

Pruijs, J. E. H., van Tol, M. J., van Kesteren, R. G., and van Nieuwenhuizen, O. (2000). Neuromuscular scoliosis: clinical evaluation pre- and postoperative. *Journal of Pediatric Orthopaedics Part B*, **9**, 217–220.

Raynor, B. L., Lenke, L. G., Kim, Y., *et al.* (2002). Can triggered electromyograph thresholds predict safe thoracic pedicle screw placement? *Spine*, **27**(18), 2030–2035.

Richter, M., Mattes, T., and Cakir, B. (2004). Computer-assisted posterior instrumentation of the cervical and cervico-thoracic spine. *European Spine Journal*, **13**, 50–59.

Rosner, M. K., Polly, Jr., D. W., Kuklo, T. R., Ondra, S. L. (2003). Thoracic pedicle screw fixation for spinal deformity. *Neurosurgical Focus*, **14**, e7.

Shi, Y. B., Binette, M., Martin, W. H., Pearson, J. M., and Hart, R. A. (2003). Electrical stimulation for intraoperative evaluation of thoracic pedicle screw placement. *Spine*, **28**(6), 595–601.

Shufflebarger, H. L., Geck, M. J., and Clare, C. E. (2004). The posterior approach for lumbar and thoracolumbar adolescent idiopathic scoliosis: posterior shortening and pedicle screws. *Spine*, **29**(3), 269–276.

Suk, S. I., Lee, S. M., Chung, E. R., *et al.* (2003). Determination of distal fusion level with segmental pedicle screw fixation in single thoracic idiopathic scoliosis. *Spine*, **28**(5), 484–491.

Yukawa, Y., Kato, F., Ito, K., *et al.* (2009). Placement and complications of cervical pedicle screws in 144 cervical trauma patients using pedicle axis view techniques by fluoroscope. *European Spine Journal*, **18**, 1293–1299.

11 Cranial nerve monitoring

Jaime R. López

Even though sensory and motor functions are properties of different cranial nerves, the term neurophysiologic intraoperative monitoring (NIOM) of cranial nerve activity typically refers to the monitoring of cranial nerve motor function. This is the case because motor function is a property of the majority of the cranial nerves. Unfortunately, at present there are no effective methods of monitoring the sensory properties of cranial nerves I (olfaction) and II (vision). Effective monitoring of cranial nerve VIII (auditory) function can be achieved using brainstem auditory evoked potentials (BAEPs). Chapter 9 provides an excellent and detailed review of BAEP monitoring, while this chapter will focus on the IOM of cranial nerve (CN) motor function.

Intraoperative monitoring of cranial nerve motor function is primarily utilized for those cranial nerves whose function is exclusively motor or has a motor component. The rationale, as well as the stimulation and recording techniques employed, is similar to that used for assessing the functional integrity of other motor peripheral nerves. The primary difference is in the placement of the recording electrodes and the neural structures that are to be stimulated. Cranial nerve dysfunction is not solely confined to a cranial nerve but can also involve the nucleus corresponding to the specific nerve. Thus, a special application of cranial nerve monitoring is the functional assessment and monitoring of cranial nerve nuclei (CNN). This chapter reviews the techniques used to monitor CN III (oculomotor), CN IV (trochlear), CN V (trigeminal), CN VI (abducens), CN VII (facial), CN IX (glossopharyngeal), CN X (vagus), CN XI (accessory), and CN XII (hypoglossal). In addition, the CNN monitoring technique used to map specific brainstem motor nuclei is reviewed and described.

Techniques utilized

Electromyography

As with peripheral nerve monitoring, electromyography (EMG), which is based on the principle of recording electrical potentials generated in the muscle fibers, can be used to monitor motor activity. This can be accomplished by using surface, subcutaneous, or intramuscular recording electrodes and placing them in the muscles innervated by that specific cranial nerve (Table 11.1). Surface and subcutaneous recording electrodes are generally considered less desirable because the electrodes may not penetrate into the muscle, and distant motor unit potential (MUP) activity may be missed. It may also be difficult to identify the specific muscle generating EMG activity.

In contrast, intramuscular electrodes allow for "near-field" recordings, since the recording electrodes are placed as near as possible to the signal generator, thereby improving the signal-to-noise ratio (Schlake *et al.*, 2001b). The two main types of intramuscular electrodes currently used are conventional subdermal EEG needle and wire electrodes, with the specific type and length of the needle electrodes used dependent upon the location and depth of the target muscle. Although subdermal EEG electrodes may not be suitable for recording the EMG activity of deeply situated muscles, these electrodes are adequate for relatively superficial muscles, such as facial muscles, the tongue, the trapezius, and the muscles of the hands and feet.

Intramuscular fine wire electrodes, also known as "hookwire" electrodes, have the advantage that they are thinner than needle electrodes, can be placed into both deeply situated and superficial muscles, and are easier to place into smaller muscles. The wire electrodes have 2–3 mm uninsulated ends, which are hooked onto the bevel of conventional hypodermic needles. The hypodermic needle is used to introduce the wire into the muscle and the needle is subsequently removed, leaving the wires in the muscle. The length of the hypodermic needle used as an introducer can be varied, according to the depth of the muscle. After the wire is introduced into the muscle, the external portion of the wire electrode is taped to the skin to prevent dislodging and is usually very stable, even during patient positioning.

Table 11.1 Cranial nerves and the corresponding muscles useful in IOM using EMG

Cranial nerve	Muscles
III	Inferior rectus, inferior oblique
IV	Superior oblique
V	Masseter, temporalis
VI	Lateral rectus
VII	Frontalis, orbicularis oculi, nasalis, orbicularis oris, mentalis
IX	Lateral soft palate
X	Vocal cords, Cricothyroid
XI	Trapezius or sternocleidomastoid
XII	Tongue or genioglossus

The recording parameters for the surface, subcutaneous, or intramuscular electrodes used during IOM are the same as those typically used for conventional outpatient needle EMG examination. The gain is set between 50 and 200 μV, low frequency filter between 20 and 30 Hz, and the high frequency filter at 10 kHz. The sweep speed is usually set at 200 ms/division for recording continuous EMG, and 2–5 ms/division for recording electrically evoked EMG potentials. Electromyograph recordings can be made from multiple muscles simultaneously with the activity observed on an oscilloscope. It is also necessary to monitor audio feedback to aid in distinguishing between EMG activity and electrode or electrical artifacts. In addition, the audio feedback can alert the surgeon to possible nerve stimulation resulting from surgical manipulation.

A detailed discussion of the different types of electrical artifacts and their likely causes is beyond the scope of this review. Nonetheless, the reader is reminded that commonly encountered operating room equipment, such as microscopes, patient and blood warmers, electrocautery, ultrasonic aspirators, surgical beds, anesthesia machines, and compression stockings, can contribute to the electrical noise interference encountered in the operation room. These should be investigated and as many as possible excluded prior to alerting the surgeon of the presence of EMG activity.

Different types of EMG activity can be encountered during IOM. Motor unit potentials (MUPs), fibrillation potentials (which remain present during anesthesia), and neurotonic discharges can be encountered when monitoring EMG activity (Harper, 1998). However, the most important of these are the neurotonic discharges (Daube and Harper, 1989). Neurotonic discharges can be distinguished by the frequency and pattern of discharge, and by the relationship of the activity to mechanical or metabolic stimulation of the neural structure at risk (Harper, 1998). Motor unit potentials have a frequency of 10–15 Hz and a continuous semi-regular pattern (Harper, 1998). Fibrillation potentials have a frequency of 1–5 Hz with a regular continuous pattern. In contrast, neurotonic discharges have discharge frequencies ranging between 50 and 200 Hz and can appear in short bursts or long trains, with the appearance and sound of high-frequency discharges similar to end-plate spikes, myotonia, and neuromyotonia. These discharges are generated when the axon membrane is depolarized, which can occur with mechanical, metabolic, or thermal stimulation, as well as ischemic or traumatic nerve injury (Møller, 1994; Harper, 1998). Unfortunately, severing of a cranial nerve may not produce any EMG activity or neurotonic discharges (Møller, 1994; Crum and Strommen, 2007).

Electrical stimulation

As with EMG recording, the concept of electrical stimulation of motor cranial nerves is no different from that of stimulating a peripheral nerve and can be achieved either with monopolar or bipolar stimulation. Bipolar stimulation refers to the situation when both the cathode and anode are directly on the nerve, whereas monopolar stimulation occurs when the cathode is directly on the nerve and the anode is placed some distance away from the nerve. An advantage of bipolar stimulation is that it provides a focal stimulus, reducing current spread to adjacent neural structures, but current shunting can occur, which can lead to submaximal stimulation. Monopolar stimulation reduces the probability of current shunting but the chances of activating other neural structures, via current spread, are increased.

Although both monopolar and bipolar stimulation techniques can produce good results, monopolar stimulation is the method most widely used, especially for the stimulation of cranial nerves and brainstem structures, since the areas stimulated are usually small and can be located deep within the small surgical opening.

In these situations, it is easier to use one electrode to stimulate rather than try to place two electrodes directly on the nerve.

In addition, electrical stimulation can be applied using either constant-current or constant-voltage. Møller (1994) has advocated the use of constant-voltage stimulation when applying electrical stimulation intracranially. The rationale is that there is little variation in electrode impedance and that shunting of current by cerebrospinal fluid (CSF), irrigation fluids, and surrounding tissues plays a much more important role. Therefore, constant-voltage stimulation is more likely to deliver the same amount of current through the tissue regardless of shunting.

Unfortunately, not all commercially available IOM equipment has the capacity, without modification, to deliver both constant-current and constant-voltage stimulation, or deliver low-level constant-voltage stimulation. Regardless, effective safe electrical stimulation can be performed using constant-current stimulation and many groups use this technique and are able to achieve excellent results. A more important issue is to determine the most appropriate level of stimulation to use, which will vary dependent on the different surgical contexts (Yingling and Ashram, 2005).

Special considerations in IOM

The application of EMG in the operating room (OR) differs from that practiced in the neurodiagnostic laboratory in that the effects of anesthesia and of the patient's blood pressure and temperature must all be considered. Unlike in the neurodiagnostic laboratory, sampling of different muscles cannot be performed once the electrodes are fixed in place, since they usually cannot be moved to another muscle. Furthermore, due to esthesia, muscle activity cannot be assessed under volitional effort and the anesthetic regimen itself may affect muscle function. In addition, the amount of recording artifact generated by electrical noise is usually much worse than that seen in the outpatient laboratory setting. The operating room also presents a very dynamic situation, which requires that the clinical neurophysiologist and the IOM team monitor many different parameters. These include monitoring and recording any changes in the IOM parameters and how they relate to the surgical events, and monitoring any changes in the anesthetic regimen, as well as the patient's blood pressure and temperature. All these can influence the IOM results and need to be factored into the interpretation of the findings (Blair *et al.*, 1994).

Anesthesia

As neuromuscular blocking (NMB) agents will affect EMG activity, it is recommended that the patient be free of their effects during the time when EMG monitoring is being performed. Short-acting NMB agents can be used at the time of anesthesia induction, with their effects wearing-off by the time the critical stages of surgery take place. Constant infusion of a short-acting NMB agent titrated to a level of 50% reduction of the baseline CMAPs recorded over the hypothenar muscle with successful IOM of EMG activity has been reported (Lennon *et al.*, 1992; Kizilay *et al.*, 2003). In addition, since it appears that NMB agents often have more effect on limb rather than cranial muscles, motor cranial nerve EMG activity can usually still be monitored in the presence of NMB agents (Harper, 1998). Inhaled anesthetic agents have little or no effect on EMG monitoring.

Monitoring of oculomotor function

The extraocular muscles controlling eye movement are innervated by the oculomotor (CN III), trochlear (CN IV), and abducens (CN VI). In certain skull-based, brainstem, anterior fossa, cavernous sinus, and orbital surgical procedures, oculomotor function may be at risk and, therefore, IOM of the oculomotor system is vital to preserve neural function. Diplopia can result from injury to one or more of the oculomotor nerves. If the diplopia cannot be corrected, the patient will probably have to wear an eye patch, which will render the patient monocular with the loss of stereoscopic vision and reduction of the peripheral visual field. Obviously, this can have a serious adverse impact on a person's quality of life. Thus, preserving normal oculomotor function is of critical importance. Monitoring EMG activity of the extraocular muscles may help in preserving oculomotor function.

Eye movements can be assessed and monitored invasively via intramuscular electrodes or non-invasively using surface electrodes placed on the skin around the eyeball (Fukaya *et al.*, 1999; Sasaki *et al.*, 2002). Recording electrodes are most commonly placed in the medial rectus (MR), superior rectus (SR), superior oblique (SO), and lateral rectus (LR) muscles. Although there appears to be a consensus that the preferred method for obtaining EMG recordings

Table 11.2 Summary of stimulation and recording techniques used by different authors for invasive monitoring of oculomotor function

Author	Number of patients	Recording electrode type	Stimulation	Stimulus duration	Stimulus intensity	Frequency rate
Sekiya *et al.*, 1993	18	Ring electrode	Bipolar	100–200 µs	7.0 mA (maximum)	1–2 Hz
Sekiya *et al.*, 2000	12	Spring electrode Needle electrode		100 µs	1.0 mA	2–3 Hz
Alberti *et al.*, 2001	10	Intraorbital eye muscle monopolar electrode	Bipolar	200 µs	Started at: 0.05 mA (nuclear regions); 0.5 mA (peripheral sites)	3 or 30 Hz
Schlake *et al.*, 2001b	18	Single-shafted bipolar needle electrode	Bipolar	10 µs	0.5 mA	30 Hz

from extraocular muscles is via intramuscular electrodes, both invasive and non-invasive techniques will be reviewed in this chapter.

A number of authors have reported results obtained using various types of intramuscular electrodes and stimulation techniques (Sekiya *et al.*, 1993; Schlake *et al.*, 2001b; Sekiya *et al.*, 2000; Alberti *et al.*, 2001) (Table 11.2).

Sekiya *et al.* (1993) stimulated the intracranial oculomotor nerves electrically in 18 patients and were able to record compound muscle action potentials (CMAP) using a custom ring electrode that they had developed. The electrode required placement by an ophthalmologist and was sutured into the muscle. Recording electrodes were placed in the MR, SR, SO, and LR muscles. In two patients, they placed small needle electrodes in the upper eyelid to record EMG responses from the levator palpebrae superioris (LPS), which is innervated by the superior ramus of the oculomotor nerve. Using either microsurgical bipolar forceps or a custom-made bipolar stimulator and the custom ring electrode, they were able to record from all muscles but had difficulty distinguishing the EMG activity between the SO and SR muscles. They concluded that monitoring of MR EMG activity provided a more reliable indicator of oculomotor nerve function than that from the SR.

In a different study, Sekiya *et al.* (2000) compared two different methods for recording EMG from extraocular muscles in 12 patients. The first method used a spring electrode, which replaced the custom ring suture electrode from their previous study (Sekiya *et al.*, 1993), placed in the SR or MR muscle with a reference electrode on the forehead. In the second method, a needle electrode was inserted into the space just beneath the roof of the orbit and above the eyeball. Both recording methods resulted in equally robust CMAPs. Sekiya and coworkers postulated that the recordings from the second method probably represented the summed potentials from the SR and LPS muscles. They also concluded that in certain situations, such as when an ophthalmologist is unavailable, needle placement into the superior intraorbital space might be a useful alternative to using the spring electrode. A different interpretation of their findings suggests that the "indirect recording" technique from the superior supraorbital space could be used to monitor extraocular muscle activity as an alternative to placing electrodes directly into the SR or MR muscles.

Schlake *et al.* (2001b) used a custom-made single-shafted bipolar needle electrode to record intramuscular EMG from selected extraocular muscles in 18 patients undergoing surgical resection of various skull base tumors. Using the novel approach of ultrasound (US) to visualize the needle tip within the muscle in the orbital space, they were able to obtain EMG recordings from the MR, SR, and LR muscles. After insertion of an eyelid retractor, a conventional orbital ultrasound probe was used to place the needle electrodes through the use of an orbital ultrasound

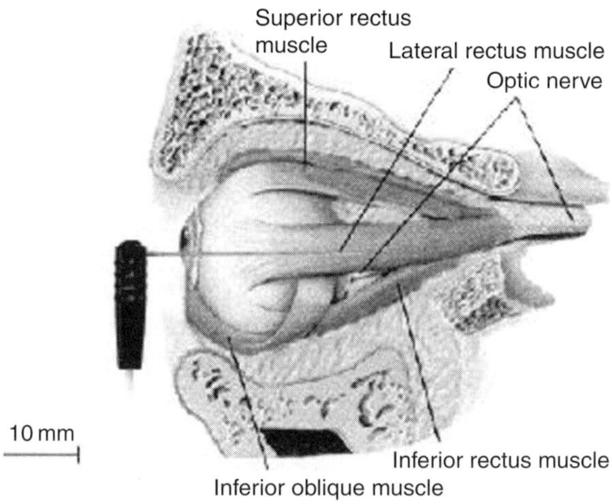

Figure 11.1 Example of the anatomical position of the single-shafted bipolar needle in the lateral rectus muscle (Schlake *et al.*, 2001b).

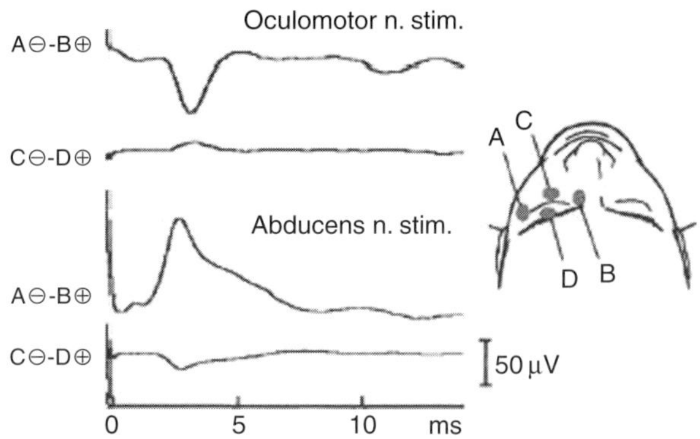

Figure 11.2 Example of electrooculographic responses after intracranial stimulation of the left oculomotor and abducens nerves (Fukaya *et al.*, 1999).

technique. The probe emitted a focused sound beam at a frequency of 10 MHz with its focal zone coinciding with the posterior wall of the eye and the anterior orbit. Once the electrodes were in place, the eyelid retractor was removed and the electrodes were fixed in position using adhesive tape (Figure 11.1). The advantages offered by their technique over using anatomical landmarks for needle placement included exact determination of needle placement, as well as a more controlled and safer insertion of the electrode. They also felt that the use of bipolar recordings derived from their custom-made electrode improved noise rejection characteristics and spatial selectivity of the receptive fields over conventional monopolar recordings. Using this technique, they were able to show that the CMAP latencies were similar to those reported in other studies but that the amplitudes were on average 3 to 20 times higher than those values calculated from previous studies. These differences in amplitude, they postulated, were due to achieving true "near-field" responses.

Alberti *et al.* (2001) used a conventional neuronavigation system for anatomically correct intraorbital eye muscle electrode placement in ten patients. Prior to brain imaging, the neuronavigation system was used to determine placement of the skin fiducials, with at least one skin marker placed close to the orbit, either at the frontal process, the zygomatic bone, or the supraorbital margin of the frontal bone. Neuronavigation was then used for insertion of monopolar needle electrodes into the LR, inferior rectus (IR) and SO muscles. The electrode was deemed in position when the tip of the electrode was in the belly of the muscle and the main part of the shaft was in the orbital fat. Reference electrodes were placed on the contralateral forehead and the ground was placed on the nasion. The authors reported that successful monitoring of CN III, IV, and VI was safely accomplished without additional resources required, compared with routine set-up.

Several authors have also reported on the application of non-invasive methods using surface recordings and electrooculography (EOG) rather than EMG to monitor oculomotor nerve function (Fukaya *et al.*, 1999; Sasaki *et al.*, 2002). Electrooculography is based on the electrical potential difference between the cornea and the retina. Fukaya *et al.* (1999) reported the use of EOG to identify eye movements consistent with oculomotor and abducens nerve activity in eight patients with skull-based tumors (Figure 11.2). In their technique, the channel 1 active electrode was placed on the right side of the eye with the reference on the left side and the channel 2 active electrode was placed on the upper side of the eyeball with the reference on the lower side. Monopolar stimulation using rectangular impulses of 0.2 ms duration at a rate of 3 Hz and a maximum intensity of 3 mA were used. After stimulation, movement of the eyeball toward the right induced a positive wave, while movement toward the left generated a negative wave.

To evaluate further the utility of non-invasive IOM of the extraocular muscles, the origins of the surface-recorded evoked potentials (EP) generated by intracranial stimulation of the oculomotor nerve were evaluated by Sasaki *et al.* (2002). Using a canine model with the retina removed, they found that the surface recorded EPs were almost identical to the simultaneously recorded intramuscular bipolar EMG recordings. Based on this finding, the authors concluded that the surface recorded EPs elicited by electrical stimulation of the oculomotor nerves originate from the extraocular muscles and are not derived from the retina or eye.

At our institution, we use hookwire electrodes to record EMG activity from the different extraocular muscles (Figure 11.3). A pair of wire electrodes is placed percutaneously via a hypodermic needle, into the IR, the SO, and the LR muscles to monitor CNs III, IV, and VI, respectively. A transeyelid technique similar to that described by Møller (1994) and Rivero *et al.* (1995) is typically used (Figure 11.5).

For CN III monitoring, the needle is placed in the infraorbital space slightly medial to the infraorbital foramen, which can easily be palpated, in order to avoid the infraorbital nerve. The needle is aimed slightly away from the eye toward the orbital bone and the globe is gently pushed up as the needle is introduced. This allows recordings from the IR and inferior oblique (IO) muscles.

For EMG monitoring of the trochlear nerve, wire electrodes are placed in the SO muscle. The needle is introduced medial to the supraorbital foramen directing it toward the orbital bone, which can be palpated and used as a landmark, and directed medially into the superior intraorbital space. For the LR, the needle is placed lateral to the eye, again aiming it slightly away from the globe toward the orbital bone and gently pushing the eye medially. Following introduction, the needle is then removed and the wires are secured in place with strong adhesive tape (Figures 11.4 and 11.5).

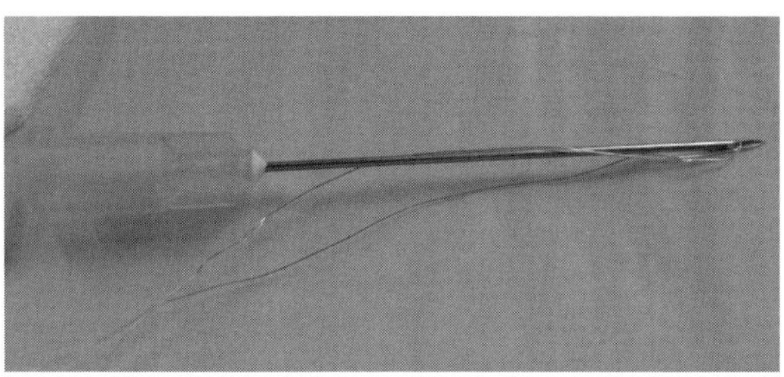

Figure 11.3 Paired hookwire electrodes for EMG (Medtronic Xomed SKU 82–26326).

In addition, to differentiate between extraocular and facial muscle activity, recordings are also taken from the lateral canthus region of the orbicularis oculi, using that muscle as a control. Monopolar stimulation with a duration of 0.05 ms, a rate between 3–5 Hz and an intensity usually not exceeding 5 mA is used. Stimulation intensity begins at 0.1 mA and is increased by 0.1 mA increments as necessary.

We have also used constant-voltage stimulation, using similar settings to those described for constant-current stimulation with stimulation gradually increased by 0.1 V increments to a maximum level of 4–5 V. Figure 11.6 shows an example of neurotonic discharges and a CMAP recorded from the inferior rectus and inferior oblique complex. We have found this technique to generate good EMG potentials from the IR, inferior oblique, and LR muscles; however, obtaining adequate EMG potentials from the SO muscle is much more challenging.

With respect to safety, little is discussed in the literature regarding adverse effects of needle placement into the extraocular muscles, other than subconjunctival hemorrhages (Rivero *et al.*, 1995). However, insertion of intramuscular electrodes into extraocular muscles is no trivial matter and it is recommended that they be placed by an experienced physician. In over 10 years of using the above-mentioned recording technique, we have had only one patient suffer a significant adverse event, a retrobulbar hemorrhage in a child,

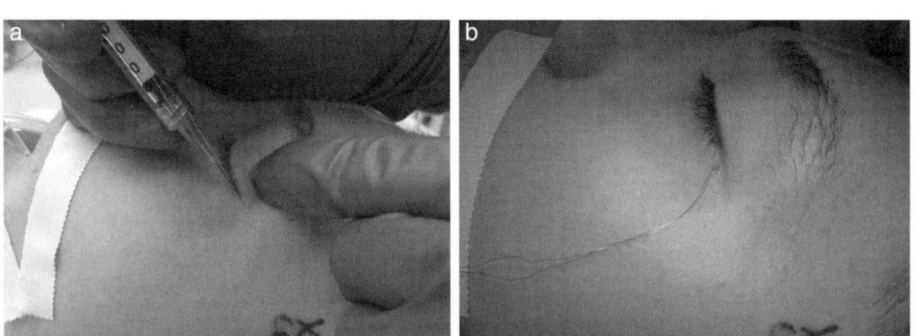

Figure 11.4 The percutaneous placement of hookwire electrodes in the lateral rectus muscle using a 25-gauge needle attached to a tuberculin syringe.

which occurred shortly after placing the wire electrodes into the lateral rectus. Fortunately, delay in surgery was the only intervention required and the patient recovered completely without any visual disturbance or lasting effects (Figure 11.7).

Monitoring of CN V

The trigeminal nerve is a mixed nerve containing motor and sensory fibers. It provides sensory input from the face and also motor innervation to the muscles of mastication. In regard to the IOM of trigeminal nerve function, most strategies have focused on monitoring motor function. Even though sensory input is a significant component of the trigeminal nerve, attempts to monitor sensory function have been met with limited success. Intraoperative trigeminal somatosensory evoked potentials (SEPs), as a means for assessing sensory function, are technically difficult to obtain owing to the stimulus artifact and the short latency of the cortical evoked potentials. Stechison (1993) concluded that the technique he tested, using peripheral stimulation at the face and central stimulation at the trigeminal nerve root entry zone, provided the ability to monitor the integrity of the extra-axial trigeminal nerve but not the central trigeminal circuitry. It is likely that monitoring of trigeminal nerve function will continue to focus on the motor component while the monitoring of sensory function will depend on future techniques that can provide reliable, robust, and accurate results.

A basic principle of monitoring trigeminal nerve motor function is similar to that of other motor cranial nerves. Recording EMG activity and electrically evoked CMAPs from trigeminal nerve innervated muscles yields good results. The temporalis and

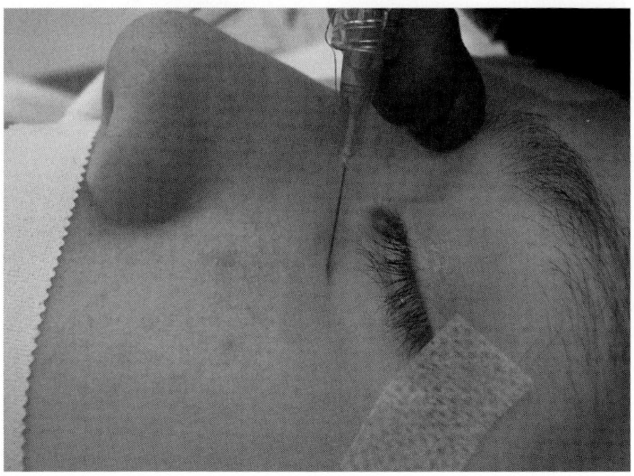

Figure 11.5 The percutaneous placement of hookwire electrodes in the IR and inferior oblique (IO) muscles.

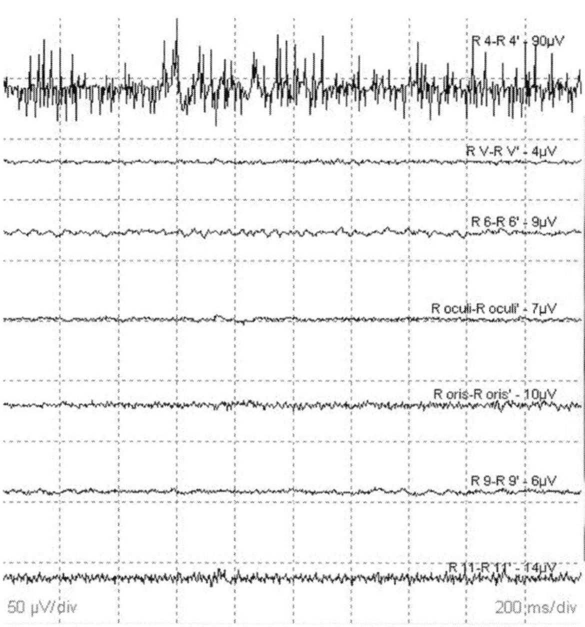

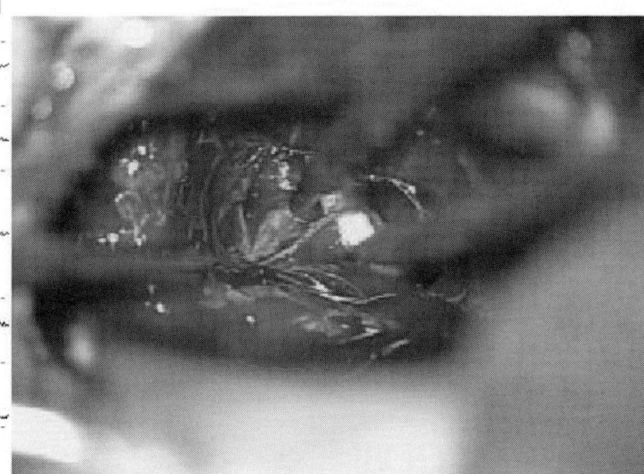

Figure 11.6 Neurotonic discharges recorded using wire electrodes placed in the right superior oblique muscle during resection of a petrous meningioma.

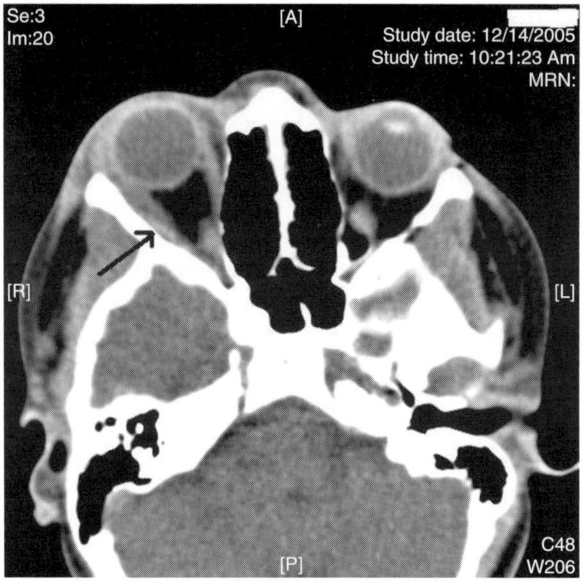

Figure 11.7 Orbital CT scan performed after attempted placement of hookwire electrodes in the right lateral rectus muscle of an 8-year-old female with a fourth ventrical tumor. A small hemorrhage in the LR muscle (arrow) can be seen. The patient developed right eye proptosis shortly after the wires were introduced into the LR muscle and a retrobulbar hemorrhage was suspected.

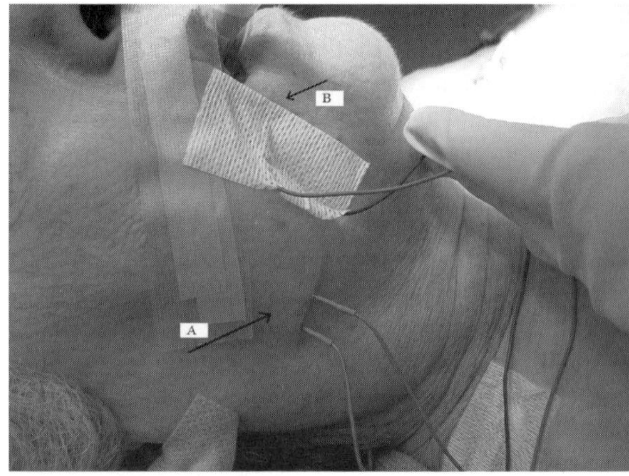

Figure 11.8 Example of paired subdermal needle electrode placement into: (A) masseter muscle (trigeminal nerve); (B) orbicularis oris muscle (facial nerve).

masseter muscles are the most commonly used muscles to monitor trigeminal nerve EMG activity. Standard intramuscular, subdermal, and wire electrodes can be used to monitor EMG activity (Figure 11.8). Because of the proximity of the masseter to facial muscles innervated by the facial nerve, care must be taken to minimize crosstalk and not confuse the origin of the EMG activity. In addition, intracranial stimulation of structures near the trigeminal nerve may activate nearby nerves, such as the facial nerve, which can sometimes be confused with a trigeminal response. One method of distinguishing the origin of the evoked CMAP is by reviewing the onset latency. Facial nerve stimulation generally produces a CMAP with an onset latency between 6 and 8 ms. Trigeminal nerve stimulation typically yields a CMAP onset latency shorter than that of the facial nerve, ranging between 3.5 and 5 ms (Yingling and Ashram, 2005).

Monitoring of CN VII

Facial nerve function can easily be monitored using similar recording and stimulation techniques to those described for other motor cranial nerves. From a historical perspective, the first published account of cranial nerve monitoring was by Dr. Fedor Krause (1912), who monitored facial nerve function through the use of intraoperative visual confirmation of facial movement. As described in Chapter 12, he reported that in 1898 he had applied electrical stimulation to the facial nerve trunk, during a cochlear nerve section procedure, which resulted in contractions of the face.

Similar techniques were used during the ensuing decades of the twentieth century by other surgeons, until 1966 when Parsons described the use of electrically evoked EMG responses, without simultaneous recording of continuous free-running EMG activity, following facial nerve stimulation during parotid surgery. Subsequently, Delgado *et al.* (1979) and Sugita *et al.* (1979) separately described the use of electrically evoked EMG monitoring after intracranial facial nerve stimulation during removal of acoustic neuromas. Møller and Jannetta (1984) described a technique of facial nerve stimulation and simultaneous monitoring of EMG activity in patients undergoing resection of medium and large-sized acoustic tumors. In 1986, Harner *et al.* described the use of continuous monitoring of EMG activity and characterized neurotonic discharges in a study of 32 surgical procedures in and around the temporal bone.

Electromyography of facial nerve function has now become routine in many different types of surgical procedures where the facial nerve may be at risk of injury. However, one of its most important and common applications is in acoustic neuroma surgery, where its use is critical in attempting to preserve postoperative facial function. In addition, the use of facial nerve EMG in acoustic neuroma surgery has contributed to anatomic preservation of the facial nerve in greater than 95% of cases (Kartush and Lundy, 1992).

Unfortunately, anatomic preservation does not guarantee preserved postoperative facial nerve function (Moskowitz and Long, 1991). Lalwani *et al.* (1994) reported a series of 129 patients who underwent surgical resection of acoustic neuromas using IOM. Anatomic preservation of the facial nerve occurred in 99.2% of patients, and 90% of patients had grade 1 or grade 2 facial nerve function one year following surgery. They also found that tumor size was an important factor in long-term facial function. In addition to tumor size, the proximal facial nerve stimulation threshold at the end of surgical tumor removal was predictive of long-term facial nerve function. Their results showed that 98% of patients with stimulation threshold of 0.2 V had grade 1 or grade 2 facial function one year later, compared with 50% of those with stimulation threshold levels between 0.21 and 0.6 V.

Analysis of other electrophysiologic parameters has been studied as a possible means of predicting postoperative facial function. Taha *et al.* (1995) were able to show the positive predictive value of proximal-to-distal amplitude ratios in determining postoperative facial function.

Similarly, using proximal-to-distal amplitude ratios and proximal electrical stimulation threshold levels, Isaacson *et al.* (2003) successfully employed a mathematical model to predict the probability of good postoperative facial function, finding their methods to be statistically significant. In 2005, these same investigators published a second study that confirmed their earlier findings using the same statistical equation and electrophysiologic parameters. They also found that immediate postoperative facial nerve function was predictive of final facial nerve function. Their study showed a 100% sensitivity, 60% specificity, 96% positive predictive value, and 100% negative predictive value in patients whose immediate postoperative facial nerve function was House–Brackmann Grade I–III. They reported that patients who could close their eye completely on initial postoperative evaluation had a 96% chance of a good final facial nerve outcome without evidence of synkinesis. The results of these studies suggest that assessing the functional state of the facial nerve after tumor removal using either threshold stimulation levels or an analysis of CMAP amplitudes is predictive of long-term facial function. Being able to identify those patients with likely poor facial nerve function permits early rehabilitative treatment.

Facial nerve stimulation and recording

Direct electrical stimulation of structures in the surgical field helps to localize viable neural tissue and to identify and differentiate between different cranial nerves (Harper, 2004). In general, a monopolar stimulator is used when screening the surgical area of interest while recording a CMAP (Figure 11.9). Direct nerve stimulation, when the nerve is visible, can be done with either monopolar or bipolar electrical stimulation. This can be done at different areas along the course of the nerve and it should also be performed proximal to the injured area near the brainstem, in order to test central-to-peripheral continuity of the facial nerve.

Stimulation can be done using a variety of different parameters, many of which were discussed earlier in this chapter. Unfortunately, there is no clear consensus on which techniques are superior. Harper (2004) recommends using very low constant-current stimulation intensities ranging over 1–5 mA with stimulation duration of 0.05–0.1 ms. The stimulation intensity should be gradually increased until a maximum response is recorded, which in normal nerves occurs at 2–5 mA or 10–20 V (Harper and Daube, 1998). Although higher stimulation intensities may be required for damaged or distant nerves, stimulation intensities greater than 5 mA carry the risk of current spread and possible unintended activation of adjacent nerves (Harper, 2004). However, it has been our experience that lower stimulation intensities, usually between 0.05 and 1 mA or 0.1 and 1 V are sufficient to activate the facial nerve during acoustic neuroma surgery if a longer stimulation duration, such as 200 μs, is used. Yokoyama *et al.* (1991) studied the use of monopolar low constant-current stimulation during acoustic neuroma surgery and reported that facial nerve activation reliably occurred at stimulation intensities 0.5–0.6 mA and 0.1 ms stimulation duration.

Recording of CMAPs can be done using surface EEG disk, subdermal needle, or hookwire electrodes placed in muscles innervated by the facial nerve. At a minimum, two facial muscles need to be monitored. Electromyographs should be obtained by monitoring at least one upper and one lower facial muscle, such as the orbicularis oculi and orbicularis oris, respectively. Yingling and Ashram (2005) prefer using monopolar

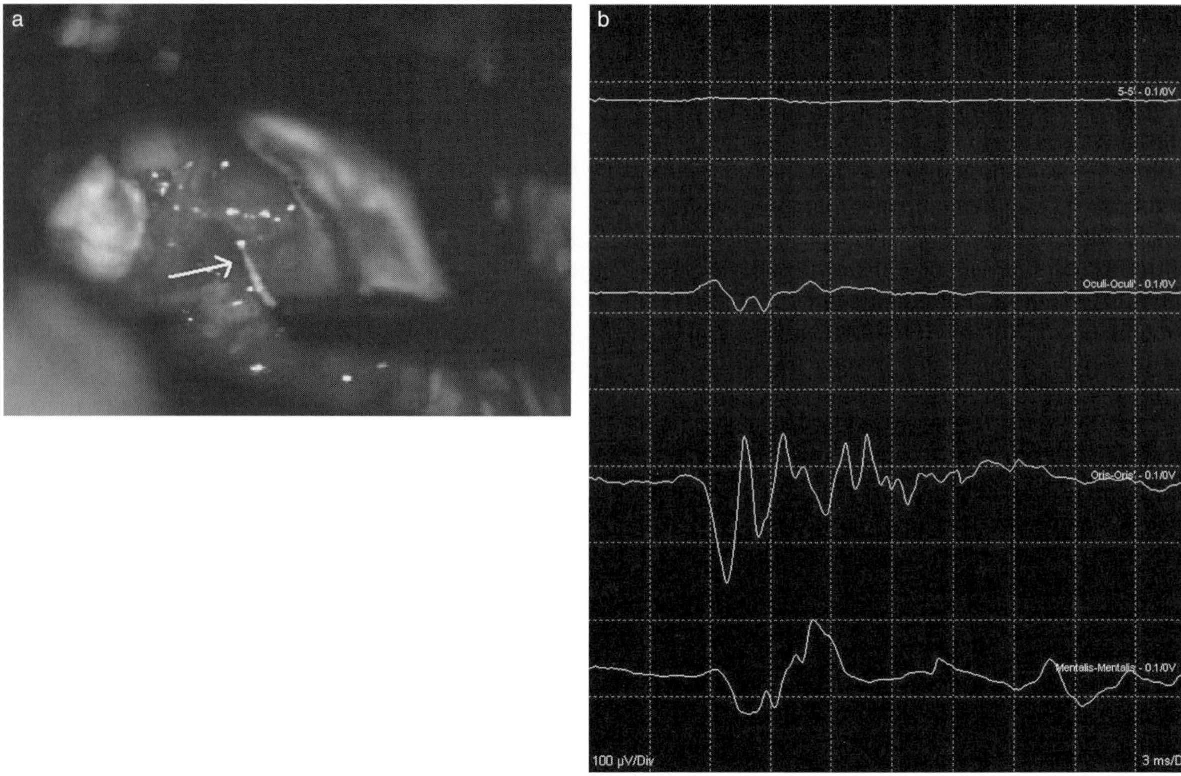

Figure 11.9 Intracranial facial nerve monopolar stimulation. (a) View from the operating microscope. The arrow points to the wire tip of the monopolar stimulating electrode. (b) Evoked CMAP recordings from facial innervated muscles, orbicularis oculi, orbicularis oris, and mentalis (traces 2–4, respectively). Note the lack of a CMAP from the masseter muscle at trace 1.

constant-voltage stimulation with stimulation duration of 0.2 ms at a rate of 5 to 10 Hz. When using these parameters, the threshold for an evoked CMAP ranged between 0.05 and 0.2 V.

The use of constant-voltage stimulation is felt to reduce current spread compared with constant-current stimulation, especially when the stimulating electrode is in contact with fluid. However, Yokoyama *et al.* (1991), during acoustic neuroma surgery, were able to demonstrate that a stimulating electrode, which was insulated except for an area 0.5 mm from the tip, placed in cerebrospinal fluid produced 1 mm of current spread when stimulating at 0.6 mA, and 5 mm when the intensity was increased to 1.8 mA. This spatial resolution of 1 mm is similar to that reported for constant-voltage stimulation at the suprathreshold level (Yingling and Ashram, 2005).

The important point to be made here is that both constant-current and constant-voltage stimulation can be used to stimulate the facial nerve in a safe, effective, and accurate manner that allows identification and localization within the operative field. In addition, the responses obtained after acoustic neuroma resection can provide useful predictive information regarding postoperative facial function.

Other facial nerve monitoring techniques

Facial nerve motor-evoked potentials

Recently, the use of transcranial motor-evoked potentials (TcMEPs) has been studied as a method of monitoring the facial nerve function in skull base surgery (Akagami *et al.*, 2005; Dong *et al.*, 2005; Fukuda *et al.*, 2008). Facial nerve motor-evoked potentials (FNMEPs) could be readily recorded from the contralateral orbicularis oris muscle (Akagami *et al.*, 2005; Dong *et al.*, 2005) and from the orbicularis oculi and oris muscles (Fukuda *et al.*, 2008).

Transcranial electrical stimulation was delivered via scalp electrodes at positions C3 or C4 and Cz with the cathode always at Cz (Akagami *et al.*, 2005; Fukuda *et al.*, 2008). Dong *et al.* (2005) stimulated using two different scalp montages using Cz as reference: 1 cm anterior to C1 and Cz; or 1 cm anterior to C3 and C4. The results of these studies showed that a ratio of greater than 50% of the final FNMEP amplitude (compared with the initial baseline amplitudes) predicts satisfactory postoperative facial function. In other words, higher ratios seem to correlate with better postoperative facial function.

Akagami *et al.* (2005) found a strong correlation between FNMEPs and facial nerve CMAPs. However, a study comparing sensitivity and specificity between FNMEPs and CMAPs does not yet exist and is needed to address the utility and the role of FNMEPs in monitoring facial nerve function. Nonetheless, FNMEPs are a promising new modality, which may provide different and useful information when monitoring facial nerve function. Similarly, transcranial electrical stimulation may prove useful in monitoring of other motor cranial nerves.

IOM of facial nerve function using F waves

Intraoperative recordings of facial nerve F waves recorded from nasal muscles have been studied as an effective measure of facial nerve function in surgical resection of cerebellopontine angle (CPA), cerebellar, and brainstem tumors (Wedekind and Klug, 1996; 2001a). Peripheral stimulation produces antidromic excitation that reaches and activates small numbers of motor neurons, which then generate orthodromic excitation, producing a late muscle response resulting in an F wave. The production of F waves depends on and reflects the functional integrity of the entire motor axon between its axon hillock and the motor end plate (Wedekind and Klug, 1996). Diminished function of the nerve along its course should then be reflected in the F wave.

Wedekind and Klug (2001a) compared IOM of facial nerve function using nasal muscle F waves and EMG activity in 33 patients with CPA tumors. Their results demonstrated that a permanent loss of nasal muscle F waves indicated a severe dysfunction of the facial nerve and was linked to a poor outcome. A transient loss of the F wave was detected more frequently in patients with moderate outcomes.

Unfortunately, the intraoperative use of facial nerve F waves is limited by its sensitivity to anesthesia and its absence in some healthy adults (Wedekind and Klug, 1998; 2001b).

Lateral spread reflex

Microvascular decompression of the facial nerve for treatment of hemifacial spasm places the facial nerve at potential risk for injury since the surgical treatment involves separating the nerve from an adjacent vascular structure and placing a soft implant between the structures to cushion the arterial pulsations against the nerve. Thus, IOM of facial nerve function is performed to reduce the risk of facial nerve injury. In addition to the possibility of facial nerve injury, postoperative hearing loss is another potential complication of microvascular decompression and is thought to result from excessive retraction or manipulation of CN VIII (Danneman *et al.*, 2008). These cases are typically monitored using multimodality IOM of BAEPs and facial nerve EMG activity.

However, hemifacial spasm is a condition associated with hyperactivity and hyperexcitability of the facial nucleus (Møller, 1999). It has been observed that in patients with this disorder an action potential in one branch of the facial nerve following an antidromic stimulus of a different facial nerve terminal branch can induce an action potential in a separate, different facial nerve branch (Harper, 2004).

The term "lateral spread reflex" (LSR) is used to describe this electrophysiologic phenomenon, which is thought to be a result of ephaptic transmission between facial axons at the site of the vascular compression or hyperexcitability of the pontine facial nucleus (Nielsen, 1985; Møller, 1999; Harper, 2004).

Studies have shown that elimination or reduction of the size of the LSR is a good predictor of clinical response to microvascular decompression (Møller and Jannetta, 1987; Shin *et al.*, 1997; Mooij *et al.*, 2001; Kong *et al.*, 2007; Neves *et al.*, 2009). However, others have questioned whether intraoperative monitoring of the LSR is a reliable predictor of postoperative clinical outcome (Hatem *et al.*, 2001; Joo *et al.*, 2008). Hatem *et al.* (2001), in their study of 33 patients with hemifacial spasm, question the prognostic value of the persistence of the LSR because the majority of their patients with LSR persistence after microvascular decompression still achieved a clinical cure.

In a separate study, Joo *et al.* (2008) concluded that hemifacial spasm clinical outcome does not always correlate with intraoperative EMG abnormality and that the prognostic value of IOM of the LSR, with

Table 11.3 Summary of the lateral spread stimulation and recording techniques used to monitor CN VII microvascular decompression surgical procedures

	Stimulation	Recording
Site	Marginal mandibular nerve Temporal	Mentalis muscle
Stimulation parameters	10–20 mA; 0.1 ms duration; or 4–20 V	Orbicularis oris muscle
Recording parameters		Sweep speed: 5–ms/div Sensitivity: 5–500 μV Filters: 5 Hz to 10 kHz

Adapted from Hatem *et al.*, 2001 and Yingling and Ashram, 2005

respect to long-term clinical results, is questionable. However, review of their data shows that 11/32 (34.3%) patients with LSR persistence had mild hemifacial spasm at discharge and 18.7% (6/32) had mild hemifacial spasm at six months. In contrast, in patients where the intraoperative LSR disappeared, 5 of 40 (12.5%) patients had mild hemifacial spasm at discharge and 4 of 40 (10%) had mild hemifacial spasm at six months. Although there is still some debate over the prognostic value of intraoperative LSR changes, the bulk of the clinical literature lends support to it as a valuable marker for adequate nerve decompression and an aide in predicting clinical outcome. In general, it is reasonable to conclude that the intraoperative elimination of the LSR predicts clinical cure of the hemifacial spasm but that LSR persistence does not exclude the possibility of postoperative benefit (Harper, 2004). The LSR IOM technique is summarized in Table 11.3.

One of the barriers for the routine use of IOM is cost. To evaluate this, Wilson *et al.* (2003) studied the cost-effectiveness of IOM of facial function in patients undergoing middle ear or mastoid surgery. Their results strongly favored use of IOM for all patients undergoing middle ear or mastoid surgery and they estimated that monitoring added approximately $222.73 to $528.00 to the total cost of surgery. However, in estimating the overall cost they used the hourly pay rate of an audiologist and not physician professional charges or the CPT IOM code, which would raise the cost. Their overall conclusion was that facial nerve IOM is cost-effective and its routine use to reduce the risk of iatrogenic injury should be adopted during otologic surgery.

In conclusion, IOM of facial nerve function can be used in a variety of different surgical procedures, alters the surgical management of many cases and can provide important prognostic information regarding postoperative facial function. It has proven to be particularly useful in CPA surgery, especially in resection of acoustic neuromas. Furthermore, in 1991 the National Institute of Health in their consensus development conference statement on acoustic neuroma (The Consensus Development Panel, 1994) stated:

> The benefits of routine intraoperative monitoring of the facial nerve have been clearly established. This technique should be included in surgical therapy for vestibular schwannoma. Routine monitoring of other cranial nerves should be considered.

Monitoring of lower cranial nerve function

The lower cranial nerves, which include CN IX (glossopharyngeal), CN X (vagus), CN XI (accessory), and CN XII (hypoglossal), may be at risk for injury in certain types of surgical procedures. These include surgeries in the region of the cerebellopontine angle, petroclival area, rhomboid fossa, brainstem, jugular foramen, and foramen magnum, and especially those procedures involving large tumors of the skull base (Schlake *et al.*, 2001a). Injury to one or more of these nerves can lead to dysphagia, dysarthria, dysphonia, dysgeusia, and possibly an increased risk of aspiration (Schlake *et al.*, 2001a). It may be difficult to differentiate between neural and non-neural tissue when a space-occupying mass is distorting the anatomy. In these situations, electrical stimulation of structures within the surgical field may assist in identifying viable cranial nerve fibers. Of course, the ability to differentiate neural tissue will be limited to the nerves being monitored. In addition to identifying nerves within the surgical field, continuous EMG monitoring using similar techniques to those previously described for peripheral and other cranial nerves can be used to detect changes consistent with manipulation or injury. As is the case with other nerves, the appearance of neurotonic discharges is considered to be the most important continuous, free-running EMG finding. Table 11.4 summarizes the stimulation parameters for CNs IX–XII.

Glossopharyngeal nerve (CN IX)

The glossopharyngeal nerve primarily contains sensory and autonomic fibers with only a small motor component, which supplies the stylopharyngeus

Table 11.4 Summary of stimulation protocols described in different studies of NIOM of the lower cranial nerves

Author	CN	Type of stimulation	Stimulation duration	Frequency rate	Stimulation intensity
DeMonte *et al.*, 1994	IX	Single stimulus	Not reported	Not reported	0.05 mA or 0.2–0.5 V
Romstöck *et al.*, 2000	VI–VII and IX–XII	Concentric, bipolar, constant-current	200 μs	30 Hz	0.05–0.5 mA
Schlake *et al.*, 2001a	IX and X	Bipolar, constant-current	100 μs	30 Hz	0.05 or 0.5 mA
Thomusch *et al.*, 2002	X	Monopolar	200 μs	3 or 30 Hz	0.05 to 0.5 mA
Tamano *et al.* 2004	CNN: vagal trigone in the medulla	Single stimulus	100 μs	1 Hz	0.8 to 3 mA
Snyder and Hendricks, 2005	X	Monopolar	Not reported	Not reported	1 mA
Husain *et al.*, 2008	IX and X	Not reported	50 μs	Not reported	0.5 mA
Topsakal *et al.*, 2008	X, XI, XII	Monopolar	50 μs	4.7 Hz	0.1 to 1 mA

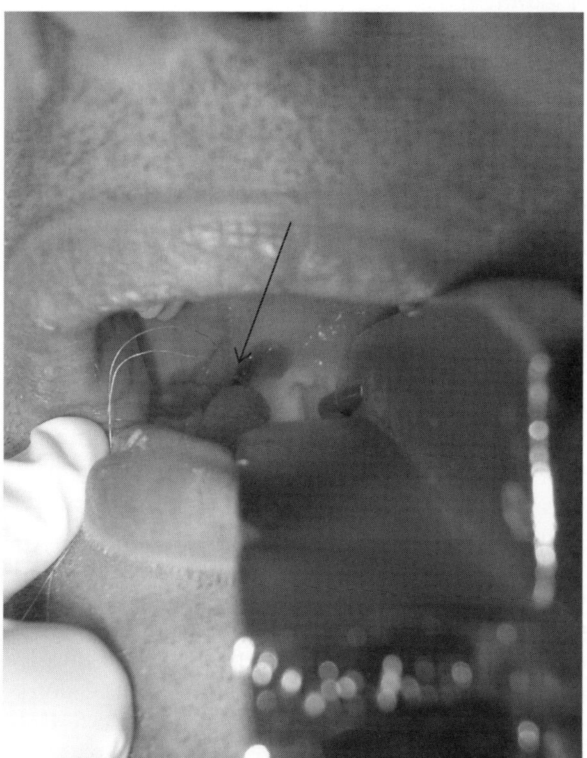

Figure 11.10 Hookwire (paired wire) electrodes placed, under direct laryngoscopy, in the right soft palate. The arrow points to the two wire electrodes and their attachment in the soft palate.

muscle. However, a number of authors have described EMG monitoring of CN IX, which can be accomplished by placing recording electrodes in the lateral aspect of the soft palate (Møller, 1994; Schlake, *et al.*, 2001a).

In addition, to differentiate glossopharyngeal from vagally innervated laryngeal muscle activity, it is recommended that CN IX function be monitored in conjunction with vagal nerve monitoring (DeMonte *et al.*, 1994). Direct nerve stimulation can be done using methods similar to those previously described for other cranial nerves. Schlake *et al.* (2001a) used bipolar constant-current stimulation and applied square-wave pulses at a frequency of 30 Hz with a duration of 100 μs and an intensity of either 0.5 or 0.05 mA for intracranial nerve or brainstem stimulation, respectively. However, DeMonte *et al.* (1994) recommend using a single stimulus instead of a train of stimuli. They also recommended initiation of the stimulus at the lowest intensity available, 0.05 mA, or 0.2 to 0.5 V if using constant-voltage stimulation as side effects such as bradycardia and hypotension may result from electrical stimulation of CN IX due to inhibitory effects exerted on the carotid sinus and carotid body.

Husain *et al.* (2008) describe the use of a non-invasive, adhesive surface electrode mounted on the

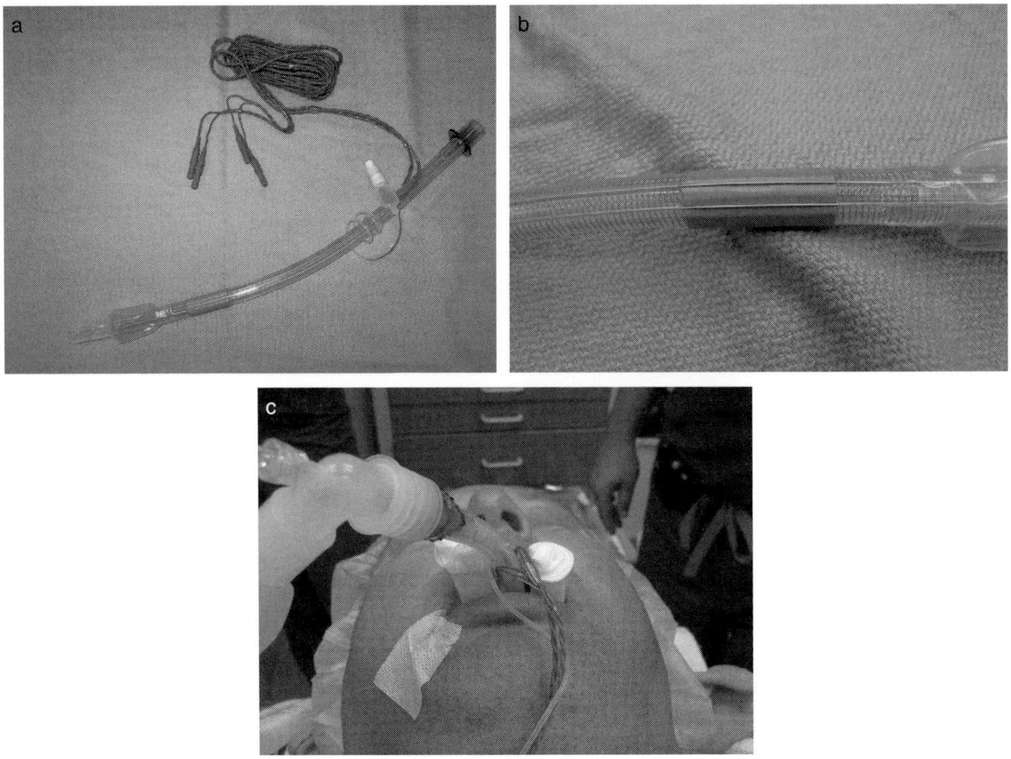

Figure 11.11 (a) Medtronic NIM EMG endotracheal tube; (b) close-up view of exposed recording wires embedded in the endotracheal tube; (c) example of proper placement of the Medtronic NIM EMG endotracheal tube after intubation.

cuff of a laryngeal mask airway to record and monitor electrically evoked CMAPs after stimulation of the glossopharyngeal nerve in a 30 year-old patient undergoing resection of a cerebellopontine angle mass. Their report did not indicate whether they were able to record any neurotonic discharges or other non-electrically evoked EMG activity during continuous, free-run EMG monitoring.

At our institution we use fine wire electrodes ("hookwire" electrodes) to record EMG activity from the lateral soft palate. A needle, attached to a syringe, is used to introduce the wires into the soft palate. This requires the assistance of an anesthesiologist who maintains the patient's mouth open and, using the laryngoscope, provides a light source to be able to visualize the soft palate and posterior pharynx (Figure 11.10).

In regard to nerve stimulation, we have obtained good results using a single stimulus technique similar to that described by DeMonte *et al.* (1994). In addition, using a 3–5 Hz stimulation mode yields good results without encountering any significant adverse effects related to the stimulation.

Vagus nerve (CN X)

The vagus nerve is also a mixed, motor, sensory, and autonomic nerve and the motor component, which innervates the vocal cords and laryngeal muscles, can also be monitored using EMG. Both invasive and non-invasive techniques have been described to monitor EMG activity from the vocal cords. Insertions of needle electrodes into the vocal cords by direct laryngoscopy or intraoperatively into the vocalis muscles or thyroarytenoid through the cricothyroid ligament are examples of invasive techniques (Snyder and Hendricks, 2005; Petro *et al.*, 2006).

Unfortunately, needle insertion into the vocal cords carries the risk of vocal cord hemorrhage or edema and placement into the false cords may also cause edema (DeMonte *et al.*, 1994). A non-invasive alternative is to use a commercially available endotracheal tube (ETT) that has embedded, but externally exposed, bipolar surface electrodes, which rest against the vocal cords once the ETT is properly positioned. Similar recordings can be obtained using commercially available adhesive surface electrodes, which are

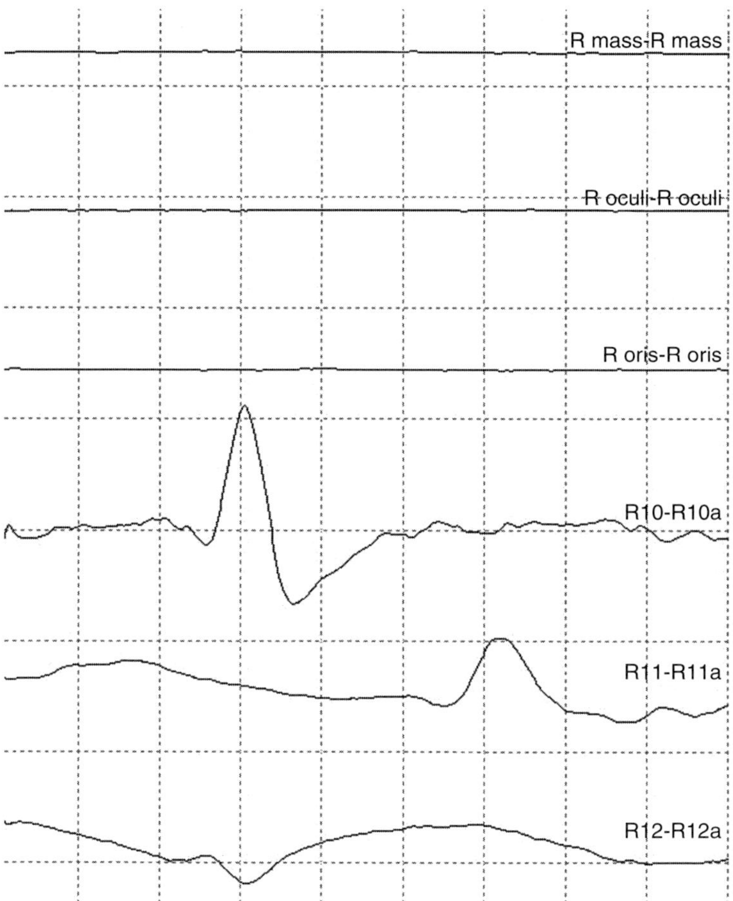

Figure 11.12 CMAP (trace 4) surface recordings from the right vocal cord using a wire embedded endotracheal tube after monopolar stimulation of the right vagus nerve in a patient undergoing microvascular decompression for glossopharyngeal neuralgia. The stimulation intensity was 0.4 mA and the duration was 0.1 ms. Recordings were obtained using a sweep speed of 3 ms/division and a gain of 100 μV/division.

attached onto a standard ETT. These methods require special care to insure that the exposed wires embedded or attached to the ETT are placed up against the vocal cords. This contact is best achieved by having the anesthesiologist directly observe that the exposed wires on the endotracheal tube are resting against the vocal cords. In addition, good contact can be demonstrated by showing low impedance for each wire.

Another non-invasive method is the use of a monometric balloon, which is placed between the vocal cords with video endoscopy to enable observation of vocal cord movement (DeMonte *et al.*, 1994; Tamano *et al.*, 2004).

Jackson and Roberson (1999) compared three techniques of monitoring vagal nerve function using laryngeal EMG in 13 out of 17 patients undergoing lateral skull base surgery who completed the study. Electromyographs were recorded using three different electrodes: endolaryngeal surface electrodes mounted on an ETT; endoscopically placed endolaryngeal thyroarytenoid intramuscular hookwire electrodes; and percutaneously placed cricothyroid intramuscular hookwire electrodes. They concluded that the surface-mounted ETT electrodes were the easiest to place, had a low false-positive response rate, but demonstrated the lowest sensitivity when compared with the intramuscular electrodes. They also found the percutaneous intramuscular electrode technique to be the least reliable. Overall, they preferred the endolaryngeal intramuscular electrode technique because of its highest relative sensitivity and reliability. However, with only a total of 13 patients studied, it is not possible to definitively declare any of the techniques studied to be clearly superior.

Because of the risk associated with the needle insertion and the fact that the expertise of an

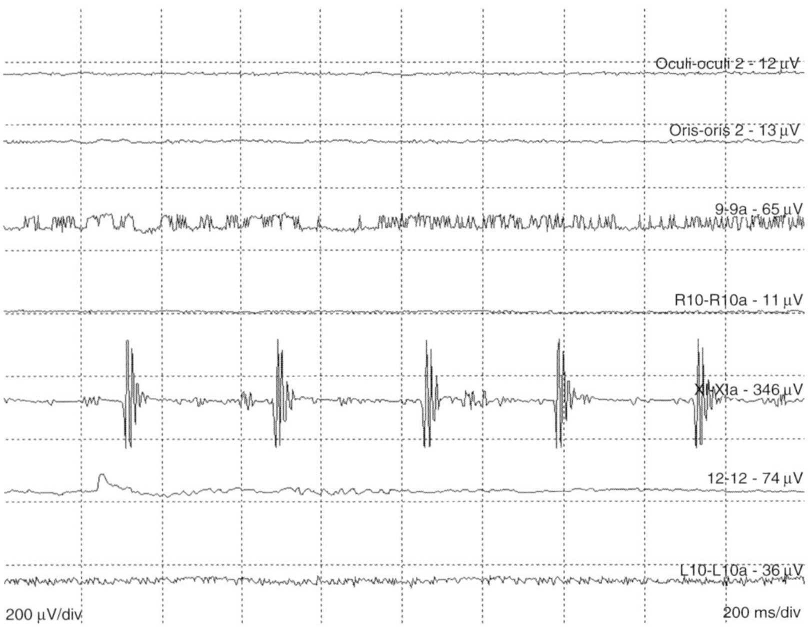

Figure 11.13 Neurotonic discharges recorded from the right trapezius muscle (CN XI, trace 5) during resection of a right glomus jugulare tumor. Small amplitude EMG activity is also recorded from the right soft palate (CN IX, trace 3).

otolaryngologist is often required, the non-invasive techniques are generally preferred. Regardless of the type of electrode used, a variety of stimulation techniques have been reported (DeMonte *et al.*, 1994; Romstöck *et al.*, 2000; Schlake *et al.*, 2001a; Thomusch *et al.*, 2002; Tamano *et al.*, 2004; Snyder and Hendricks, 2005; Husain *et al.*, 2008; Topsakal *et al.*, 2008;) (Table 11.4).

At our institution, we have used both direct needle insertion into the vocal cords and wire embedded endotracheal tubes to record EMG activity. We have found the endotracheal tube method of recording surface EMG to yield adequate results, and now use this technique almost exclusively (Figure 11.11).

Stimulation of the vagus nerve can result in adverse effects similar to those seen after stimulation of the glossopharyngeal nerve and, thus, care should be exercised when stimulating this nerve. Figure 11.12 shows an example of a CMAP recorded from the right vocal cord after stimulation of the right vagus nerve.

Accessory nerve (CN XI)

The spinal motor division of the accessory nerve can be monitored using EMG by placing recording needle electrodes into the corresponding muscles, the sternocleidomastoid and trapezius. Placement of the electrodes is easy and straightforward, since both muscles are superficial and are easily accessible. When monitoring the trapezius, it is important that the recording electrodes are placed in the lower portion of the muscle, since the upper part of the trapezius has mixed innervation from the C1 and C2 spinal roots (Schlake *et al.*, 2001a).

The stimulation techniques used are similar to those described for CNs IX and X. However, stimulation should be initiated at low levels since higher stimulus intensities are likely to produce strong shoulder movement and head turning, which can lead to excessive patient movement (DeMonte *et al.*, 1994). In addition, because electrical stimulation of the nerve bypasses the feedback from muscle spindles and tendon organs that normally limit the force of muscle contraction, muscle and tendon injuries may occur if the stimulation intensities are set too high (Møller, 1994; Schlake *et al.*, 2001a). An example of neurotonic discharges recorded from the right trapezius muscle is shown in Figure 11.13.

Hypoglossal (CN XII)

Injury to CN XII can cause dysarthria, and dysphagia. Fortunately, hypoglossal motor nerve function can be monitored using EMG. This can be easily performed by inserting needle or hookwire electrodes into the ipsilateral tongue or genioglossus muscle. Some authors advocate against submental percutaneous placement into the genioglossus because of possible injury to vascular and neural structures in that area

(DeMonte *et al.*, 1994). However, we have experience recording from the tongue as well as the genioglossus and have noticed no differences in adverse effects or quality of EMG recordings.

Recording and stimulation techniques are similar to those described for the other lower CNs (Romstöck *et al.*, 2000; Schlake *et al.*, 2001a). In a study reviewing intraoperative monitoring of lower cranial nerves in 123 cases, Topsakal *et al.* (2008) reported that the hypoglossal nerve was monitored in 67% (83/123) of the cases, whereas the vagus nerve was monitored in 30% (37/123) and the spinal accessory nerve in 96% (118/123) of the cases, respectively. In their study, IOM of CN XII was performed using continuous EMG monitoring after placement of needle electrode pairs in the lateral aspect of the anterior third of the tongue, and by CMAP recordings after direct electrical nerve stimulation. In correlating neurophysiologic findings with postoperative neurological status, their results of hypoglossal nerve monitoring showed a true positive rate of 50%, a false-positive rate of 43.6%, a true negative of 56.3%, and a false-negative rate of 50%. Overall, the authors concluded that the use of lower cranial nerve IOM reduced cranial nerve morbidity in high-risk patients compared with lower risk patients. They also felt that the immediate feedback obtained with IOM may help in preventing surgical injury to the cranial nerves being monitored and that IOM can help in identifying the course of a nerve in patients with severely distorted anatomy.

Overall, the available clinical evidence indicates that IOM of lower cranial nerves can be performed using a combination of continuous, free-running EMG activity to monitor neurotonic discharges resulting from nerve manipulation or injury and electrically evoked CMAPs to identify CN fibers in the surgical field, or to predict postoperative neurologic outcome. Table 11.4 summarizes the stimulation protocols of the different studies of lower cranial nerve IOM presented.

Cranial nerve nuclei

As mentioned previously, CNN monitoring represents a unique application of IOM. Although cranial nerve function is monitored in many different types of surgical procedures, the monitoring of CNN is performed less frequently and only in specific types of neurosurgical procedures where damage to the CNN is a possibility.

Because of the risk of injuring the tightly packed CNN and neural tracts within the rhomboid fossa and brainstem, surgical treatment of intra-axial brainstem lesions carries a substantial risk of postoperative morbidity (Lang *et al.*, 1991) and, historically, neurosurgeons considered this area to be a "no man's land" with most lesions being inoperable (Baker, 1965; Katsuta *et al.*, 1993).

However, since the study by Lassiter *et al.* (1971), surgical treatment of intra-axial brainstem lesions has been considered a viable option. In addition, continued improvement of technical imaging, and microsurgical techniques in magnetic resonance imaging (MRI), have allowed better definition of the pathological anatomy. In addition, the increased use of IOM with SEPs, brainstem BAEPs, and motor-evoked potentials (MEPs) have allowed for increased success in the treatment of brainstem lesions (Heffez *et al.*, 1990; Strauss *et al.*, 1994; Steinberg *et al.*, 2000).

Unfortunately, conventional neurophysiologic techniques are of limited use when attempting to resect intra-axial brainstem lesions surgically because they cannot be used to identify and map functional structures on the brainstem surface. This can result in neurologic deficits as a consequence of mechanical or vascular injury to neural pathways and nuclei during surgical dissection of brainstem surface structures prior to the resection of the intrinsic lesions (Strauss *et al.*, 1994). Therefore, the use of neurophysiologic techniques has been advocated to assist in identifying neural structures, such as cranial nerve nuclei, thus allowing the surgeons to avoid these structures during surgery (Eisner *et al.*, 1995).

Anatomy of the rhomboid fossa

To plan the neurophysiologic techniques to use for mapping and monitoring of the brainstem properly, an understanding of the anatomy of the rhomboid fossa is critical.

Anatomical landmarks of this area have been described by Lang *et al.* (1991), who reported that the striae medullares could not be considered anatomical landmarks because of their highly variable location on the surface, in addition to the fact that no striae were observed in 14% of their sample. They also found that the broadest area of the rhomboid fossa frequently does not correspond to the midpoint of its length. This has important implications because the facial colliculus is normally located above the striae, and, therefore, if the striae are not present, visual identification of the facial colliculus becomes very difficult (Strauss *et al.*, 1994).

Several other neural pathways are located in close proximity to the facial colliculus including the abducens nucleus and the medial longitudinal fasciculus (MLF) as well as the parapontine reticular formation (PPRF), which is ventral and medial to the facial colliculus and the laterally located vestibular nuclei (Strauss *et al.*, 1994). Injury to this region may cause facial nerve and abducens nucleus dysfunction and can affect eye movements and vestibular function. Thus, identification of the facial colliculus is crucial (Lang *et al.*, 1991).

The ability to identify functional neural structures precisely is further complicated by the distortion and displacement produced by intrinsic brainstem space occupying lesions (Strauss *et al.*, 1994; Eisner *et al.*, 1995; Morota *et al.*, 1996). Owing to the limitations of BAEPs and SEPs, when used in combination they provide electrophysiologic coverage of only 20% of the brainstem cross-sectional area between the pontomedullory and pontomesencephalic junctions at any given level (Lang *et al.*, 1991; Strauss *et al.*, 1994). The addition of TcMEPs provides additional brainstem coverage but still falls far short of complete or comprehensive coverage. Furthermore, these conventional techniques are not useful in identifying surface structures prior to dissection. Therefore, the need for a method to identify functional structures on the surface of the rhomboid fossa is vital. Such a technique, using electrical stimulation of brainstem surface structures, was first described by Fahlbusch and Strauss in 1991 (Lang *et al.*, 1991; Katsuta *et al.*, 1993). Subsequently, similar techniques have been described by other authors (Rusyniak *et al.*, 1992; Katsuta *et al.*, 1993; Strauss *et al.*, 1993; Eisner *et al.*, 1995; Morota *et al.*, 1995; Morota *et al.*, 1996; Morota and Deletis 2006; Chang *et al.*, 1999), which can be combined with conventional sensory and motor IOM techniques.

Intraoperative monitoring of CNN

Simultaneous multimodality IOM of BAEPs, SEPs, and TcMEPs should be performed in conjunction with continuous EMG monitoring. This may present some difficulty if one is using older IOM equipment, because of the limited number of channels. However, newer equipment typically has a sufficient number of electrode inputs and channels to be able to monitor multiple, bilateral cranial nerves as well as many, if not all, of the evoked potentials mentioned.

The purpose of electrical stimulation of brainstem structures is to identify cranial motor nuclei and their motor tracts. The nucleus of greatest interest is the facial colliculus, not only because it lies above and adjacent to important neural structures, as mentioned previously, but also because it lies only 0.25 mm beneath the ependyma. In addition, the hypoglossal nucleus is also reported to be situated 0.5 to 2.6 mm deep from the surface (Strauss *et al.*, 1994). Direct electrical stimulation at relatively low stimulation intensities can be used to localize both of these nuclei.

Although the intent, when using these techniques, is to activate the CNN, Strauss *et al.* (1999) postulated that the peripheral motor neuron rather than the CNN itself is the likely site of activation. They deduced that because the triggered EMG responses can be obtained with low level stimulation intensities, as low as 0.1 mA; and stimulus durations of 20–50 μs (Chang *et al.*, 1999), these responses correlate best with time constants known to activate peripheral neural structures. Shannon *et al.* (1997), in reviewing the literature, reported that the time constant for stimulating myelinated central nervous system axons ranges from approximately 50–100 μs to 200–300 μs for large cortical neurons; and 1–10 ms for other neurons and their dendrites. Strauss *et al.* (1999) determined that peripheral motor neuron axons are the probable site of intraoperative electrical stimulation, based on estimates of the stimulation required to activate neural structures within the brainstem and comparing these with measured motor-threshold stimulation intensities.

Although the literature suggests that either monopolar or bipolar stimulation is useful and provides adequate stimulation to activate the motor nuclei and tracts, a study by Strauss *et al.* (1993) reported that bipolar electrodes were difficult to use for mapping and that a monopolar handheld probe defined the stimulation area more easily. They also reported that in cases using constant-current stimulation, shunting effects were a result of the presence of cerebrospinal fluid but that a preliminary review of their data revealed no differences between constant-current and constant-voltage stimulation. In our experience, we have also found no appreciable differences between the two.

Prior to stimulating any neural structures, the stimulation system should be tested to see if electrical current is being delivered. A simple and easy method for doing this is to stimulate a muscle adjacent to the surgical site, whereby the surgeon should be able to observe muscle contraction in response to the stimulation. However, keep in mind that muscle requires higher stimulation intensities than those used for direct nerve stimulation. To reduce current spread and prevent activation of other more distant brainstem regions, the surgeon should take special care to keep the area to be stimulated clear of blood and fluids.

As the goal of brainstem stimulation is to activate motor nuclei or pathways, recordings need to be obtained from muscles corresponding to the motor CNN and tracts. Because of the complex structures of the rhomboid fossa, different sets of muscle recordings have been reported by the different groups, with all groups reporting electromyographs from at least two facial nerve innervated muscles. The types of recording electrodes that can be used include surface disk or disposable electrodes, which are commonly used for diagnostic nerve conduction studies, conventional subdermal needle electrodes, or intramuscular wire (hookwire) electrodes.

In regard to safety, experimental studies demonstrate that current density, charge per phase, and charge density per phase are important variables in determining the potential for neural damage when studying the effects of electrical stimulation on the brain (Agnew and McCreery, 1987; McCreery *et al.*, 1990; Harnack *et al.*, 2004).

Agnew and McCreery (1987) reported that neural damage occurred in brain tissue immediately subjacent to the stimulating electrodes when the charge density at the surface of the electrodes exceeded approximately 40 μC/(cm^2 ph) and the charge per phase (ph) was greater than approximately 0.4 μC/ph, at a pulse frequency of 50 Hz. They also found that charge density correlated more closely with the severity of neural damage when using small electrodes. They also postulated that the possible mechanisms of neural damage included charge transfer across an electrode–tissue interface, such as electrochemically produced toxic products, passage of current through tissue causing neuronal hyperactivity, or power dissipation leading to thermal injury.

Some studies have reported on their levels of charge densities delivered during brainstem stimulation. Strauss *et al.* (1999) reported a charge density of 0.59 μC/(cm^2 ph) using constant-voltage stimulation and charge densities up to 0.0014 μC/(cm^2 ph) when using constant-current stimulation. Chang *et al.* (2000) later reported on an average stimulation intensity of 3.4 mA and a pulse duration of 0.02 ms. Based on the surface area of the stimulating electrode, they calculated the charge per phase as 0.068 μC/ph and the charge density per phase was 17.4 μC/(cm^2 ph).

These levels, reported by both groups, are well below the safety limits and, clinically, patients seem to tolerate this type of electrical stimulation well, with only scattered reports of ventricular arrhythmia in patients with a known history of cardiac arrhythmia (Strauss *et al.*, 1999). Nevertheless, as the parasympathetic dorsal motor nucleus of the vagus is lateral to the hypoglossal trigone, patients need to be monitored for possible cardiovascular side effects during brainstem stimulation (Strauss *et al.*, 1999). Fortunately, no patient at our institution has experienced side effects as a result of this type of brainstem stimulation. During the stimulation protocol we carefully monitor a patient's blood pressure and heart rate and, to date, have not encountered any alterations in blood pressure or heart rate. It is important to emphasize that we are careful to use higher stimulus intensities only on tissues that appear to be tumor, necrotic, or non-viable.

Stimulation and recording techniques

Several authors have reported different stimulation parameters, using both bipolar and monopolar techniques, with a variety of electrodes utilized for EMG. The following will be a summary of the various methods described.

Rusyniak *et al.* (1992) reported using monopolar, constant-current stimulation with an intensity of 0.1 mA with stimulation applied for approximately 1.0 s. They estimated that when using these stimulation parameters the current spread from a monopolar electrode was 0.5 mm. Electromyography responses were recorded from electrodes placed in the face and tongue with gross motor responses noted, presumably from visual inspection of the face for movement.

Katsuta *et al.* (1993) used bipolar silver-ball electrodes for electrical stimulation of the facial colliculus with recording of CMAPs from surface electrodes placed on the ipsilateral orbicularis oculi and oris muscles. Square-wave pulses of 0.2 ms and an intensity of 1 mA, with a frequency of 1 Hz, were applied and

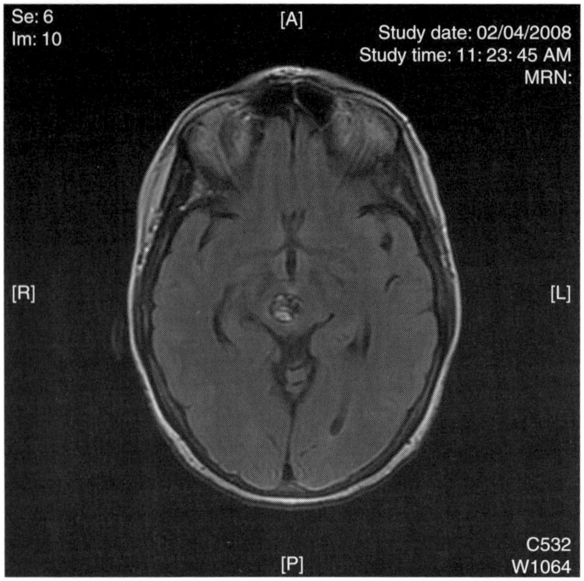

Figure 11.14 A preoperative MRI scan shows the midbrain cavernous malformation, which lies primarily on the right.

the EMG signal was amplified with a bandpass filter of 10 to 3000 Hz with no averaging required.

Strauss *et al.* (1993) used both monopolar and bipolar, as well as constant-current and constant-voltage, stimulation. Stimulation intensities varied between 0.1 and 2 mA and 0.1 and 1 V; stimulation duration was between 50 and 400 μs, and single rectangular stimuli, with levels up to 10 Hz, were used. When using monopolar stimulation, a handheld probe, which was insulated, except at the tip, served as the cathode and a needle electrode placed in a wound muscle acted as the anode stimulus. Pulse duration was 100 μs for constant-voltage stimulation. Monopolar stainless steel needles were used for placement of electrodes. For facial nerve recordings, active electrodes were placed in the orbicularis oris and oculi muscles with an indifferent electrode placed at the frontal midline. For hypoglossal recordings, the active electrode was placed in the genioglossus muscle.

A combination of mapping and continuous monitoring was employed by Eisner *et al.* (1995). They used stimulation forceps (GK 675, Aeskulap, Germany) to apply constant-current, square-wave impulses at 0.1 to 3 mA with a repetition rate of 4.7 Hz and custom-made, non-insulated, steel-needle recording electrodes. Active and reference electrodes were placed in the muscles that corresponded to the nerves being monitored, including the superior rectus, medial rectus, superior oblique, masseter, orbicularis oculi, orbicularis oris, trapezius, velum palatinum, and lingual muscles. In addition to recording triggered EMG responses, they monitored continuous EMG activity with auditory feedback to the surgeon via a loudspeaker.

Morota *et al.* (1995) reported using monopolar stimulation and applying one- to two-second trains of stimuli of 0.2 ms duration delivered at 4 Hz. Distinct and stable muscle responses were obtained at stimulation intensities of 1.5 to 2.0 mA. Subsequently, the intensity was reduced and the threshold level that was established (0.3–2.0 mA) was used for mapping. Custom-made, 30-gauge, stainless steel, Teflon-coated wires with a 2 mm bare hook tip encased in a 27-gauge needle were used as their recording electrodes. Facial nerve responses were recorded from the orbicularis oris and oculi, cranial nerve IX and X activity was monitored using the posterior pharyngeal wall, and hypoglossal nerve responses were recorded from the lateral tongue. Four averaged EMG responses were recorded using epoch lengths of 20 ms, amplified 10 000 times, filtered between 50 and 2133 Hz.

In 1999, Chang *et al.* reported using stimulation intensities as high as 10.2 mA with stimulus duration varying from 0.02 to 0.05 ms in patients undergoing resection of pontine cavernous malformations. After starting at zero, stimulation was increased at 0.1 mA increments until a threshold response was identified. Stimulation was then increased until a maximum compound motor action potential (CMAP) or spread to other nuclei was observed. Intramuscular hookwire electrodes were used with recordings obtained from the masseter, orbicularis oculi, and orbicularis oris muscles. The active and reference electrodes were placed in the same muscle and spaced a few millimeters apart. In addition to recording electrically triggered compound action potentials, neurotonic discharges from the muscles monitored was also monitored continuously. In the few patients where high stimulus intensities were reported, stimulation levels above 4 mA were utilized either on what was felt to be necrotic tissue or in regions adjacent to areas of hemorrhage; no adverse outcomes were reported.

The differences in the techniques highlights that there is no single method which is clearly superior. Adequate and reliable recordings can be obtained using monopolar, bipolar, constant-current or constant-voltage stimulation. Furthermore, stimulation

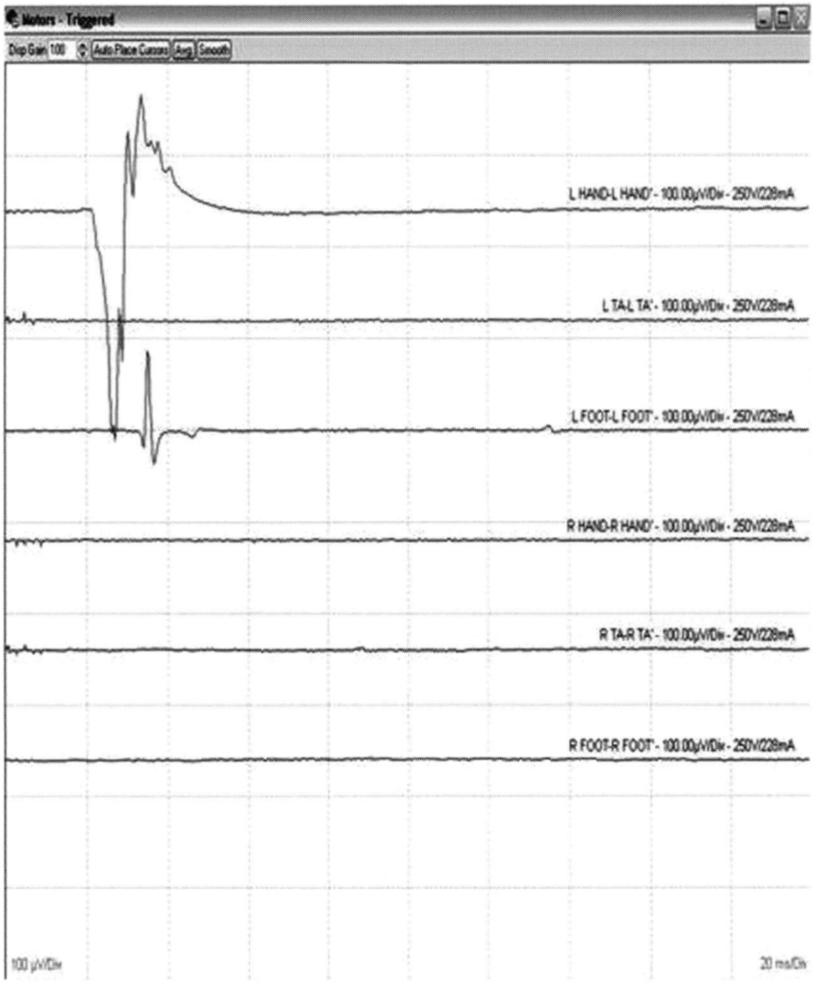

Figure 11.15 Traces 1 to 3 show pre-resection TcMEPs: EMG recordings from left dorsal interosseus, left tibialis anterior, and left abductor hallucis muscles, respectively. Traces 4–6 show TcMEPs using reverse stimulation polarity, showing evoked myogenic potentials from the right dorsal interosseus, tibialis anterior, and abductor hallucis muscles).

can be either non-recurrent with only single stimuli, or recurrent with frequencies reported as high as 10 Hz.

Anesthesia in IOM of CNN

Different anesthetic protocols can be used successfully during these procedures, but as with the monitoring of cranial nerves, the most important aspect is the use of muscle relaxants. It is preferred that the neuromuscular blocking agents are not used during the time that EMG monitoring is to take place. These are also likely to have adverse effects on TcMEPs, which are frequently used as part of a multimodality approach when monitoring these cases. The following is the typical anesthetic protocol used at our institution:

1. Induction with a short-acting barbiturate, an intravenous narcotic agent, and a short-acting neuromuscular blocking agent (NMBA)
2. Thereafter, anesthesia is usually maintained with a combination of isoflurane, with a maximum end-tidal not exceeding 0.6%, and a maximum of 50% nitrous oxide. In addition, very short-acting intravenous narcotics, such as remifentanyl, are used after induction to supplement the inhaled agent. NMBAs are not used after induction.

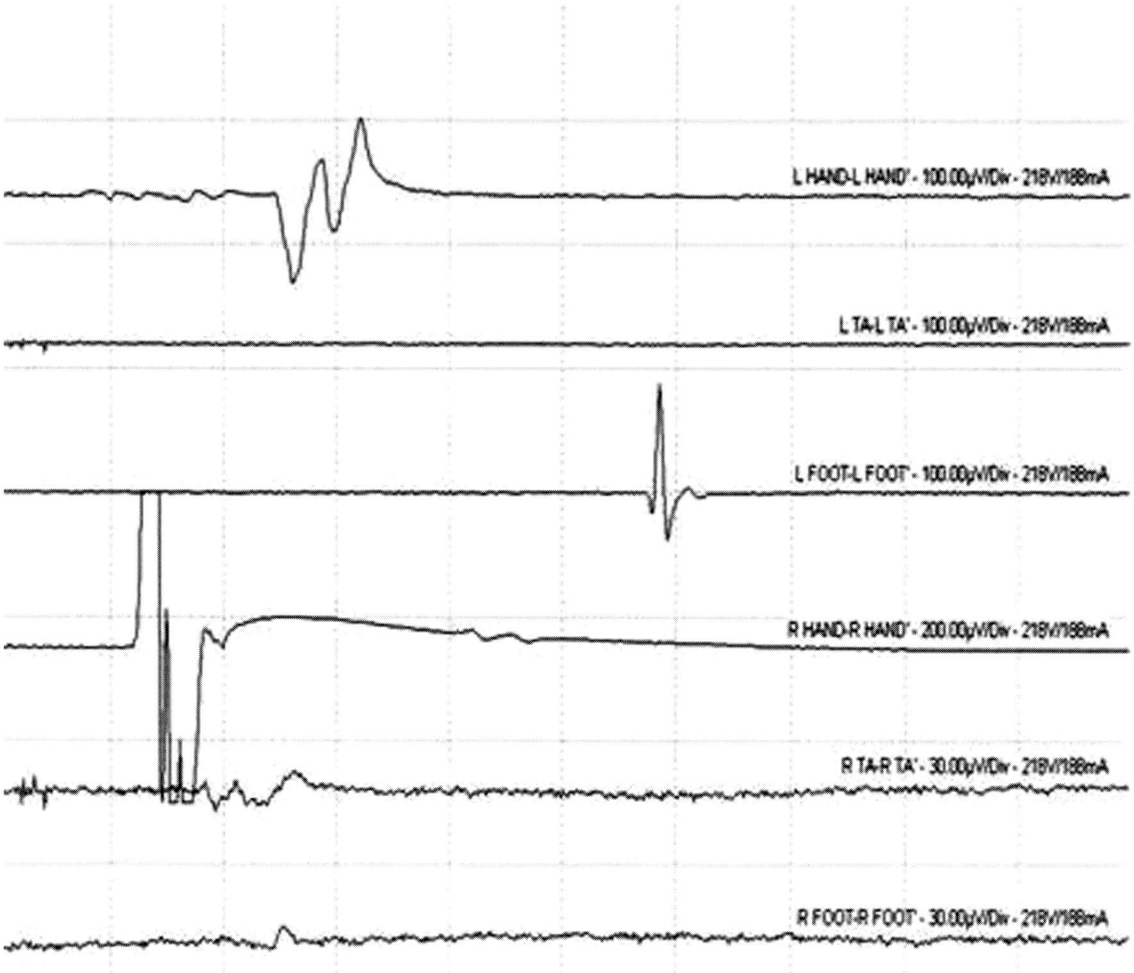

Figure 11.16 TcMEPs using reverse stimulation polarity, as compared to Figure 11.15, showing evoked myogenic potentials from the right dorsal interosseus, tibialis anterior, and abductor hallucis muscles (Traces 4–6, respectively).

Cranial nerve nuclei monitoring protocol used at our institution

Cranial nerve nuclei monitoring and brainstem motor nuclei mapping are performed in conjunction with SEPs generated at the median and posterior tibial nerve, bilateral BAEPs, and TcMEPs with muscle recordings obtained from all four limbs. This type of multimodality monitoring allows for monitoring of the intrinsic brainstem auditory pathways, somatosensory, and motor tracts passing through the brainstem, as well as CN and CNN motor activity.

The recording technique for mapping and monitoring of motor CNN involves placing needle or "hookwire" electrode pairs in the muscles corresponding to the CNN at risk for injury. We have found that the needle and hookwire electrodes seem to provide the best responses without decay in the signal characteristics, which can occur with lengthy surgical cases. In addition, when properly secured, these electrodes will easily stay in place, and the biggest technical problem we have encountered is that the electrodes may inadvertently be pulled out during the positioning of the patient. For this reason, we usually wait until the patient is placed in the headholder device while supine and before being turned onto the prone position prior to placing the electrodes into the masseter, orbicularis oculi, orbicularis oris, trapezius, and genioglossus muscles. Unfortunately, this cannot be done for electrodes inserted into the lateral soft palate (CN IX), which require placement with the

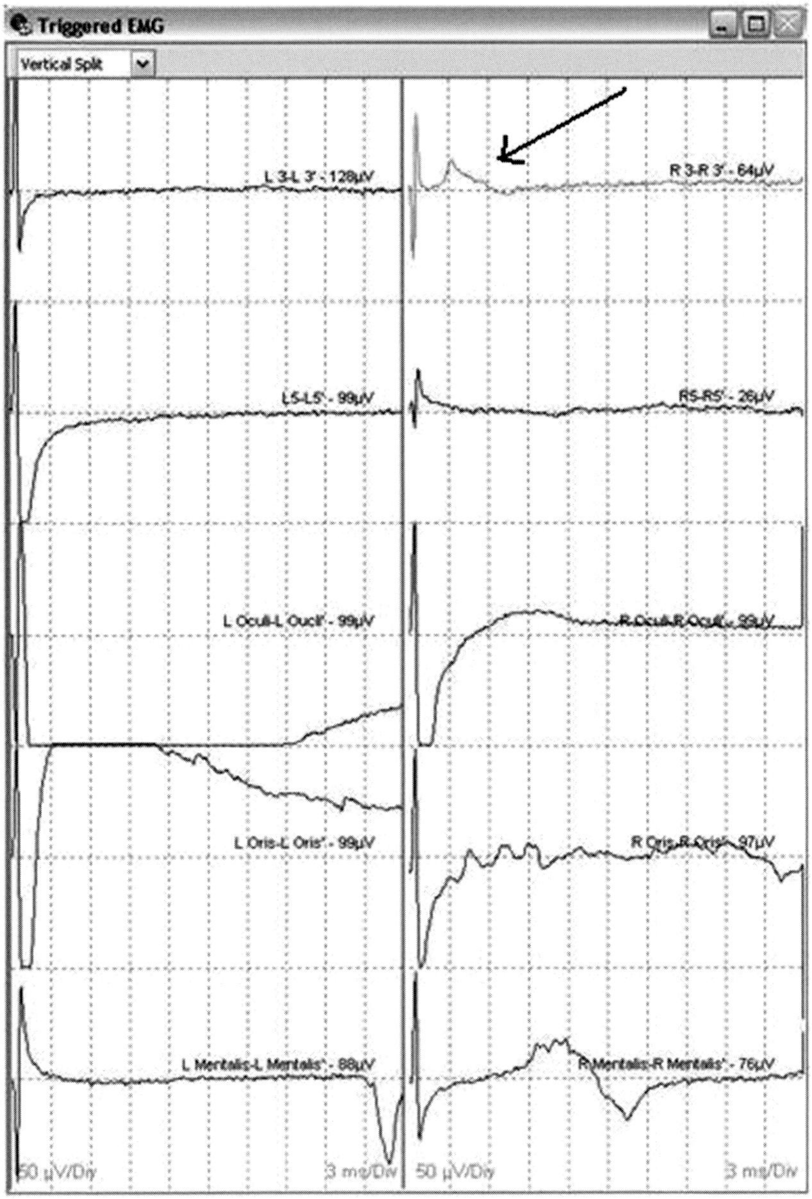

Figure 11.17 Pre-resection electrically evoked CNIII CMAP (trace 1 on right) after stimulation of the right dorsal-lateral midbrain.

aid of a laryngoscope. As stated previously, we place these electrodes with the assistance of the anesthesiologist, who will hold the patient's mouth open, move the tongue to the side, and provide a light source with the laryngoscope. We also prefer to insert wire electrodes into the extraocular muscles, if we plan to monitor CNs III, IV, or VI, before placement of the headholder.

We usually monitor a minimum of eight muscles of three bilateral nerves: masseter (CN V), orbicularis oculi and oris (CN VII), and trapezius (CN XI). The trapezius serves mainly as a control muscle and is useful in differentiating neurotonic discharges from electrical noise. In addition, stimulation of the spinal accessory nerve (if exposed) can be used to test the stimulation and recording set-up before the start of the brainstem mapping session. This will ensure that the stimulation and recording capabilities of the equipment and the electrodes are functioning properly.

Baseline EMG recordings from appropriate muscles are made prior to the start of surgery, and monitoring is

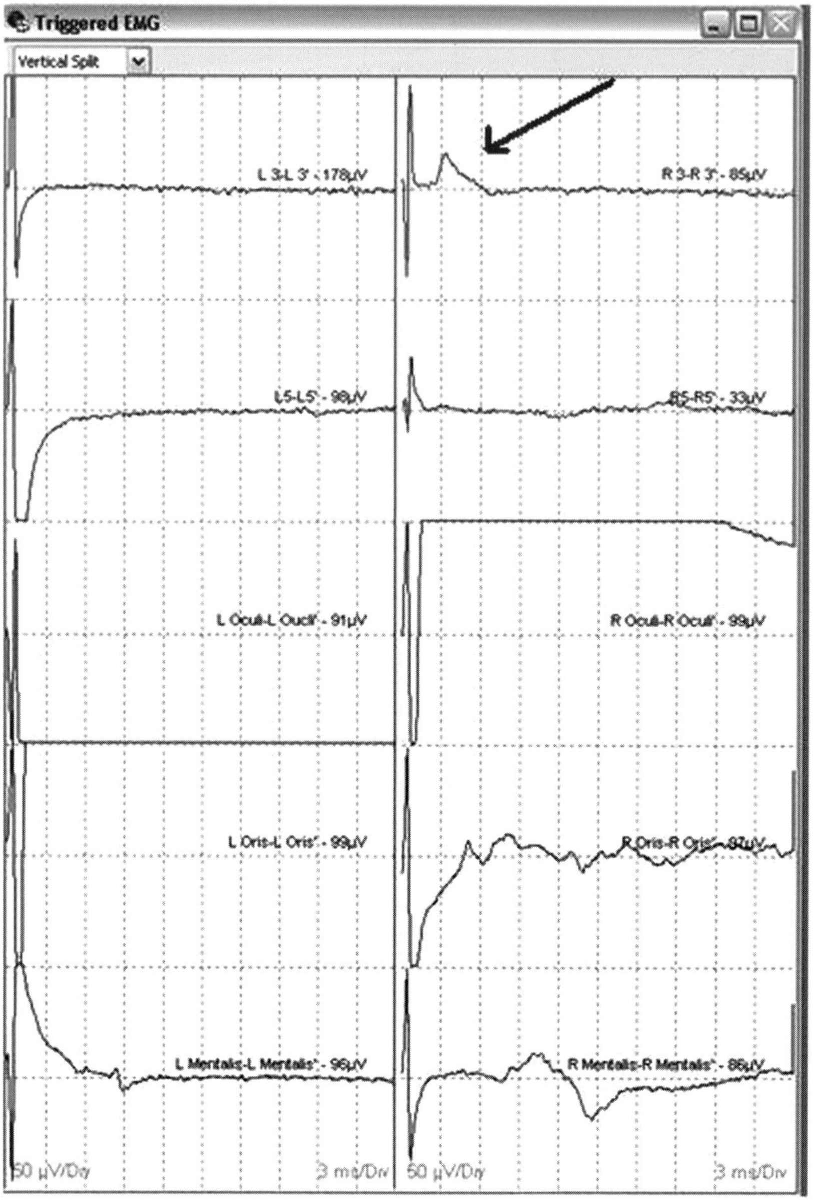

Figure 11.18 Post-resection electrically evoked CN III CMAP showing no change in morphology or amplitude.

continuous until surgical closure. At intervals during the surgery, when desired by the operating surgeon, a small sterile monopolar electrode probe is placed on the brainstem in areas felt to correspond to non-diseased CNN, and a small electrical stimulus is applied.

Stimulation intensity begins at zero, and is increased at 0.1-mA (or 0.1- V) increments until a threshold response is identified. Stimulation is then increased until a maximum CMAP is observed or EMG activity from another motor nucleus is triggered. This maximal stimulation level is then used to search for electrically excitable tissue in the surgical target area. By observing the stimulus intensities required for a threshold response or a maximum CMAP, the approximate locations of the CNN are determined. Triggered CMAPs are usually obtained with stimulus intensities below 3 mA and stimulus durations of 0.02–0.1 ms delivered as a single pulse with a rate of 1.1 Hz.

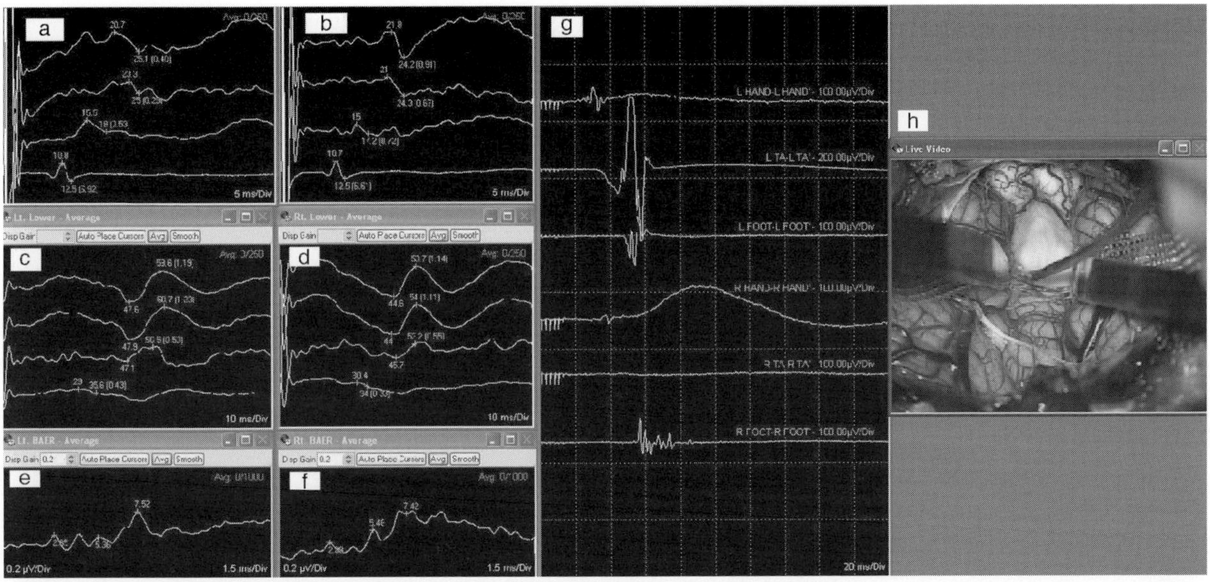

Figure 11.19 Baseline, bilateral median and posterior tibial nerve SEPs, BAEPs, and TcMEPs. (a) Baseline SEPs after left median nerve stimulation. Traces 1–2: scalp recorded cortical SEPs; trace 3: cervical spine (C7) recorded SEP; trace 4: Erb's point recorded SEP. (b) Baseline SEPs after right median nerve stimulation. Traces 1–2: scalp recorded cortical SEPs; trace 3: cervical spine (C7) recorded SEP; trace 4: Erb's point recorded SEP. (c) Baseline SEPs after left posterior tibial nerve stimulation. Traces 1–3: scalp recorded cortical SEPs; trace 4: cervical spine (C7) recorded SEP. (d) Baseline SEPs after right posterior tibial nerve stimulation. Traces 1–3: scalp recorded cortical SEPs; trace 4: cervical spine (C7) recorded SEP. (e, f) Baseline BAEPs after left and right ear stimulation. (g) TcMEPs: Traces 1–3: EMG recordings from left dorsal interosseus, left tibialis anterior and left abductor hallucis muscles, respectively. Traces 4–6 show EMG recordings obtained from the same muscles on the right. (h) resection of brainstem cavernous malformation.

Although this slow stimulation frequency rate reduces the overall number of stimuli to the brainstem, this technique requires that the surgeon hold the stimulating electrode at the site of interest for a few seconds to acquire responses. If the surgeon prefers to search for excitable areas by quickly moving the probe across the brainstem, then the stimulation frequency should be increased and we have found that stimulation frequencies between 5.1 and 7.1 Hz provide good results. There are several commercially available monopolar stimulating electrodes that function well. However, the one preferred by us is a monopolar stimulating probe with a ball end, attached by a small flexible wire to the handle (Medtronic Xomed Surgical Products; SKU 82–25251).

Interpretation of the electrophysiologic data

As one cannot obtain preoperative or pre-incision data regarding the characteristic responses from brainstem stimulation, each patient must serve as their own control. If the brainstem lesion lies primarily on one side, one should stimulate the opposite side and record the triggered EMG response, which will provide an estimate of the stimulus intensity needed to obtain the EMG signal, as well as some of the characteristics of the recorded potential, such as latency and amplitude. Mapping of the brainstem should be performed prior to initiating the resection of the lesion. Once a CMAP is obtained, the stimulus intensity should be adjusted to threshold and maximum response levels and the latencies and amplitudes recorded for both levels.

Although a critical change in the triggered EMG potentials during or following resection has not been defined, Chang *et al.* (1999) reported that one patient developed a transient VI and VII cranial nerve palsy after post-resection stimulation revealed an 80% amplitude reduction in the recorded CMAP when compared with the pre-resection baseline. In our experience, we have found that patients do not develop new cranial nerve deficits if the post-resection triggered CMAPs remain unaltered from the baseline. In contrast, if higher stimulus intensities are required to obtain the CMAP or if the CMAP amplitudes are smaller after post-resection stimulation, there is a higher probability of developing a new transient or permanent cranial

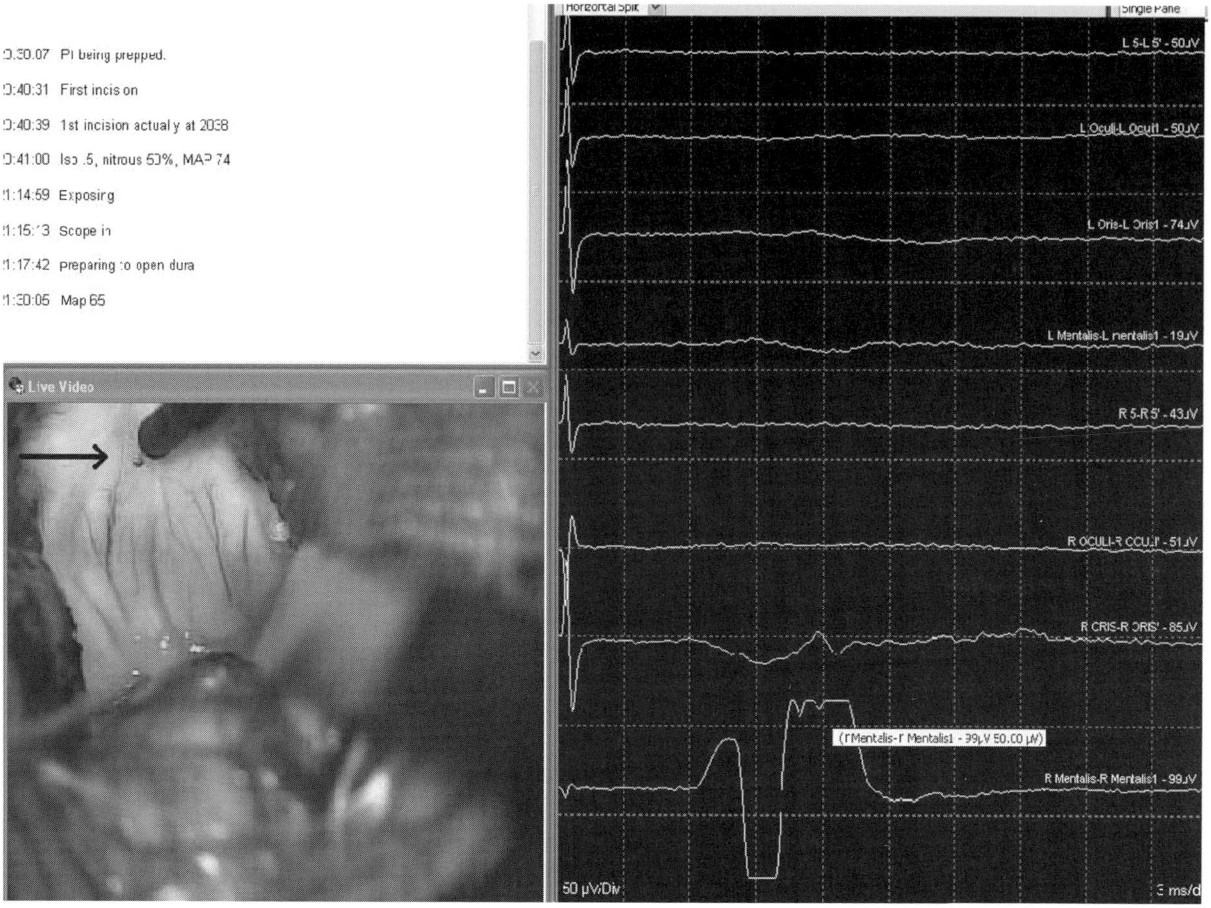

Figure 11.20 Mapping of the floor of the fourth ventricle. Electrical stimulation was performed using a monopolar electrode (arrow). Note the triggered large CMAP recorded from the right mentalis (trace 8) and a small CMAP from the right orbicularis oris (trace 7).

nerve deficit, or a worsening of preexisting cranial nerve deficits. However, we have found that the probability of a permanent deficit is highest in those patients where the CMAP cannot be registered after resection. We have not observed any significant changes in latencies secondary to CNN injury.

Unfortunately, the literature does not provide clear criteria on how to interpret neurotonic discharges. In our practice, we use a strategy similar to that described by Eisner *et al.* (1995) with standard audible and video-displayed EMG to monitor neurotonic discharges in addition to triggered CMAPs during brainstem mapping. Electromyographic activity of short duration, which coincided with the start and end of a surgical manipulation and was similar to the EMG bursts typically seen during facial nerve monitoring, was defined as "contact activity." They also described "pathological spontaneous activity" (PSA) and classified it as "slight PSA" if it lasted for a few seconds or "extreme PSA" if it persisted for several hours after the surgical manipulation had been terminated.

Using these methods, the surgeon's dissection strategy was guided and modified based on the amount of audible EMG produced by surgical manipulation. One of their patients developed transient followed by gradually permanent EMG discharges in the lingual muscles and suffered severe bilateral and permanent tongue weakness. In the remaining 15 patients the surgeons modified their dissection strategy by reducing pressure or traction on tissue, or by changing the direction and location of dissection based on the occurrence of EMG activity. They felt that the alterations in surgical management that were a response to the EMG findings ultimately resulted in EMG activity of shorter duration and lower amplitude. In this group of patients, new postoperative deficits were mild, transient, or absent.

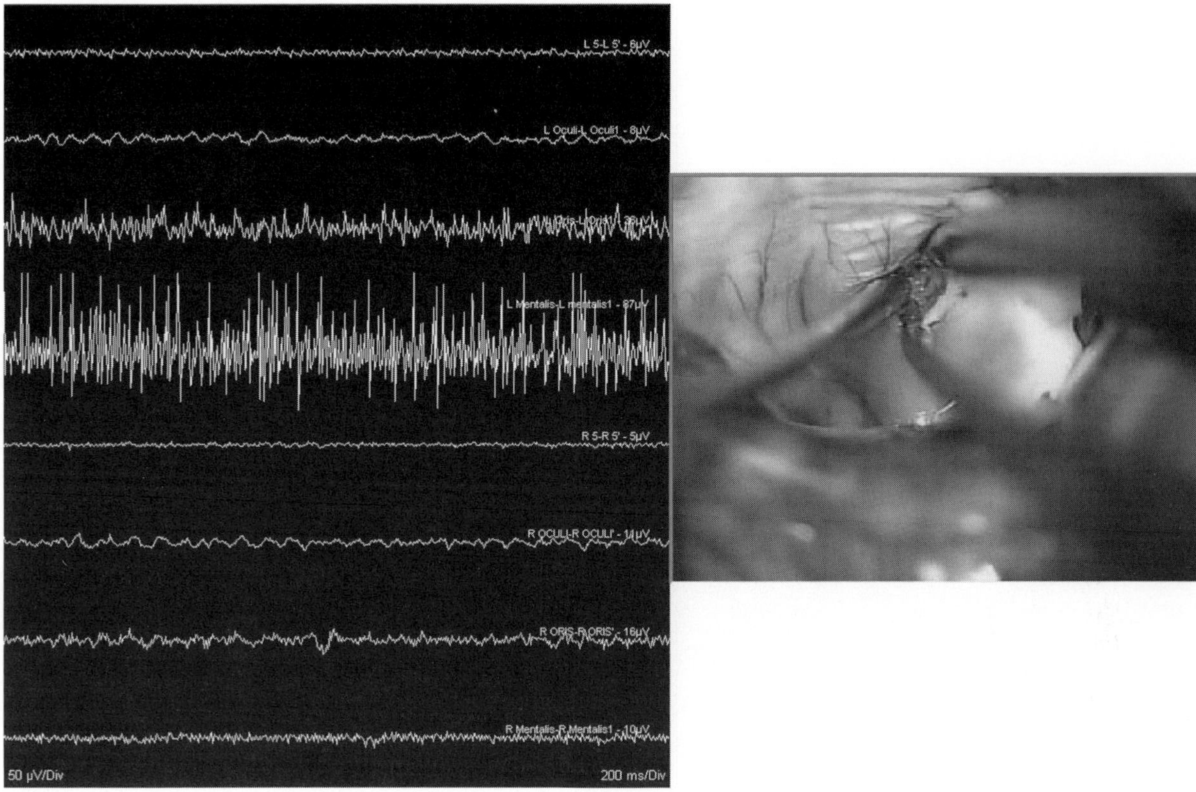

Figure 11.21 Neurotonic discharges noted primarily from the left mentalis (trace 4) and a lesser degree from the left orbicularis oris (trace 3) during the initial resection of the cavernous malformation.

Conclusion

Our ability to identify different neural structures and assess their functional state during a variety of surgical procedures has improved and evolved over the past two decades. The literature shows that we now possess excellent electrophysiologic techniques to identify cranial nerves within the operative field, even when they are not visible. This can be done using electrical stimulation to activate the nerve and generate a CMAP.

The presence of unchanged CMAP amplitudes after cranial nerve stimulation at the end of the operative procedure provides reassurance that there remain a sufficient number of functioning motor axons, predictive of good postoperative motor function. In many instances, the intraoperative results obtained alter the surgical management, as has been the case with acoustic neuroma surgery. In addition, the dynamic functional state of cranial nerves can also be continuously assessed by monitoring EMG activity. This can provide real-time feedback and guidance to the surgeon.

The basic principles of EMG used to monitor cranial nerves have also been applied to IOM of CNN. Electrical stimulation of brainstem structures is the only method currently available to identify CNN and CN motor pathways and is now considered a safe and reliable technique that can guide neurosurgeons during certain types of brainstem surgery. Although the current literature is compelling and is suggestive that IOM of CNs and CNN is useful, and can alter surgical management, there is still a relative shortage of clinical outcome studies and further research in this area is needed.

In closing, it is important to note that effective IOM of CNs and CNN requires a detailed understanding of CN function, neuroanatomy, neurologic and neurosurgical disorders, proper electrical stimulation and recording techniques, and the possible adverse effects of CN and brainstem stimulation. When these criteria are fulfilled, IOM of CNs and CNN is a vital and safe adjunct to surgery where these structures are at risk for injury, and many surgeons consider its use indispensable.

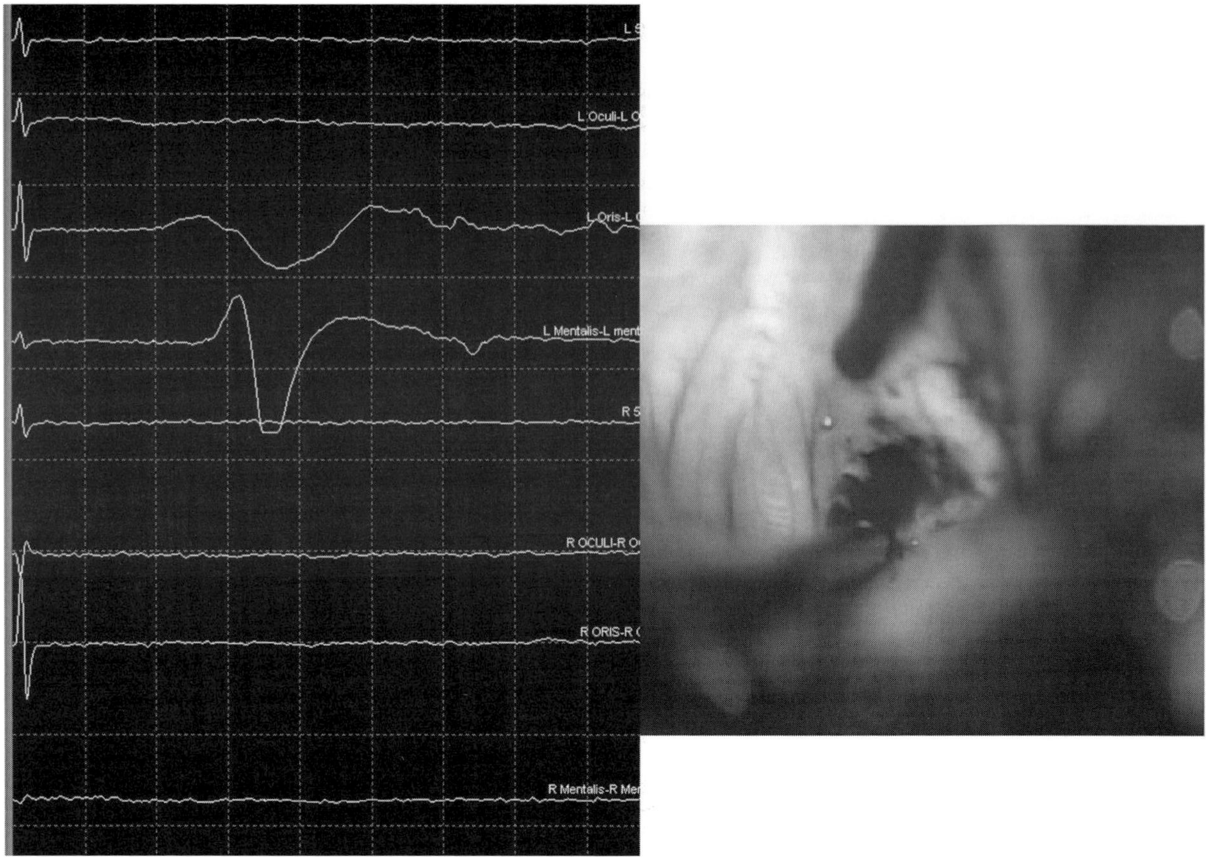

Figure 11.22 Post-resection mapping of the floor of the fourth ventricle and electrical stimulation using a monopolar electrode. Note the triggered CMAP recorded from the left mentalis (trace 4) and a small CMAP from the left orbicularis oris (trace 3).

Case examples for Chapter 11

Case example 1

A 65-year-old female with midbrain cavernous malformation developed a mild right ptosis and transient diplopia after a midbrain hemorrhage (see Figure 11.14).

Multimodality IOM was performed, using the following modalities: median and posterior tibial nerve SEPs, TcMEPs, bilateral BAEPs, continuous, free-running EMG activity of bilateral CNs III, V, and VII, and direct CN III electrical stimulation (see Figures 11.15 to 11.17).

There were no significant changes in the SEPs, MEPs, or BAEPs during the surgical procedure. The patient developed slight worsening of her right ptosis but no diplopia or other neurologic symptoms or new neurologic findings on physical examination.

Case example 2

A 20-year-old female underwent resection of a brainstem cavernous malformation after experiencing a brainstem hemorrhage. Prior to surgery, the patient had persistent left facial weakness but no other abnormalities on neurologic examination.

Intraoperative monitoring, using multimodality recordings, included the following modalities: bilateral median and posterior tibial nerve SEPs, TcMEP, bilateral BAEPs, continuous, free-running EMG activity of bilateral CNs V and VII, and CNN mapping of the floor of the fourth ventricle (see Figures 11.19–11.24).

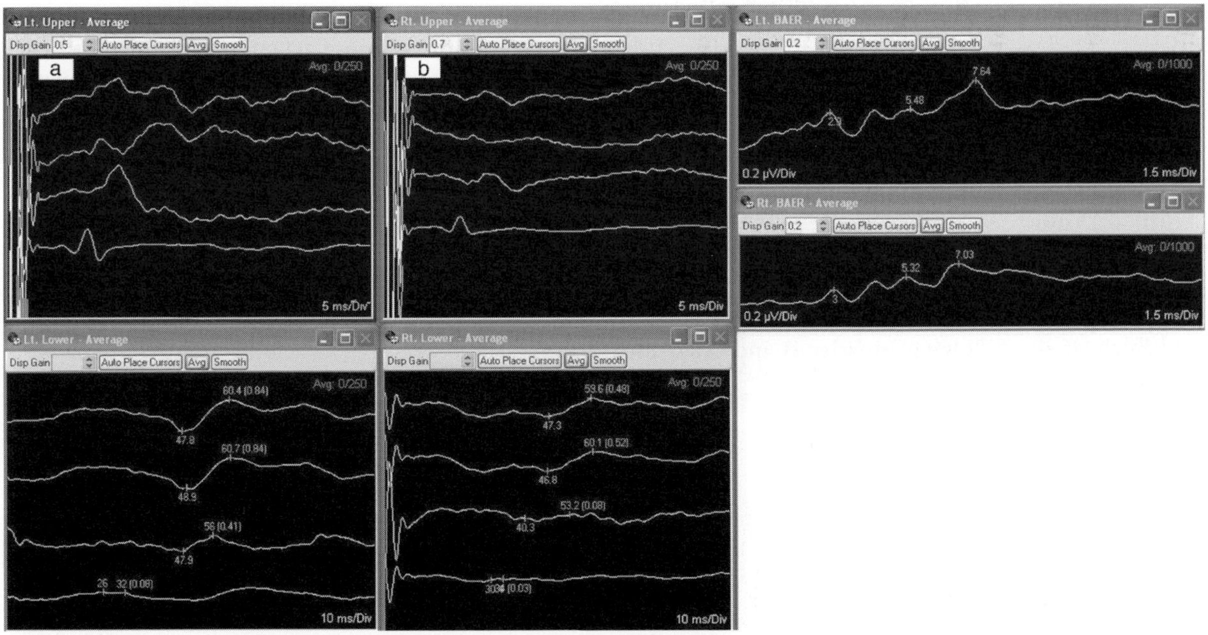

Figure 11.23 Loss of right median nerve scalp recorded cortical SEPs (panel b; traces 1–2). Left median nerve scalp recorded cortical SEPs (panel a; traces 1–2) show no significant changes from baseline.

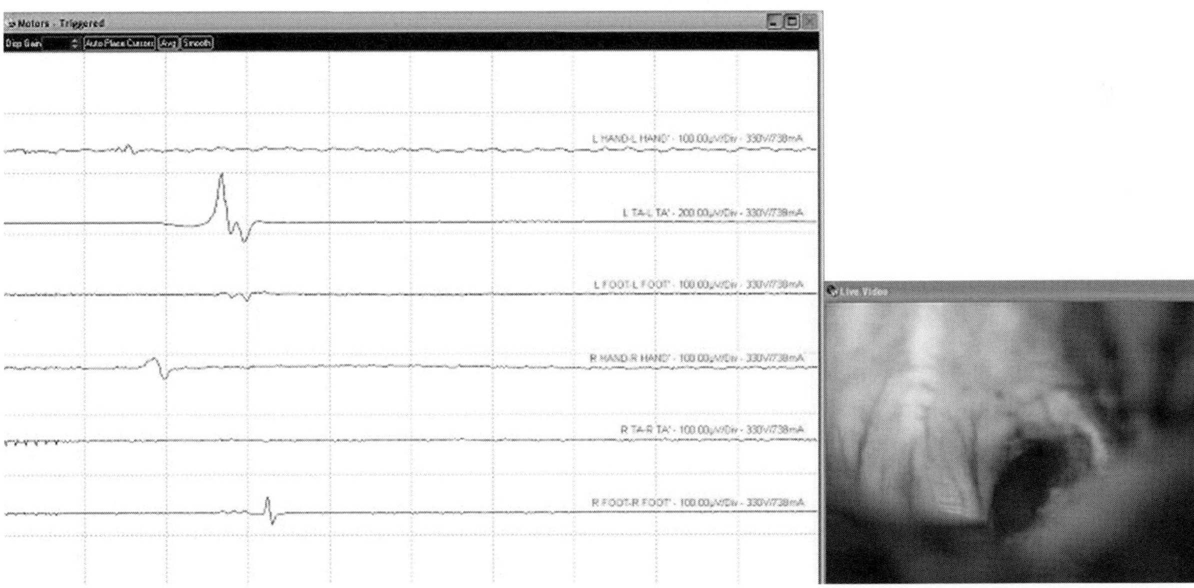

Figure 11.24 Post-resection TcMEP shows no significant changes from baseline. The picture on the right demonstrates the post-resection cavity in the brainstem.

Postoperatively, the patient developed numbness in the right side of her body. Her neurologic examination showed a moderate right-sided hemisensory loss and a left CN VI palsy. Her facial strength remained unchanged after surgery. Intraoperative monitoring accurately predicted the new sensory loss with preserved extremity strength and no change in her preexisting facial weakness. The CN VI palsy could not have been predicted by IOM because that nerve had not been directly monitored.

Questions

1. The facial colliculus is an important structure to identify on the surface of the brainstem. It can easily be electrically stimulated owing to its location, as it lies approximately how far beneath the ependyma?
   (a) 0.05 mm
   (b) 0.15 mm
   (c) 0.25 mm
   (d) 0.35 mm
   (e) 0.50 mm
2. Which of the following muscles is indicated for EMG of CN V?
   (a) Orbicularis oris
   (b) Orbicularis oculi
   (c) Lateral rectus
   (d) Masseter
   (e) Trapezius
3. High levels of inhaled anesthetic agents are contraindicated during EMG?
   (a) True
   (b) False
4. EMG monitoring of CN III function requires placing electrodes in which of the following muscles?
   (a) Orbicularis oculi
   (b) Inferior rectus
   (c) Lateral rectus
   (d) Masseter
   (e) Temporalis
5. Neurotonic discharges can be most easily recorded using which of the following electrodes?
   (a) Intramuscular needle electrodes
   (b) Subdermal needle electrodes
   (c) Surface disk electrodes
   (d) All of the above
   (e) None of the above
6. The extraocular muscles that control eye movement are innervated by three cranial nerves, including:
   (a) CN I (olfaction)
   (b) CN IV (trochlear)
   (c) CN V (trigeminal)
   (d) CN VII (facial)
   (e) CN XI (accessory)
7. Electrical stimulation of CN IX can result in which of the following side effects?
   (a) Head rotation
   (b) Hypothermia
   (c) Excessive shoulder movement
   (d) Hypotension
   (e) Seizures
8. Which of the following is most likely to generate a maximum response when stimulating a normal intracranial facial nerve using 0.05 ms stimulation duration?
   (a) 0.5 mA
   (b) 1.0 mA
   (c) 1.5 mA
   (d) 2.0 mA
   (e) All of the above
9. Effective electrical stimulation and activation of cranial nerves have been reported using which of the following?
   (a) Monopolar stimulation
   (b) Bipolar stimulation
   (c) Constant-current stimulation
   (d) Constant-voltage stimulation
   (e) All of the above
10. The intraoperative use of facial nerve F waves is limited because of which of the following?
    (a) Their absence in some healthy adults
    (b) Excessive facial movement
    (c) Interference with continuous monitoring of EMG activity
    (d) Difficulty placing recording electrodes
    (e) The need to perform peripheral stimulation

References

Agnew, W. F. and McCreery, D. B. (1987). Considerations for safety in the use of extracranial stimulation for motor-evoked potentials. *Neurosurgery*, **20**, 143–147.

Akagami, R., Dong, C. C. J., and Westerberg, B. D. (2005). Localized transcranial electrical motor-evoked potentials for monitoring cranial nerves in cranial base surgery. *Operative Neurosurgery*, **57**, 78–85.

Alberti, O., Sure, U., Riegel, T., and Bertalanffy, H. (2001). Image-guided placement of eye muscle electrodes for intraoperative cranial nerve monitoring. *Neurosurgery*, **49**, 660–664.

Baker, G. S. (1965). Physiologic abnormalities encountered after removal of brain tumors from the floor of the fourth ventricle. *Journal of Neurosurgery*, **23**, 338–343.

Blair, E. A., Teeple, E., Sutherland, R. M., Shih, T., and Chen, D. (1994). Effect of neuromuscular blockade on facial nerve monitoring. *American Journal of Otology*, **15**, 161–167.

Chang, S. D., López, J. R., and Steinberg, G. K. (1999). Intraoperative electrical stimulation for identification of cranial nerve nuclei. *Muscle & Nerve*, **22**, 1538–1543.

Chang, S. D., López, J. R., and Steinberg, G. K. (2000). Reply: identification of cranial nerve nuclei. *Muscle & Nerve*, **23**, 1446.

Crum, B. A. and Strommen, J. A. (2007). Peripheral nerve stimulation and monitoring during operative procedures. *Muscle & Nerve*, **35**, 159–170.

Danneman, M., Lega, B. C., Suki D., Harper, R. L., and Yoshor, D. (2008). Microvascular decompression for hemifacial spasm: long-term results from 114 operations performed without neurophysiological monitoring. *Journal of Neurosurgery*, **109**, 410–415.

Daube, J. R. and Harper, C. M. (1989). Surgical monitoring of cranial and peripheral nerves. In *Neuromonitoring in Surgery*, ed. J. E. Desmedt. Amsterdam: Elsevier, pp. 115–138.

Delgado, T. E., Buchheit, W. A., Rosenholtz, H. R., and Chrissian, S. (1979). *Neurosurgery*, **4**, 418–421.

DeMonte, F., Warf, P., and Al-Mefty, O. (1994). Intraoperative mapping of the lower cranial nerves during surgery of the jugular foramen and lower clivus. In *Intraoperative Monitoring Techniques in Neurosurgery*, ed. C. M. Loftus and V. C. Traynelis. New York: McGraw-Hill, pp. 205–212.

Dong, C. C. J., MacDonald, D. B., Akagami, R., *et al.* (2005). Intraoperative facial motor-evoked potential monitoring with transcranial electrical stimulation during skull base surgery. *Clinical Neurophysiology*, **116**, 588–596.

Eisner, W., Schmid, U. D., Reulen, H.-J., *et al.* (1995). The mapping and continuous monitoring of the intrinsic motor nuclei during brain stem surgery. *Neurosurgery*, **37**, 255–265.

Fukaya, C., Katayama, Y., Kasai, M., Kurihara, J., and Yamamoto, T. (1999). Intraoperative electrooculographic monitoring of oculomotor nerve function during skull base surgery. Surgery technical note. *Journal of Neurosurgery*, **91**, 157–159.

Fukuda, M., Oishi, M., Takao, T., Saito, A., and Fujii, Y. (2008). Facial nerve motor evoked potential monitoring during skull base surgery predicts facial nerve outcome. *Journal of Neurology, Neurosurgery, and Psychiatry*, **79**, 1066–1070.

Harnack, D., Winter, C., Meissne, W., *et al.* (2004). The effects of electrode material, charge density and stimulation duration on the safety of high-frequency stimulation of the subthalamic nucleus in rats. *Journal of Neuroscience Methods*, **138**, 207–216.

Harner, S. G., Daube, J. R., and Ebersold, M. J. (1986). Electrophysiologic monitoring of facial nerve during temporal bone surgery. *Laryngoscope*, **96**, 65–69.

Harper, C. M. (1998). Cranial nerve monitoring. In 1998 *American Association of Electrodiagnostic Medicine Annual Meeting, Course F: Update in Intraoperative Monitoring Syllabus*, pp. 23–31.

Harper, C. M. (2004). Intraoperative cranial nerve monitoring. *Muscle & Nerve*, **29**, 339–351.

Harper, C. M. and Daube, J. R. (1998). Facial nerve electromyography and other cranial nerve monitoring. *Journal of Clinical Neurophysiology*, **15**, 206–216.

Hatem, J., Sindou, M., and Vial, C. (2001). Intraoperative monitoring of facial EMG responses during microvascular decompression for hemifacial spasm. Prognostic value for long-term outcome: a study in a 33-patient series. *British Journal of Neurosurgery*, **15**, 496–499.

Heffez, D. S., Zinreich, J., and Long, D. M. (1990). Surgical resection of intrinsic brain stem lesions: an overview. *Neurosurgery*, **27**, 789–798.

Husain, A. M, Wright, D. R., Stolp, B. W., Friedman, A. H., and Keifer, J. C. (2008). Neurophysiological intraoperative monitoring of the glossopharyngeal nerve: technical case report. *Neurosurgery*, **63**: 277–278. Discussion, 278.

Isaacson, B., Kileny, P. R., and El-Kashlan, H. K. (2003). Intraoperative monitoring and facial nerve outcomes after vestibular schwannoma resection. *Otology and Neurotology*, **24**, 812–817.

Isaacson, B., Kileny, P. R., and El-Kashlan, H. K. (2005). Prediction of long-term facial nerve outcomes with intraoperative nerve monitoring. *Otology and Neurotology*, **26**, 270–273.

Jackson, L. E. and Roberson, J. B. (1999). Vagal nerve monitoring in surgery of the skull base: a comparison of efficacy of three techniques. *American Journal of Otology*, **20**, 649–656.

Joo, W. I., Lee, K. J., Park, H. K., Chough, C. K., and Rha, H. K. (2008). Prognostic value of intraoperative lateral spread response monitoring during microvascular decompression in patients with hemifacial spasm. *Journal of Clinical Neuroscience*, **15**, 1335–1339.

Kartush, J. M. and Lundy, L. B. (1992). Facial nerve outcome in acoustic neuroma surgery. *Otolaryngologic Clinics of North America*, **25**, 623–647.

Katsuta, T., Morioka, T., Fujii, K., and Fukui. M. (1993). Physiological localization of the facial colliculus during direct surgery on an intrinsic brain stem lesion. *Neurosurgery*, **32**, 861–863.

Kizilay, A., Aladag, I., Cokkeser, Y., *et al.* (2003). Effects of partial neuromuscular blockade on facial nerve monitorization in otologic surgery. *Acta Oto-laryngologica*, **123**, 321–324.

Kong, D.-S., Park, K., Shin, B.-G., Lee, J. A., and Eum, D.-O. (2007). Prognostic value of the lateral spread response for intraoperative electromyography monitoring of the facial musculature during microvascular decompression for hemifacial spasm. *Journal of Neurosurgery*, **106**, 384–387.

Krause, F. (1912). *Surgery of the Brain and Spinal Cord Based on Personal Experience*. Vol. II. New York: Rebman Company.

Lalwani, A. K., Butt, F. Y., Jackler, R. K., Pitts, L. H., and Yingling, C. D. (1994). Facial nerve outcome after acoustic neuroma surgery: a study from the era of cranial nerve monitoring. *Otolaryngology and Head and Neck Surgery*, **111**, 561–570.

Lang, J., Ohmachi, N., and Lang Sen, J. (1991). Anatomical landmarks of the rhomboid fossa (floor of the fourth ventricle), its length and width. *Acta Neurochirurgica*, **113**, 84–90.

Lassiter, K. R. L., Alexander, E., Davis, C. H., and Kelly, D. L. (1971). Surgical treatment of brain stem gliomas. *Journal of Neurosurgery*, **34**, 719–725.

Lennon, R. L., Hosking, M. P., Daube, J. R., and Welna, J. O. (1992). Effect of partial neuromuscular blockade on intraoperative electromyography in patients undergoing resection of acoustic neuromas. *Anesthesia and Analgesia*, **75**, 729–733.

McCreery, D. B., Agnew, W. F., Yuen, T. G. H., and Bullara, L. (1990). Charge density and charge per phase as cofactors in neural injury induced by electrical stimulation. *IEEE Transactions on Rehabilitation Engineering*, **37**, 996–999.

Møller, A. R. (1994). Monitoring techniques in cavernous sinus surgery. In *Intraoperative Monitoring Techniques in Neurosurgery*, ed. C. M. Loftus and V. C. Traynelis. New York: McGraw-Hill, pp. 141–155.

Møller, A. R. (1999). Vascular compression of cranial nerves: II: pathology. *Neurological Research*, **21**, 439–443.

Møller, A. R. and Jannetta, P. J. (1984). Preservation of facial function during removal of acoustic neuromas. Use of monopolar constant-voltage stimulation and EMG. *Journal of Neurosurgery*, **61**, 757–760.

Møller, A. R. and Jannetta, P. J. (1987). Monitoring facial EMG responses during microvascular decompression operations for hemifacial spasm. *Journal of Neurosurgery*, **66**, 681–685.

Mooij, J. J. A., Mustafa, M. K., and van Weerden, T. W. (2001). Hemifacial spasm: intraoperative electromyographic monitoring as a guide for microvascular decompression. *Neurosurgery*, **49**, 1365–1371.

Morota, N. and Deletis, V. (2006). The importance of brainstem mapping in brainstem surgical anatomy before the fourth ventricle and implication for intraoperative neurophysiological mapping. *Acta Neurochirurgica*, **148**, 499–509.

Morota, N., Deletis, V., Epstein, F. J., *et al.* (1995). Brain stem mapping: neurophysiologic localization of motor nuclei on the floor of the fourth ventricle. *Neurosurgery*, **37**, 922–930.

Morota, N., Deltis, V., Lee, M., and Epstein, F. J. (1996). Functional anatomic relationship between brain stem tumors and cranial motor nuclei. *Neurosurgery*, **39**, 787–794.

Moskowitz, N. and Long, D. M. (1991). Acoustic neurinomas. Historical review of a century of operative series. *Neurosurgery Quarterly*, **1**, 2–18.

Neves, D. O., Lefaucheur, J. P., Ciampi de Andrade, *et al.* (2009). A reappraisal of the value of lateral spread response monitoring in the treatment of hemifacial spasm by microvascular decompression. *Journal of Neurology, Neurosurgery, and Psychiatry*, **80**, 1375–1380.

Nielsen, V. K. (1985). Electrophysiology of the facial nerve in hemifacial spasm: ectopic/ephaptic excitation. *Muscle & Nerve*, **8**, 545–555.

Parsons, R. C. (1966). Electrical stimulation of the facial nerve. *Laryngoscope*, **76**, 391.

Petro, M. L., Schweinfurth, J. M., and Petro, A. B. (2006). Transcricothyroid, intraoperative monitoring of the vagus nerve. *Archives of Otolaryngology – Head and Neck Surgery*, **132**, 624–628.

Rivero, A., Crovetto, L., Lopez, L., Maselli, R., and Nogues, M. (1995). Single fiber electromyography of extraocular muscles: a sensitive method for the diagnosis of ocular myasthenia gravis. *Muscle & Nerve*, **18**, 943–947.

Romstöck, J., Strauss, C., and Fahlbusch, R. (2000). Continuous electromyography monitoring of motor cranial nerves during cerebellopontine angle surgery. *Journal of Neurosurgery*, **93**, 586–593.

Rusyniak, W. G., Ireland, P. D., Radley, M. G., Pilcher, W. H., and Okawara, S.-H. (1992). Ultrasonographic and electrophysiological adjuncts to surgery within the brain stem: technical note. *Neurosurgery*, **31**, 798–801.

Sasaki, T., Suzuki, K., Matsumoto, M., *et al.* (2002). Origins of surface potentials evoked by electrical stimulation of oculomotor nerves: are they related to electrooculographic or electromyographic events? *Journal of Neurosurgery*, **97**, 941–944.

Schlake, H.-P., Goldbrunner, R. H., Milewski, C., *et al.* (2001a). Intraoperative electromyographic monitoring of the lower cranial motor nerves (LCN IX-XII) in skull base surgery. *Clinical Neurology and Neurosurgery*, **103**, 72–82.

Schlake, H.-P., Goldbrunner, R., Siebert, M., Behr, R., and Roosen, K. (2001b). Intraoperative electromyographic monitoring of extraocular motor nerves (Nn III, VI) in skull base surgery. *Acta Neurochirurgica*, **143**, 251–261.

Sekiya, T., Hatayama, T., Iwabuchi, T., and Shuji, M. (1993). Intraoperative recordings of evoked extraocular muscle activities to monitor ocular motor nerve function. *Neurosurgery*, **32**, 227–235.

Sekiya, T., Hatayama, T., Shimamura, N., and Suzuki, S. (2000). Intraoperative electrophysiological monitoring of oculomotor nuclei and their intramedullary tracts during midbrain tumor surgery. *Neurosurgery*, **47**, 1170–1177.

Shannon, R. V., Moore, J. K., McCreery, D. B., and Portillo, F. (1997). Threshold-distance measures from electrical stimulation of human brainstem. *IEEE Transactions on Rehabilitation Engineering*, **5**, 70–73.

Shin, J. C., Chung, U. H., Kim, Y. C., and Park, C. I. (1997). Prospective study of microvascular decompression in hemifacial spasm. *Neurosurgery*, **40**, 730–735.

Snyder, S. K. and Hendricks, J. C. (2005). Intraoperative neurophysiology testing of the recurrent laryngeal nerve: plaudits and pitfalls. *Surgery*, **138**, 1183–1192.

Stechison, M. T. (1993). The trigeminal evoked potential: Part II. Intraoperative recording of short-latency responses. *Neurosurgery*, **33**(4), 639–643.

Steinberg, G. K., Chang, S. D., Gewirtz, R. J., and Lopez, J. R. (2000). Microsurgical resection of brainstem, thalamic, and basal ganglia angiographically occult vascular malformations. *Neurosurgery*, **46**, 260–270.

Strauss, C., Romstock, J., Nimsky, C., and Fahlbusch, R. (1993). Intraoperative identification of motor areas of the rhomboid fossa using direct stimulation. *Journal of Neurosurgery*, **79**, 393–399.

Strauss, C., Romstock, J., and Fahlbusch, R. (1994). Intraoperative mapping of the floor of the IVth ventricle. *Intraoperative Monitoring Techniques in Neurosurgery*, ed. C. M. Loftus and V. C. Traynelis. New York: McGraw-Hill, pp. 213–218.

Strauss, C., Romstock, J., and Fahlbusch, R. (1999). Pericollicular approaches to the rhomboid fossa. Part II. Neurophysiological basis. *Journal of Neurosurgery*, **91**, 768–775.

Sugita, K., Kobayashi, S., Mutsuga, N., *et al.* (1979). Microsurgery for acoustic neurinoma – lateral position and preservation of facial and cochlear nerves. *Neurologia Medico-chirurgica*, **19**, 637–641.

Taha, J. M., Tew, J. M., and Keith, R. W. (1995). Proximal-to-distal facial amplitude ratios as predictors of facial nerve function after acoustic neuroma excision. *Journal of Neurosurgery*, **83**, 994–998.

Tamano, Y., Ujiie, H., Kawamata, T., and Hori, T. (2004). Continuous laryngoscopic vocal cord monitoring for vascular malformation surgery in the medulla oblongata: technical note. *Neurosurgery*, **54**, 232–235.

The Consensus Development Panel (1994). National Institutes of Health consensus conference statement on acoustic neuroma, December 11–13, 1991. *Archives of Neurology*, **51**, 201–207.

Thomusch, O., Sekulla, C., Walls, G., Machens, A., and Dralle, H. (2002). Intraoperative neuromonitoring of surgery for benign goiter. *American Journal of Surgery*, **183**, 673–678.

Topsakal, C., Al-Mefty, O., Bulsara, K. R., and Williford, V. S. (2008). Intraoperative monitoring of lower cranial nerves in skull base surgery: technical report and review of 123 monitored cases. *Neurosurgical Review*, **31**, 45–53.

Wedekind, C. and Klug, N. (1996). F-wave recording from nasal muscle for intraoperative monitoring of facial nerve function. *Zentralblatt für Neurochirurgie*, **57**, 184–189.

Wedekind, C. and Klug, N. (1998). Nasal muscle F-wave for peri- and intraoperative diagnosis of facial nerve function. *Electromyography and Clinical Neurophysiology*, **38**, 481–490.

Wedekind, C. and Klug, N. (2001a). Recording nasal muscle F waves and electromyographic activity of the facial muscles: a comparison of two methods used for intraoperative monitoring of facial nerve function. *Journal of Neurosurgery*, **95**, 974–978.

Wedekind, C. and Klug, N. (2001b). A normative study on human facial F waves. *Muscle & Nerve*, **24**, 900–904.

Wilson, L., Lin E., and Lalwani, E. (2003). Cost-effectiveness of intraoperative facial nerve monitoring in middle ear or mastoid surgery. *Laryngoscope*, **113**, 1736–1745.

Yingling, C. D. and Ashram, Y. A. (2005). Intraoperative monitoring of cranial nerves in neurotologic surgery. In *Cummings: Otolaryngology: Head and Neck Surgery*, 4th edn. Philadelphia: Mosby, pp. 2542–2556.

Yokoyama, T., Uemura, K., and Ryu, H. (1991). Facial nerve monitoring by monopolar low constant current stimulation during acoustic neuroma surgery. *Surgical Neurology*, **36**, 12–18.

12 Peripheral nerve monitoring: basics and indications

Jaime R. López

Introduction

Neurophysiologic intraoperative monitoring (IOM) of surgical procedures involving the peripheral nerves has become more common over the past four decades, owing to the recognition that routinely used clinical neurophysiologic techniques in the outpatient setting can also be employed in the operating room. Neurophysiologic techniques, such as direct nerve stimulation and recording of compound nerve action potentials (CNAPs), somatosensory evoked potentials (SEPs), and electromyography (EMG) allow not only the functional assessment of peripheral nerves but also evaluation of the peripheral-to-central neural pathways.

Depending on the technique used and the method utilized, IOM of the peripheral nervous system can provide valuable information, which may alter the course of the surgical procedure. In specific cases, such as in the surgical treatment of peripheral nerve injury in continuity, the monitoring of intraoperative CNAPs can be critical in ultimately determining the proper treatment course to take. In other instances, continuous recording of EMG can provide real-time data, which may alter surgical management. Moreover, these techniques are commonly applied using a multimodality approach, which allows for a more comprehensive and dynamic functional assessment of the peripheral nervous system region in question. When correctly used, IOM can achieve the following (Crum and Strommen, 2007; Holland, 2002; Kliot and Slimp, 1994):

1. Identify abnormal nerves or specific areas of abnormality within a nerve.
2. Define the severity of nerve injury.
3. Localize neural tissue or specific peripheral nerves.
4. Provide warning of possible injury to a nerve.
5. Aid in determining the most appropriate surgical treatment, such as decompression, neurolysis, or neuroma resection.
6. Assist in determining the probability of root avulsion.
7. Identify appropriate regions for nerve biopsy.
8. Provide prognostic information.

To achieve the above, the technologists, physicians, surgeons, and anesthesiologists all need to function as a team. It is very important that there be clear preoperative communication between the surgical and IOM teams in regard to the patient's diagnosis, neurologic status, and preexisting neurologic deficits, and the planned surgical procedure. This will help determine the neural territory that requires physiologic assessment or that is at risk for injury. In addition, the results of preoperative electrodiagnostic studies should also be made available to the IOM team prior to surgery. This information will allow the IOM team to properly devise an appropriate monitoring or testing strategy. Furthermore, after reviewing the preoperative data the IOM team, if appropriate, may suggest further preoperative electrodiagnostic testing prior to surgery.

Communication is equally important during surgery. Although the anesthesiologists are well aware of the planned surgical procedure, they may not be clear on some of the limitations imposed by IOM, such as the use of neuromuscular blocking agents. Thus, it is imperative to discuss the planned IOM protocol and its possible implication for the anesthetic regimen prior to the induction of anesthesia. This level of communication between the IOM service, anesthesiologists, and surgeons must continue during the surgical procedure. Any changes in anesthetic agents or significant changes in their levels need to be reported to the IOM team. Nonetheless, the IOM staff should be trained to review the anesthetic levels and the patient's blood pressure independently at regular intervals and these should be recorded in the IOM flow sheet. Any significant changes in the IOM parameters have to be communicated to the surgeons and anesthesiologists immediately.

The goal of this chapter is to present the physiologic basis of peripheral nerve monitoring, to review the different IOM techniques useful in assessing the peripheral nervous system, to discuss the different surgical procedures where IOM may be useful, and, lastly, to provide case examples.

History

Intraoperative monitoring in peripheral nerve surgery has been in use for approximately 40 years. Kline and DeJonge (1968) were the first to report on the use of electrically evoked nerve action potentials (CNAPs) to evaluate peripheral nerve injuries. This technique has subsequently been studied and further described, and its use has been reported in the surgical management of peripheral nerve injuries (Kline *et al.*, 1969; Kline and Happel, 1993; Tiel *et al.*, 1996).

Even though we perceive of IOM of the peripheral nervous system as a relatively recent development, in actuality the first published description of IOM was by Dr. Fedor Krause, a neurosurgeon, who during a cochlear nerve section on 14 July 1898 noted that electrical stimulation, or "unipolar faradic irritation of the remaining nerve-trunk with the weakest possible current ... resulted in contractions of the right facial region, especially the orbicularis oculi ..." (Krause, 1912). Essentially, what Dr. Krause was performing was an assessment of the functional integrity of the facial nerve by producing electrically evoked myogenic responses during surgery.

Subsequently, other surgeons used similar techniques for nerve stimulation while either directly observing the facial muscles or having an assistant manually palpate the face during dissection or electrical stimulation (Yingling and Ashram, 2005). These techniques are now routinely used in a variety of surgical procedures. However, instead of visually inspecting for muscle movement, this can be accomplished by recording the electrically triggered electromyographic potential from specific muscles. It is also important to note that direct nerve stimulation and recording of CNAPs, as a form of evaluating the functional state of peripheral nerves, was not employed prior to the late 1960s. Direct nerve stimulation, as well as many other IOM techniques, is now in common use and can provide valuable intraoperative neurophysiologic assessment of different peripheral nerves and neural pathways.

Surgical applications of peripheral nerve IOM

Although IOM can be applied in a wide variety of surgical procedures involving peripheral nerves, its most common application is probably in the setting where injury to a specific nerve, as a direct consequence of surgery, is of high risk. In these situations, nerve identification using electrical stimulation to identify nervous tissue within the operative field and recording continuous EMG activity would be useful. A representative example would be tumor resection surgery at or near peripheral nerve structures, such as a peripheral nerve or nerve root. In such a situation, electrical stimulation on and in regions adjacent to the tumor with recording of an electrically evoked Compound Muscle Action Potential (CMAP) may provide useful information about adjacent motor nerves. In addition, monitoring continuous EMG activity would provide information about manipulation, irritation, and injury of peripheral nerves (Figure 12.1). These techniques will be discussed in more detail later in this chapter.

A different scenario would be one where surgical management depends on the intraoperative functional assessment of a peripheral nerve or group of nerves. In these cases, the specific surgical intervention is directly dependent on the IOM results. For example, in the surgical treatment of traumatic nerve injury, intraoperative neurophysiologic testing of nerve function can guide the surgeon and help determine the proper surgical course. In many cases it is the vital component in the intraoperative surgical decision-making process. These cases require the intraoperative testing of nerve function, which can be achieved by assessing the CNAP of the injured nerve.

Compound nerve action potentials

The following section will provide a detailed review of the use of CNAPs in surgery. Although CNAPs play a very important and prominent role in the surgical management of traumatic nerve injury, their clinical application is essentially limited to these types of cases. Thus, it is useful to first provide a review of traumatic nerve injury, followed by a description of the CNAP technique.

Traumatic nerve injury

Management of traumatic nerve injury requires knowledge of nerve anatomy and physiology, gross anatomy of the involved region, pathology, pathophysiology, and clinical neurophysiology. In addition, an understanding of the neuropathology of peripheral nerve injury and its surgical treatment is critical. Thus, prior to describing the IOM of traumatic nerve injury

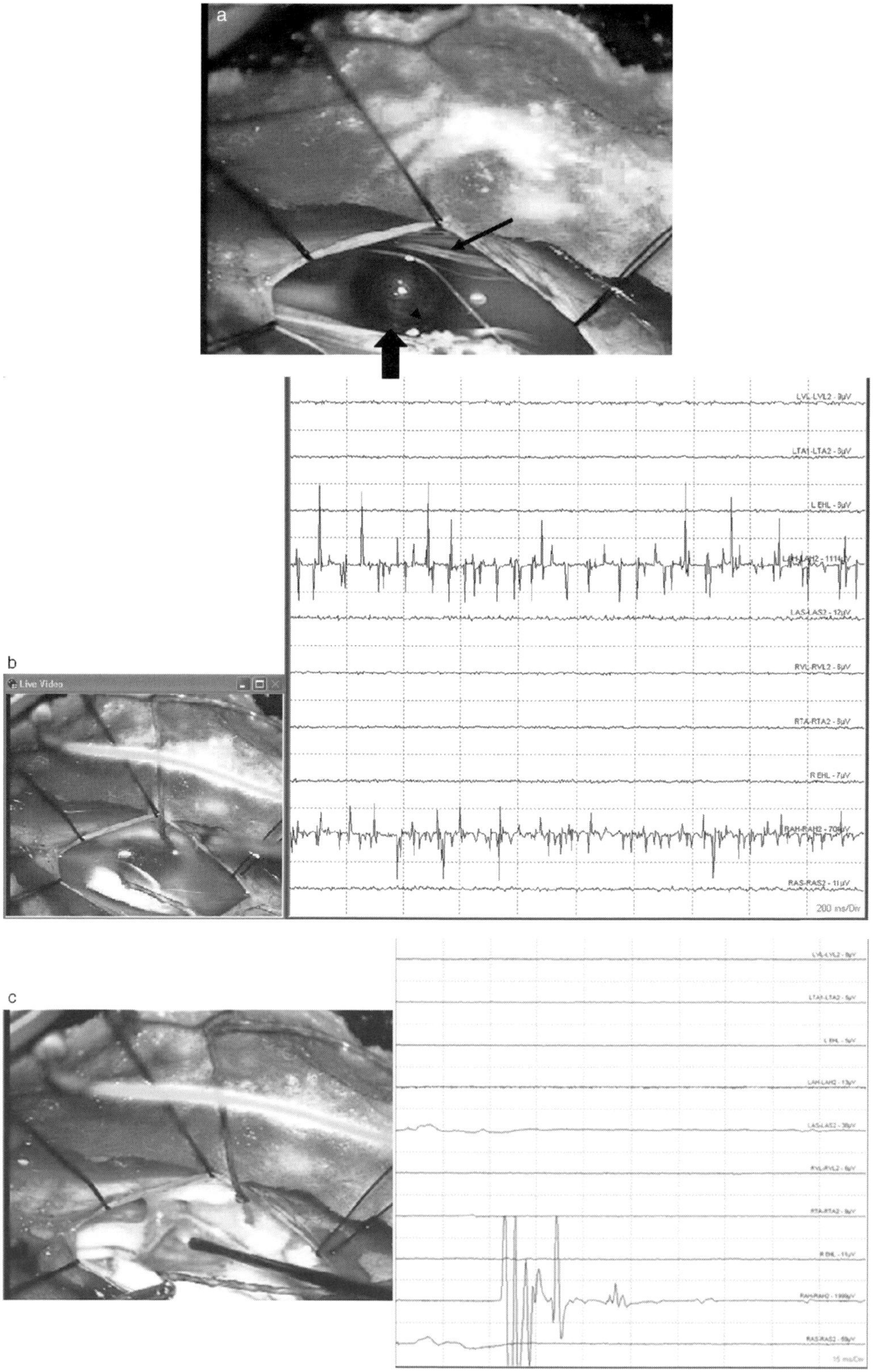

Figure 12.1 (a) A 64-year-old male with L5 intradural spinal tumor (large arrow). Note adjacent nerve roots (small arrow). (b) Manipulation-induced EMG activity from bilateral abductor hallucis (AH) muscles. (c) Monopolar stimulation of nerve roots showing an electrically triggered CMAP from the right AH after removal of tumor.

it is essential to review peripheral nerve anatomical structures and the neuropathology of peripheral nerve injury briefly.

The following is a simplified review of peripheral nerve anatomy. The anatomy of peripheral nerves reveals a compartmentalized structural organization. Individual myelinated axons and groups of unmyelinated axons are surrounded by endoneurium. These, in turn, form fascicles, which are collections of axons – endoneurium surrounded by perineurium with epineurium lying between the fascicles. Finally, a collection of fascicles forms a peripheral nerve trunk (Sunderland, 1990; Campbell, 2008). Practical knowledge of peripheral nerve anatomy is important in trying to understand and classify peripheral nerve injury. A firm grasp of these concepts allows for better preparation, planning, and execution of IOM of peripheral nerve surgery.

When trying to classify peripheral nerve injury, one must first recall and understand the two most commonly used clinical classification systems, those of Sedden and Sunderland. These classification schemes are based on the nature of the lesion and not the specific cause. In 1943, Sedden popularized Professor Henry Cohen's classification system of nerve injury, introduced in 1941, and described as neurapraxia, axonotmesis, and neurotmesis (Seddon, 1943). Later, Sunderland expanded the classification system by describing five different degrees of nerve injury (Maggi *et al.*, 2003). In Sedden's classification, neurapraxia defines the nerve as being intact but not able to transmit impulses (*apraxia*, non-action); axonotmesis indicates that the axon is damaged or destroyed (*tmesis*, a "cutting", implying a separation of related parts) while the majority of the connective tissue framework remains intact; and neurotmesis describes loss of continuity of the nerve trunk (Seddon, 1943; Campbell, 2008).

Although the Sunderland classification scheme is more complex, some consider it to be more clinically useful (Campbell, 2008). Sunderland's classification system, consisting of degrees of nerve injury, is based on the histologic structure of the nerve trunk. It defines five types of increasing nerve injury severity, as follows (Sunderland, 1990) (Figure 12.2):

1. First-degree injury: temporary conduction block but with preserved axonal continuity.
2. Second-degree injury: there is preserved continuity of the endoneurial sheath with Wallerian degeneration below the lesion. Axonal regeneration occurs within the original endoneurial tube, eventually restoring the original pattern of innervation and function is fully restored.
3. Third-degree injury: the fascicles are preserved but continuity of nerve fibers and the endoneurial sheaths is lost. Axonal regeneration may be misdirected and recovery is compromised.
4. Fourth-degree injury: there is preservation of the epineurium but discontinuity of axons, endoneurium, perineurium, and the fascicles. The traumatized nerve trunk is replaced by a barrier of fibrous tissue (neuroma), preventing axon growth. This is the type of injury that requires surgical intervention.
5. Fifth-degree injury: indicates loss of continuity of the entire nerve trunk (nerve transection).

Once a peripheral nerve (PN) injury has occurred, a plan has to be tailored for each type of nerve injury. This is where the need for and an understanding of a useful classification system becomes important. Sunderland's classification scheme lends itself well as a template for planning the most appropriate treatment, which in many instances involves surgical intervention. Cases of first-degree injury should recover without intervention because all peripheral nerve structures and elements are preserved. However, segmental demyelination is likely to produce conduction block with resultant weakness and sensory loss. These clinical neurologic deficits will remain until regeneration is restored. Recovery can take from days up to three months (Spinner and Kline, 2000; Maggi *et al.*, 2003; Campbell, 2008). In second-degree injury, regeneration should be excellent and recovery is expected to be complete because the endoneurial sheath is preserved. However, neurologic deficits will be present and may take months to resolve. Surgical treatment is usually indicated in cases of third- and fourth-degree injury, since regeneration tends to be poor. This occurs because of discontinuity of multiple nerve trunk structures, and the development of a neuroma in the case of fourth-degree injury (Sunderland, 1990; Spinner and Kline, 2000; Maggi *et al.*, 2003; Campbell, 2008).

Surgical exploration and repair of peripheral nerve trauma depends on the type of damage and is done at varying times after the injury. For example, sharp nerve transections (fifth-degree injury) which produce

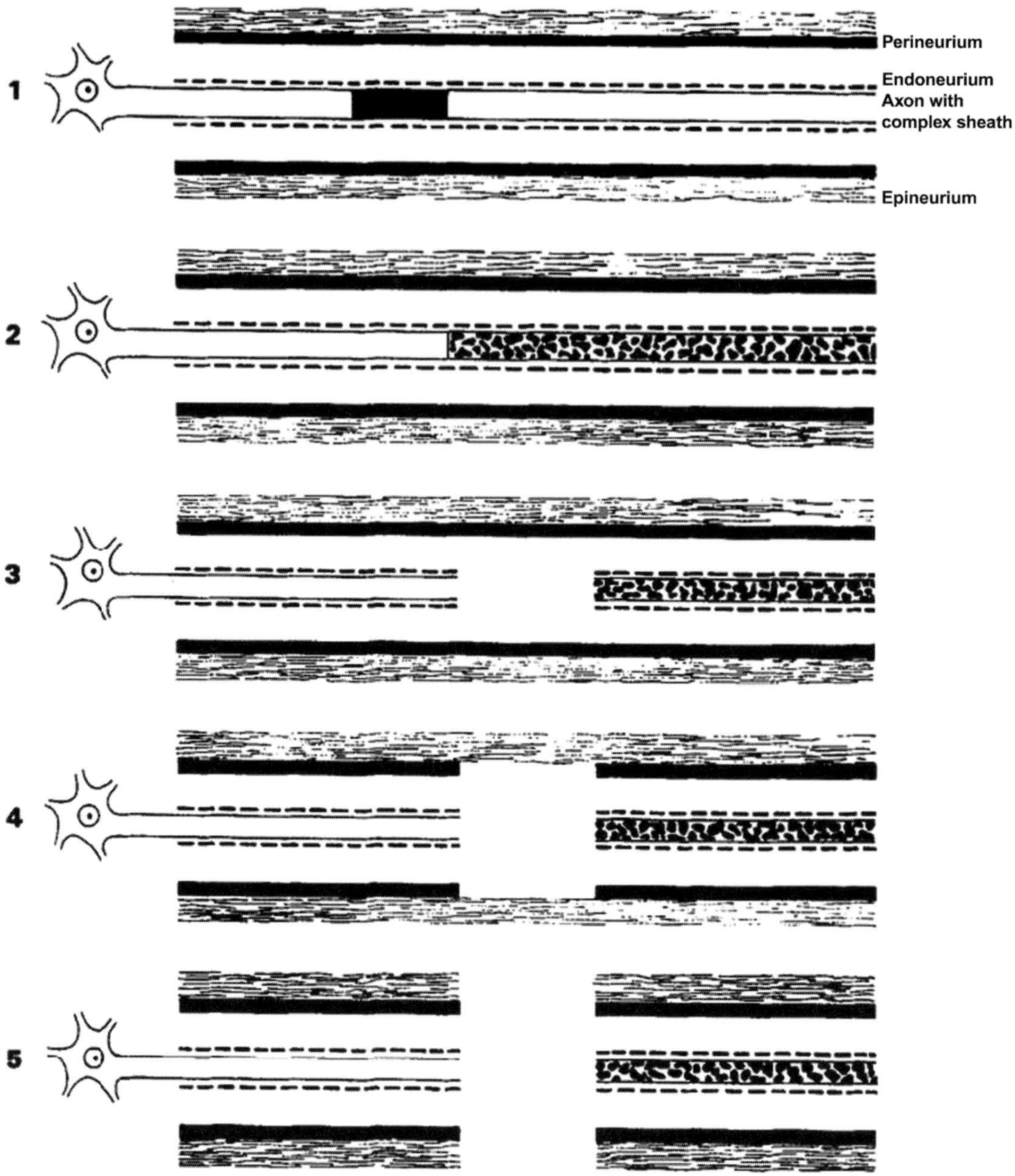

Figure 12.2 Sunderland's peripheral nerve injury; schematic representation of the five degrees of nerve injury (from Sunderland, 1990).

little trauma to the nerve stumps require early surgical treatment within 72 hours (Spinner and Kline, 2000; Campbell, 2008). However, the most common surgical approach is delayed intervention, which is done when nerve continuity is unclear or when natural recovery may provide a better outcome than surgical repair (Campbell, 2008). Another reason for delayed surgery is that blunt trauma, the major cause of peripheral nerve injuries, tends to result in lesions-in-continuity (Spinner and Kline, 2000). However, since it is unknown which specific case of peripheral nerve injury will demonstrate reinnervation, it is reasonable to wait a period of time to see if there is clinical or electrophysiologic evidence of regeneration. Spinner and Kline (2000) recommend that patients with closed nerve injuries be evaluated for clinical and electrophysiologic signs of regeneration over a period of two to five months. Patients with focal lesions-in-continuity, which can be seen in gunshot wounds, contusions, or fractures, should be surgically explored at two to three months if no signs of regeneration are seen. Patients with lengthier nerve lesions, as can be seen in stretch injuries, should be surgically explored at three to five months.

These types of cases where signs of regeneration are absent, and there appears to be a lesion-in-continuity, benefit the most from intraoperative electrophysiologic testing using CNAPs. The following are general recommendations based on intraoperative CNAP results:

1. If stimulation across the lesion reveals a reproducible CNAP, this indicates the presence of 4000 to 5000 functioning axons traversing the lesion. In this case, the recommended treatment is external neurolysis (Campbell, 2008; Spinner and Kline, 2000).
2. If a CNAP is not recorded across the lesion, then this is probably a result of one of the following (Spinner and Kline, 2000; Crum and Strommen, 2007; Campbell, 2008):
   (a) A lack of functioning axons that cross the lesion and no axonal regeneration.
   (b) That there is still the potential for axonal regeneration but that it is being compromised by internal nerve scarring. The recommended treatment in these situations is resection and nerve grafting or a transfer procedure.
   (c) The presence of a conduction block or neurapraxia.

Traumatic nerve injury IOM techniques

The specific electrophysiologic technique to use will depend on the suspected pathology, anatomic location, specific peripheral nerve or nerves involved, and the planned surgical intervention. Frequently, surgical cases require that a variety of physiologic data be monitored or obtained from one or more nerves. Therefore, multimodality recording capability is particularly important, especially when attempting to obtain different types of neurophysiologic information.

Surgical exposure of peripheral nerve structures permits direct stimulation and recording from peripheral nerves, groups of fascicles, or individual fascicles (Williams and Terzis, 1976; Spinner and Kline, 2000). Motor, sensory, and mixed nerves can be targeted for such testing. If both the stimulation and recording is done from peripheral nerves, this will yield CNAPs. When doing this, one can assess several important aspects of nerve function, such as the presence or absence of a CNAP, conduction velocity, and an estimate of the functional axons, which is reflected in the CNAP amplitude. However, recordings can also be made from a muscle innervated by the nerve being stimulated, yielding a CMAP. In many instances spinal or cortical evoked potential (EP) recordings, after peripheral nerve stimulation, can also be made from appropriately placed scalp and neck surface recording electrodes. In addition, peripheral nerve recordings can also be made in response to transcranial electrical stimulation. These are likely to be mixed-nerve evoked potentials that represent an element of continuity between the central and peripheral nervous systems.

CNAP stimulation technique

Intraoperative CNAPs can easily be monitored within the operative field. Supramaximal stimulation of motor and sensory nerves provides assessment of conduction by recording the maximum CNAP. The surgeon simultaneously places the stimulating and recording electrodes directly on the nerve; either a bipolar or a monopolar electrode may be used. However, a bipolar electrode is preferred as the cathode and anode are placed directly on the nerve; which allows the current to be better focused and minimizes unwanted current spread. A tripolar stimulator, which consists of a cathode situated between two anodes, can also be used to focus and limit current spread, and to help reduce the shock artifact (Tiel *et al.*, 1996; Crum and Strommen, 2007).

Monopolar stimulation can also be performed but this carries the risk of current spread, which, in turn, may activate a larger than intended nerve segment or entirely different neural structures. The technique is relatively straightforward and is done by placing the cathode directly on the nerve and the anode a short distance away but not in contact with the nerve being stimulated. Because the anode is at a distant location, stimulation lacks precise focus and excessive current spread with activation of adjacent neural structures can occur (Crum and Strommen, 2007). Therefore, monopolar stimulation is only recommended if bipolar stimulation is not possible.

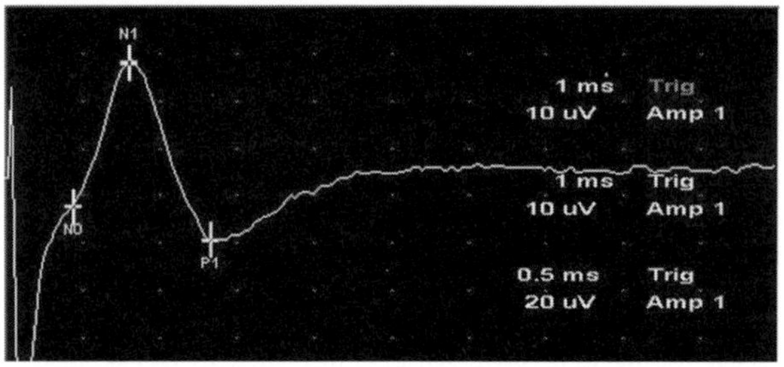

Figure 12.3 Example of peroneal nerve tripolar stimulation CNAP recorded at a 3 cm interelectrode distance between stimulating and recording electrodes. Note partial shock artifact contamination of CNAP.

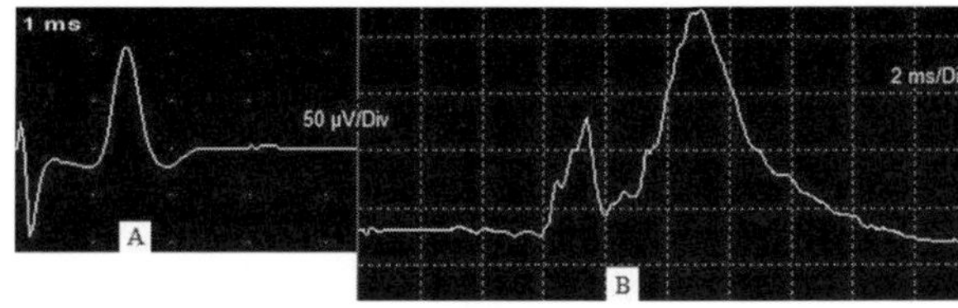

Figure 12.4 CNAPs recorded using stimulating and recording interelectrode distance of 5 (A) and 12 cm (B). Shock artifact contamination is not seen at these longer distances.

Williams and Terzis (1976) recommend that the stimulator interelectrode distance should be no less than 3 mm for smaller neural structures, such as fascicles. For larger elements, such as the brachial plexus and sciatic nerve, the distance can range from 5–7 mm (Tiel *et al.*, 1996). Stimulation of the nerve will be insufficient if the interelectrode distances are too short, which can lead to erroneously determining that the CNAP is absent.

Sufficient exposure of the nerve is also important, since CNAPs recorded from distances between the stimulating and recording electrodes less than 4 cm are likely to be obscured or contaminated by shock artifact (Tiel *et al.*, 1996) (Figures 12.3 and 12.4). To minimize current spread, the nerve should also be elevated out of the surgical field during stimulation, to avoid contact with blood and fluid within the surgical site.

Direct nerve stimulation parameters are different from those used for conventional diagnostic NCS. Since the nerve is being stimulated directly, the stimulus intensity required to depolarize a normal nerve will be less than that typically used percutaneously. Current spread is also increased with higher stimulation intensity levels and, therefore, the stimulation intensity should be kept as low as possible while still maintaining the ability to obtain CNAPs. There are other factors affecting the CNAP besides nerve injury. Hypothermia of the exposed nerve is commonly encountered in the operating room, and this should be taken into consideration when calculating conduction velocity. The temperature of the exposed nerves may reach levels as low as 20–30 °C, reducing the conduction velocity by as much as 1.5–2.0 ms per degree centigrade (Brown and Veitch, 1994). It is also common to use a tourniquet to achieve a bloodless surgical field. Unfortunately, this also affects nerve function and can produce complete conduction block in motor and sensory fibers within a period of 15–20 minutes (Yates *et al.*, 1981; Brown and Veitch, 1994). In a study by Yates *et al.* (1981), they found that ischemia distal to the applied tourniquet was the likely cause of the conduction block, while a combination of ischemia and compression contributed to more severe and faster onset of conduction abnormalities. In cases where a pneumatic tourniquet is used, it must be deflated at least 20–30 minutes before attempting to record CNAPs (Kliot and Slimp, 1994; Spinner and Kline, 2000). Although the actual tourniquet technique is widely used in upper and lower limb surgery, safe working guidelines for its use have not been clearly defined (Kam *et al.*, 2001).

Anesthetic agents typically do not interfere with recording of CNAPs, unless local or regional blocks are used that may affect the peripheral neural structures of interest. Neuromuscular blocking agents will also not interfere with CNAP recordings. However, their use is not recommended, as these agents will interfere with recordings of CMAPs. In addition, it is also important to be able to observe muscle contractions in response to nerve stimulation. This provides important information regarding the proper working state of the stimulation and recording equipment.

The following are typical parameters used for intraoperative nerve stimulation:

- Biphasic square wave pulses, with duration between 0.05 and 0.1 ms,
- Stimulus intensity of 0.2–5.0 mA or 3–15 V,
- Single stimulation or frequency rates of 1 Hz when recording CMAP and 3–10 Hz when using signal averaging.

These parameters are usually sufficient to produce supramaximal depolarization of axonal membranes of most peripheral nerves. However, higher stimulation intensities are typically needed in diseased, damaged, or scarred nerves. In these situations, stimulation levels needed to activate the nerve may be as high as 100 V but usually remain below 50 V (Tiel *et al.*, 1996). Crum and

Strommen (2007) report that they have increased the stimulus intensity as high as 25 mA if no response was initially obtained. It is also thought that this type of short-duration stimulation preferentially depolarizes motor axons. Veale *et al.* (1973) studied ulnar nerve function in seven normal human adults and found, using percutaneous stimulation, that motor fibers were more readily excitable using short-pulse stimulation of 200 μs or less, while sensory fibers were preferentially excitable using long pulses of 1 ms or more. Signal averaging is also useful when attempting to record particularly small CNAPs. In many instances, averaging two to five responses will yield acceptable results.

CNAP recording technique

Recording of CNAPs is done directly from the nerve, using either bipolar or monopolar techniques. However, the bipolar technique is preferred and most commonly used. When using the bipolar technique, the active recording electrode should be oriented toward the stimulating cathode. In the technically ideal setting, the active and reference recording electrodes require a separation by a distance equal to the wavelength of the fastest conducting fibers. Thus, they would have to be separated by a minimum of 30–60 mm to equal or exceed the most rapidly conducting nerve fibers, since the wavelength equals the spike duration (measured in ms) multiplied by the conduction velocity (measured in millimeters per millisecond) (Brown and Veitch, 1994). Since this degree of electrode separation is not practical in the surgical setting, Brown and Veitch (1994) recommend a separation of 15–20 mm as a reasonable compromise between the active and reference recording electrodes (Figure 12.5).

Although bipolar recordings are more commonly performed, monopolar recording is also possible. During monopolar recordings, the reference electrode should be placed far enough away to minimize recording of volume-conducted activity emanating from the nerve. Monopolar recordings have the advantage that they cause less distortion of the signal than bipolar recordings. Unfortunately, the impedance differences between the cathode and anode electrodes may be high enough and the distances between them great enough to degrade the common mode rejection of the amplifier significantly, leading to electrical and magnetic interference (Brown and Veitch, 1994). In general, bipolar recording of CNAPs is preferred over the monopolar technique.

Instrumentation and recording bandpass is similar to that used for conventional sensory nerve action potentials. However, it is important to understand that any recorded bioelectric signal can be altered by filtering. Fourier transformation of the CNAP indicates that most of the information describing the CNAP waveform is in the 1000 Hz range. Knowing this is essential in the proper use of filtering and filters can be set to eliminate frequencies above and below this range (Tiel *et al.*, 1996):

- High-frequency filter (low-pass) set at 2000 Hz or higher,
- Low-frequency filter (high-pass) set at 5–10 Hz or lower.

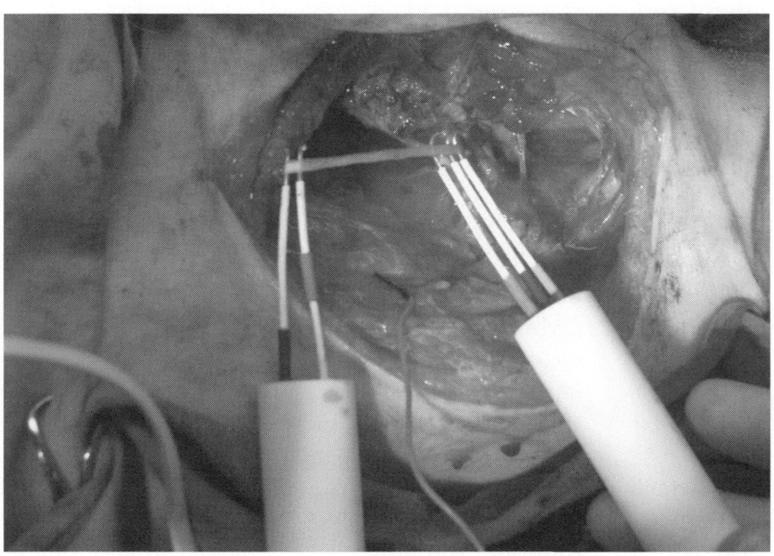

Figure 12.5 Example of CNAP bipolar recording and tripolar stimulation techniques performed in an animal model. The ground (subdermal needle) electrode is placed in the incision site between the recording and stimulating electrodes.

The CNAP amplitude also depends on the distance between the cathode and anode (usually 0.5–1.0 cm) and the density of the axons. For example, the CNAP from a normal nerve may be 1–2 mV, while that from regenerating fibers can be 10 μV or less. It is important to note that the presence of a recordable non-averaged CNAP indicates the presence of a substantial number of functioning axons in that nerve segment. Kline *et al.* (1969) studied experimental peripheral nerve injury in 12 macaque monkeys and their findings revealed that 4000–5000 moderate-diameter myelinated functional axons are needed to be able to record a non-averaged CNAP. They also found that EMG evidence of reinnervation was absent when the CNAP could not be recorded. In addition, they determined that CNAPs could be recorded weeks before EMG evidence of reinnervation could be identified.

Even though CNAP recordings are obtained directly from the nerve, the recorded amplitudes are rarely above 200 μV (Brown and Veitch, 1994; Crum and Strommen, 2007). Therefore, a reasonable gain setting for recording CNAPs is 20 to 50 μV per division. The onset latency of the CNAP will depend on the conduction velocity and on the distance between the stimulating and recording electrodes. Given the nature of these surgeries, the distances between the stimulating and recording electrodes tend to be short. Accordingly, the time base should be set with these things taken into consideration. For short distances, the time base should be set at 0.5 ms per division but for longer distances this time base should be increased. When deciding what time base to use, a simple guide is 1 ms per 5 cm distance, using a conduction velocity of 50 m/s (Crum and Strommen, 2007). Using an incorrect time base can alter the appearance of the CNAP and make it difficult to identify the onset latency accurately (Tiel *et al.*, 1996).

Knowing the distance between the stimulating and recording electrodes also allows for assessment of conduction velocity. Thus, based on the intraoperative findings of amplitude and conduction velocity a possible pathophysiologic mechanism, such as an axonal or demyelinative process, may be determined.

Clinical utility of CNAP testing in surgery

Any nerve that can be surgically exposed is a potential candidate for CNAP testing. Many clinical studies have shown the utility of intraoperative CNAP testing in guiding surgeons, enabling them to determine whether to perform neurolysis or resection in many different examples of traumatic nerve injury. For example, Kline *et al.* (1998) reported a 24-year retrospective clinical study of sciatic nerve injuries in 380 patients, where CNAP testing was used to guide intraoperative management. This technique has also been successfully used to guide surgical treatment of many different peripheral nerve injuries. Kim *et al.* (2003a) studied the surgical outcomes of 1019 patients with brachial plexus lesions. Data exist for other specific nerves, including ulnar nerves (Kim *et al.*, 2003b), posterior interosseous (Kim *et al.*, 2006a), anterior interosseous (Kim *et al.*, 2006b), amongst many others.

In summary, CNAPs can play a pivotal role in the surgical management of traumatic peripheral nerve injury because they provide crucial information about axonal function across the area of injury. This functional information can only be obtained by electrically activating the nerve during surgery. Table 12.1 provides a summary of the different degrees of nerve injury and their relationship with CNAPs and surgical treatment.

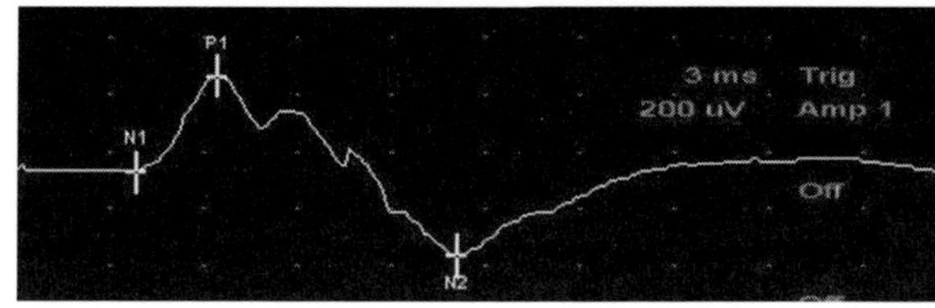

Figure 12.6 Bipolar CMAP recording from the deltoid after axillary nerve stimulation.

Compound muscle action potential

In addition to yielding CNAPs, direct nerve stimulation can also generate CMAPs, which can be recorded from muscles within or distant from the surgical site (Figure 12.6). Typically, the CMAP amplitude is larger than the CNAP amplitude since it reflects the summation of muscle fiber action potentials. In general, CMAPs are easier to record than CNAPs. These can be recorded using wire, subdermal, intramuscular, or surface electrodes. The larger recording area of surface electrodes provides a more accurate estimate of the number of functioning motor axons but the electrodes can be more easily dislodged and their impedances tend to increase as the conductive material used to interface with the skin dries. Subdermal recordings can be

Table 12.1 Summary of patterns of nerve injury and their relationship to intraoperative CNAP results and surgical treatment

Nerve injury classification						
Seddon	**Sutherland**	**Type of nerve injury**	**Spontaneous recovery pattern**	**Spontaneous rate of recovery**	**Intraoperative CNAPs**	**Surgical treatment**
Neurapraxia	First degree	Segmental demyelination	Complete	Fast, days-to-weeks	Present	None
Axonotmesis	Second degree	Axon transection with preserved endoneurium	Good	Slow (1 inch/month)	Present	Usually none
Axonotmesis	Third degree; neuroma in continuity	Axon transection, endoneurial discontinuity, perineurium and fascicles preserved	Variable, some with good recovery	Slow (1 inch/month)	Present	Neurolysis
Axonotmesis	Fourth degree; neuroma in continuity	Discontinuity of axons, endoneurium, perineurium and fascicles. Epineurium still intact	Poor to none	None or very little	CNAPs not obtainable	Nerve repair or graft
Neurotmesis	Fifth degree	Transection of entire nerve trunk	None	None	CNAPs not obtainable	Nerve repair or graft

Adapted from Campbell (2008) and Spinner and Kline (2000).

performed using conventional subdermal EEG needle electrodes, which are placed close to or into the muscle. Fine wire electrodes, a type of intramuscular electrode, are placed in the muscle using a hollow needle.

Subdermal and intramuscular electrodes record from smaller surface areas and reflect recordings obtained from smaller regions. In particular, intramuscular electrodes record from a small portion of the muscle and the responses obtained may represent activity of a fraction of the axons stimulated. An advantage of subdermal and intramuscular electrode recordings is that the reduced recording area also reduces the amount of extraneous noise introduced into the recording. Some institutions, including ours, prefer the use of subdermal EEG electrodes because they provide relatively quiet recordings, offer the ability to quantify the number of functioning axons, and are easy and convenient to place (Crum and Strommen, 2007).

Stimulation techniques are similar to those used for obtaining CNAPs and recording parameters are similar to those used for conventional diagnostic motor NCS.

Technical difficulties

There are a multitude of problems that may confound intraoperative NCS. The first step is to stimulate and record from a segment of a normal nerve. This will test the set-up and should be done prior to testing the injured nerve. Some useful strategies that can be used to reduce stimulus artifact and other problems are:

1. Increasing the stimulating–recording interelectrode distance,
2. Using the lowest stimulus intensity and duration necessary for a supramaximal response,
3. Lifting the nerve out from surrounding tissue and fluid to reduce current spread,
4. Using bipolar instead of monopolar stimulation,
5. Placing a ground between the stimulating and recording electrodes,
6. Orienting the active–reference recording electrode axis perpendicular to the cathode–anode axis to bring the signal into the range of the common mode rejection,
7. Avoiding the use of cold irrigation, since this may induce conduction block,
8. Recording CNAPs only after circulation has been restored a minimum of 20–30 minutes if a tourniquet has been used to produce a bloodless field, as tourniquet application can induce ischemia and obliterate the CNAPs.

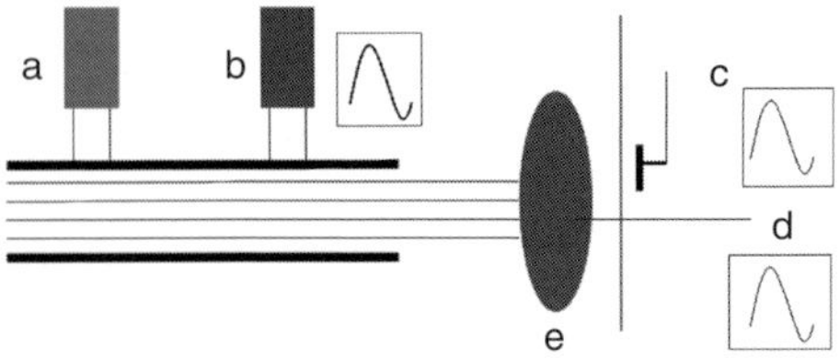

Figure 12.7 Schematic representation of direct nerve stimulation of a normal nerve with recordings obtained from several different structures: (a) bipolar stimulator; (b) bipolar CNAP recording electrode; (c) surface CMAP recording electrode; (d) intramuscular CMAP recording electrode; (e) muscle.

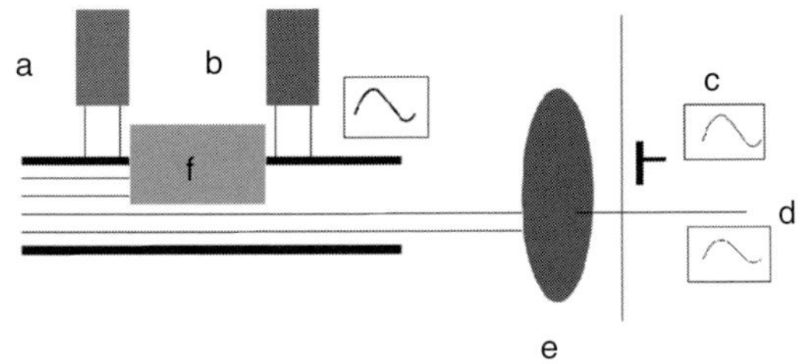

Figure 12.8 Schematic representation of direct nerve stimulation of a partial nerve injury with recordings obtained from different structures: (a) bipolar stimulator; (b) bipolar CNAP recording electrode – reduced amplitude; (c) surface CMAP recording electrode – reduced amplitude; (d) intramuscular CMAP recording electrode – reduced amplitude; (e) muscle; (f) nerve lesion.

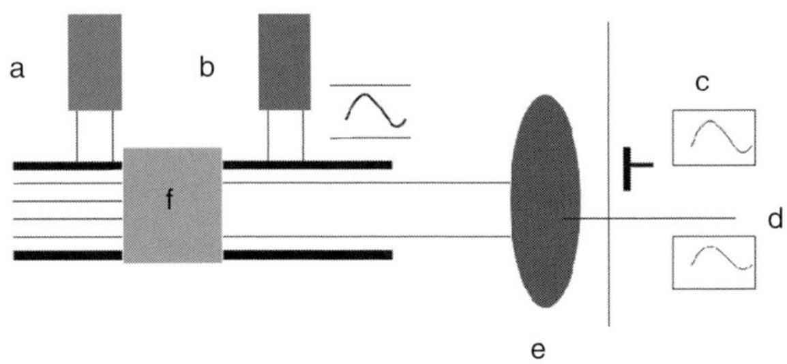

Figure 12.9 Schematic representation of direct nerve stimulation of a complete nerve injury with axonal regeneration: (a) bipolar stimulator; (b) bipolar CNAP recording electrode – reduced amplitude; (c) surface CMAP recording electrode – reduced amplitude; (d) intramuscular CMAP recording electrode – reduced amplitude; (e) muscle; (f) nerve lesion.

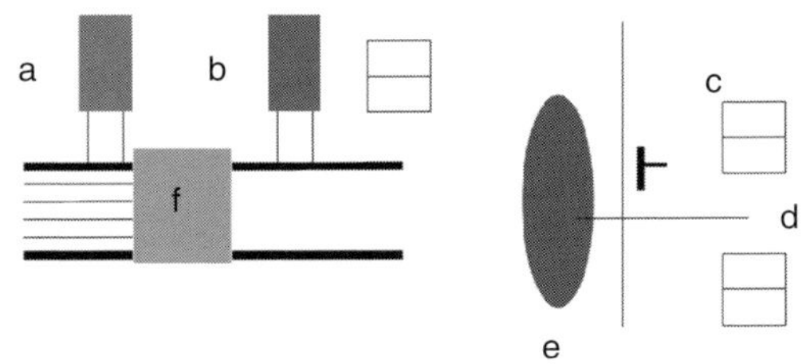

Figure 12.10 Schematic representation of direct nerve stimulation of a complete nerve injury without axonal regeneration: (a) bipolar stimulator; (b) bipolar CNAP recording electrode – no response; (c) surface CMAP recording electrode – no response; (d) intramuscular CMAP recording electrode – no response; (e) muscle; (f) nerve lesion.

Figures 12.7–12.10 are schematic representations of probable electrophysiologic results in different types of nerve injury.

Electromyography (EMG)

The rationale for the use of EMG as an IOM technique is based on the principle that EMG records muscle fiber generated electrical potentials similar to the motor unit potentials (MUPs) recorded during conventional diagnostic EMG. This can serve as a sensitive indicator of the functional state of motor axons. Thus, monitoring of continuous free-running EMG from muscles innervated by motor cranial nerves, motor nerve roots, or peripheral nerves at risk of injury during surgery can provide useful information regarding the functional status of the nerves.

Electromyographic activity is not expected to be seen in muscles at rest innervated by normal, uninjured nerves. However, a variety of different types of EMG activity can be encountered during IOM. These include MUPs, fibrillation potentials, which can remain present during anesthesia, and neurotonic discharges (Harper, 1998, see Figure 12.11).

Neurotonic discharges, defined by Daube and Harper (1989) as "distinctive discharges of a motor unit potential in rapid irregular bursts" are the most important to distinguish and identify, since they are thought to be generated when the axon membrane is depolarized, which can occur with mechanical, metabolic, or thermal stimulation, as well as ischemia, traumatic nerve injury, or saline irrigation (Møller, 1994; Harper, 1998; Crum and Strommen, 2007). Neurotonic discharges can be distinguished from other types of MUPs by the frequency and pattern of discharge, and by the relationship of their activity to mechanical or metabolic stimulation of the neural structure at risk (Table 12.2). In addition, neurotonic discharges can have different discharge frequencies and may have the appearance and sound of high-frequency discharges, similar to end-plate spikes, myotonia, and neuromyotonia (Harper, 1998).

Although neurotonic discharges are seen with traumatic nerve injury, severing or sharp transection

Table 12.2 Characteristics of EMG activity observed during intraoperative recordings

EMG activity	Frequency	Pattern
Fibrillation potentials	1–5 Hz	Regular and continuous
MUPs	10–15 Hz	Continuous and semi-regular
Neurotonic discharges	50–200 Hz	Short bursts or long trains

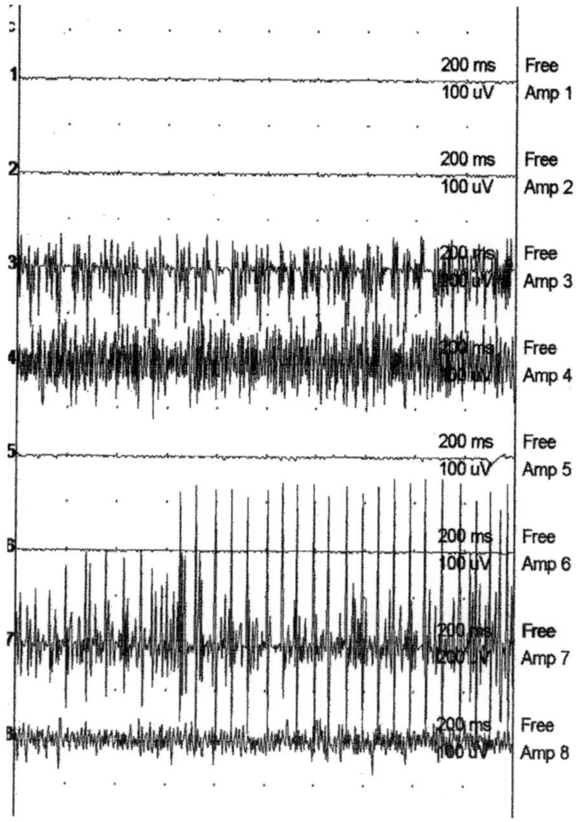

Figure 12.11 Example of neurotonic discharges.

of a nerve may not produce any EMG activity or neurotonic discharges (Møller, 1994; Crum and Strommen, 2007). It is also important to emphasize that the presence of neurotonic discharges does not always indicate axonal injury nor their absence indicate lack of damage.

For example, Gunnarsson *et al.* (2004) reported the correlation of EMG activity and postoperative neurologic deficits in a prospective series of 213 patients undergoing thoracolumbar spine surgery. They found significant electromyographic activity in 77.5% of the patients while only 14 patients (6.6%) had new postoperative neurologic symptoms. They determined that intraoperative EMG had a sensitivity of 100% and a specificity of 23.7% for the detection of a new postoperative neurologic deficit. However, it is unknown how many cases of postoperative radiculopathy were prevented as a result of the alteration of the surgical management due to EMG changes.

The recording of EMG can be performed using surface, subcutaneous, or intramuscular electrodes. Intramuscular electrodes have the advantage of improved signal-to-noise ratio and the capability for "near-field" recordings since the recording electrodes are placed in or as near as possible to the muscle producing the MUPs. In contrast, surface electrodes are generally considered less desirable because their placement on the skin is distant from where the MUPs are generated, essentially producing "far-field" recordings. Consequently, the MUP activity can be missed. In addition, in many cases when using surface electrodes, it may be difficult to identify the specific muscle generating the EMG activity. Subcutaneous electrodes have similar limitations to surface recording electrodes.

The two main types of intramuscular electrodes currently used for EMG recordings are conventional subdermal EEG needle and wire electrodes. The specific type and length of the needle electrodes used depend on the location and depth of the target muscle. In some instances, such as in deep localized muscles, subdermal EEG electrodes may not be suitable for recording EMG activity, since the length of the needle may be too short to reach the muscle. However, in the majority of cases these electrodes are adequate for most muscles, especially for relatively superficial muscles, such as those on the head, neck, hands, and feet. Intramuscular wire electrodes, also known as "hookwire" electrodes, have the advantage that they are thinner than needle electrodes and are easier to place into smaller muscles. The hookwire–needle unit consists of two wire electrodes, which have 2–3 mm uninsulated ends and are hooked into the bevel of conventional hypodermic needles. These, in turn, are used to introduce the wire into the muscle. The needles are then removed, leaving the wires imbedded in the muscle. For deeply situated muscles, longer hypodermic needles can be used as introducers (Chapter 11, Figure 11.3).

The recording parameters used are the same for all of the electrodes described and are similar to those typically used for conventional outpatient needle EMG examination:

- The gain is set to 50–200 μV,
- A low-frequency filter between 20 and 30 Hz, and a high frequency filter at 10 kHz are used,
- The sweep speed is usually set at 200 ms/division for recording continuous EMG and between 2 and 5 ms/division for recording electrically evoked EMG potentials. Of course, this latter parameter will depend on the distance between the locations of the site of nerve stimulation and the innervated muscle.

Electromyographic recordings can be made from several muscles simultaneously with the activity observed on a monitor. It is equally important to monitor the audio feedback, which can aid in distinguishing between EMG activity and electrode or electrical artifact. The audio feedback is also important for the surgeon, since this serves to indicate which surgical events are associated with the generation of EMG activity, indicating nerve stimulation.

Electrically triggered EMG

Recording a CMAP after electrical stimulation of a peripheral nerve is also possible while simultaneously recording continuous EMG activity. This can be accomplished using the same techniques as described in the CNAP section.

Technical difficulties of EMG

It is important to keep in mind that technical problems will inevitably be encountered in the operating room. However, a comprehensive discussion of the different types of electrical artifact and their likely causes is beyond the scope of this review. Nonetheless, the reader is reminded that many of the electrical devices in the operating room will generate 60 Hz interference and this can interfere and contaminate the EMG recordings. Common instruments and devices that can cause electrical interference include beds, equipment with electric motors, cautery devices, surgical microscopes, and surgical instruments. In addition, bad electrodes and faulty IOM equipment and cables can also produce artifacts. Care should be taken to differentiate EMG activity from artifacts prior to alerting the surgeon to the presence of EMG activity.

Anesthesia and EMG

Anesthetic agents are not usually a problem for IOM of the peripheral nerves. However, neuromuscular blocking agents may suppress neurotonic discharges and should be avoided or, at the very least, their use should be minimized. Nonetheless, in certain circumstances, muscle relaxation is imperative and using neuromuscular blocking agents is unavoidable. In these situations EMG monitoring can still be performed using partial neuromuscular blockade. Successful NIOM of EMG activity has been reported using a constant infusion of a short-acting NMB agent titrated to a level of 50% reduction of the baseline CMAPs recorded over the hypothenar muscle (Lennon *et al.*, 1992; Kizilay *et al.*, 2003). In addition, EMG activity can still be adequately monitored with 75% neuromuscular blockade, defined as a CMAP amplitude decrement of less than 100% over the four successive supramaximal repetitive nerve stimulations (Holland, 1998). This latter technique can easily be performed using train-of-four stimulations at 2 Hz. Neuromuscular blockade greater than 75% may lower the sensitivity of EMG monitoring, especially during attempts to identify nerve root injury (Holland, 2002).

Somatosensory evoked potentials (SEPs)

Somatosensory evoked potentials produced after direct nerve stimulation can also be obtained and can provide information regarding the peripheral-to-central nervous system continuity of sensory pathways. The primary application of SEPs, in the setting of peripheral nerve injury, is in assessing cervical nerve root avulsion. In this scenario, the nerve root, brachial plexus structure, or peripheral nerve is directly stimulated and the SEPs are recorded, using standard intraoperative SEP recording techniques, over the scalp or skin overlying the cervical spinal cord. The presence of a well-defined SEP indicates central nervous system continuity of the dorsal roots. In addition, this finding is also indicative of high likelihood of ventral root continuity (Hashimoto *et al.*, 1991; Oberle *et al.*, 2002; Harper, 2005). Unfortunately, the SEP cannot distinguish between

Table 12.3 Direct nerve SEP parameters

	Stimulation	Recording
Type	1. Bipolar (preferred) 2. Monopolar if bipolar is not possible	1. Contralateral scalp C3′/C4′–Fz, C3′–C4′ 2. High cervical spine
Electrodes	1. Hook 2. Bipolar or monopolar probes	1. Surface EEG 2. Subdermal 3. Corkscrew
Parameters	1. Intensity: 1–20 mA 2. Stimulation duration: 0.05–0.1 mS 3. Rate: Single or 1.9–4.1 Hz	1. Average: 50–250-0.1 2. Filters: 30–2000 Hz

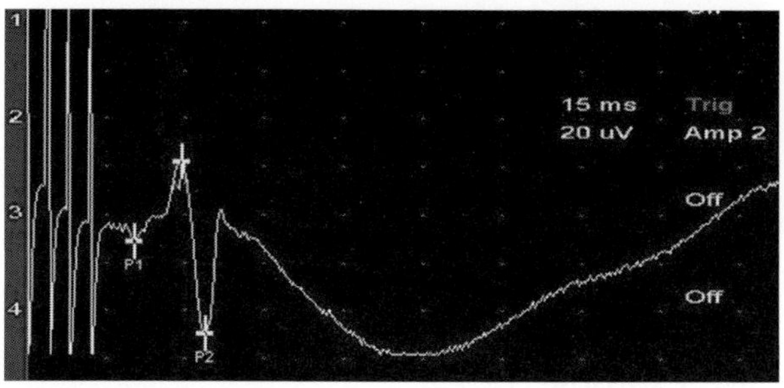

Figure 12.12 Example of a neurogenic TcMEP bipolar recording from the left nerve at the distal forearm, indicating continuity of central-to-peripheral nervous system motor pathways.

incomplete nerve root avulsion and intact roots (Hashimoto *et al.*, 1991).

Stimulation may be proximal to, distal from, or directly at the surgical site, using standard techniques. Lower stimulation intensities should be used, since the stimulation is being applied directly to the nerve (Table 12.3).

Distal peripheral nerves, most commonly median, ulnar, and tibial nerves, can also be tested for peripheral-to-central continuity. According to the surgical stimulation site, these central recordings may reflect activity that propagates through the surgical field, representing the functional state of afferent fibers. It is important to bear in mind that these recordings represent the peripheral-to-central continuity of the somatosensory–medial lemniscal pathways and do not provide any direct information regarding motor pathways. However, information about central-to-peripheral continuity of motor pathways can be obtained using TcMEPs.

Neurogenic TcMEPs

Like SEPs, the use of TcMEPs in peripheral nerve surgery is primarily in identifying nerve root avulsion, specifically an avulsion that involves the ventral root. This can be done by recording a neurogenic TcMEP in the form of a CNAP from spinal nerves or plexus structures following transcranial electrical stimulation. This is important because motor recovery is the main objective of brachial plexus surgery, and partial nerve root avulsion involving only the ventral rootlet can occur. Neurogenic TcMEPs are useful in evaluating the functional continuity of motor pathways within the spinal cord, and distal from the peripheral nerves (Figure 12.12).

In a prospective study assessing the value of computerized tomography myelography and MRI in diagnosing cervical root avulsions, Carvalho *et al.* (1997) surgically explored 135 cervical roots in 40 patients who were undergoing treatment for traumatic brachial plexus lesions. They found partial root avulsions in 11% (15/135) of the nerve roots studied. Combinations of intact ventral rootlets and avulsed dorsal rootlets or vice versa were seen. Of these, the majority, 11/15 (73%), involved ventral rootlet avulsion while the dorsal rootlet remained intact. Thus, assessing central-to-peripheral continuity of motor pathways is important in brachial plexus surgery. Although CNAP and SEP studies are useful, they are limited in that neither method provides specific functional information on the integrity of motor nerve pathways from the central corticospinal tracts, and anterior horn cell to the peripheral nerves.

Turkof *et al.* (1997) reported on the use of neurogenic TcMEPs in evaluating the status of anterior spinal roots and spinal nerves during brachial plexus surgery. They were able to report the technique and show how surgical management was altered. They concluded that neurogenic TcMEPs were an effective method of assessing the functional state of anterior motor roots and of motor fibers in exposed spinal nerves. More recently, in a case study, Burkholder *et al.* (2003) were the first to report the combined use of neurogenic TcMEPs with CNAPs and SEPs to assess the functional status of central and peripheral motor and sensory pathways successfully, for accurate localization of lesion sites.

Table 12.4 Neurogenic TcMEP parameters

	Stimulation	Recording
Site	Scalp C3–C4 Cz–Fz	Brachial plexus elements Nerve roots Spinal nerve
Electrodes	Surface EEG Subdermal EEG Corkscrew	Bipolar hook Bipolar wires
Parameters	Intensity: variable, usual range 200–750 V Stimulation duration: 0.05 ms Train: 4 Inter-stimulus interval: 2.0 ms	Same as CNAPs

Neurogenic TcMEP stimulation and recording techniques

Stimulation is done using similar techniques to those used for conventional intraoperative muscle-recorded transcranial electrical motor-evoked potentials. Stimulation is performed using scalp electrodes and recordings are obtained directly from the nerve root, brachial plexus structures or a more distal peripheral nerve. The parameters are summarized in Table 12.4.

Conclusion

Intraoperative neurophysiologic monitoring can be applied to a wide variety of peripheral nerve surgical procedures and can be used in many different ways. Since its inception, the technique of recording CNAPs has proven to be an invaluable adjunct in the surgical treatment of traumatic peripheral nerve injury, and in many instances the results of intraoperative nerve testing dictates the surgical treatment intervention. Somatosensory and neurogenic TcMEPs allow the determination of central-to-peripheral continuity of motor and sensory pathways, thus establishing the presence or absence of complete or partial nerve root avulsion. In addition, monitoring the EMG activity helps in assessing possible peripheral nerve injury during surgery and this information assists the surgeon in altering surgical management to avoid nerve damage. Furthermore, the use of these techniques in a multimodality fashion provides a greater degree of information regarding different functional aspects of the peripheral nervous system which helps to better define the type and location of the neural structures at risk, assess the severity of an underlying injury and, ultimately, prevent injury.

Case examples for Chapter 12

Case example 12.1

A 34-year-old male had a history of a gunshot wound to his left leg resulting in a left foot drop without clinical improvement over a period of greater than three months. Outpatient EMG showed absent peroneal nerve CMAPs and denervation of peroneal nerve innervated muscles, consistent with an axonal peroneal neuropathy. Intraoperative nerve exploration and direct nerve stimulation were planned.

Intraoperative nerve stimulation revealed complete nerve block between the sciatic and peroneal nerves (Figure 12.13a).

Recordings of CNAPs were also obtained between sciatic nerve segments (Figure 12.13b). This demonstrated that the equipment and electrodes were all functioning properly and that the lack of CNAP illustrated in Figure 12.13a was not due to technical factors.

Case example 12.2

A 26-year-old male with right brachial plexus injury after a motorcycle accident showed resultant right arm weakness, predominantly involving the deltoid. It was planned to explore the right brachial plexus surgically and perform intraoperative nerve conduction studies of the different brachial plexus elements.

Figure 12.14 shows recordings of CNAPs after supramaximal electrical stimulation and stimulation and recording sites with calculated conduction velocities (CVs) for each segment tested. Note the reduced CNAP amplitude and subnormal CV of the posterior cord–axillary nerve segment. These findings are consistent with partial conduction block between the posterior cord and the axillary nerve. Based on these findings, the patient was treated with neurolysis.

Case example 12.3

A 30-year-old male presented with a history of left proximal shoulder weakness, consistent with traumatic brachial plexus injury, following a motorcycle accident. The surgical plan was to perform intraoperative electrophysiologic studies and determine peripheral-to-central nervous system continuity.

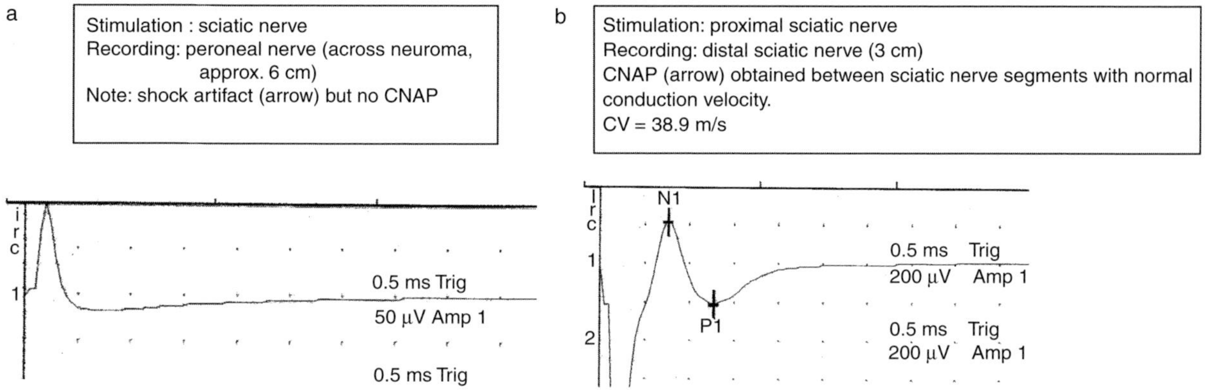

Figure 12.13 (a) Intraoperative nerve stimulation revealed complete nerve block between the sciatic and peroneal nerve; (b) CNAPs obtained between sciatic nerve segments.

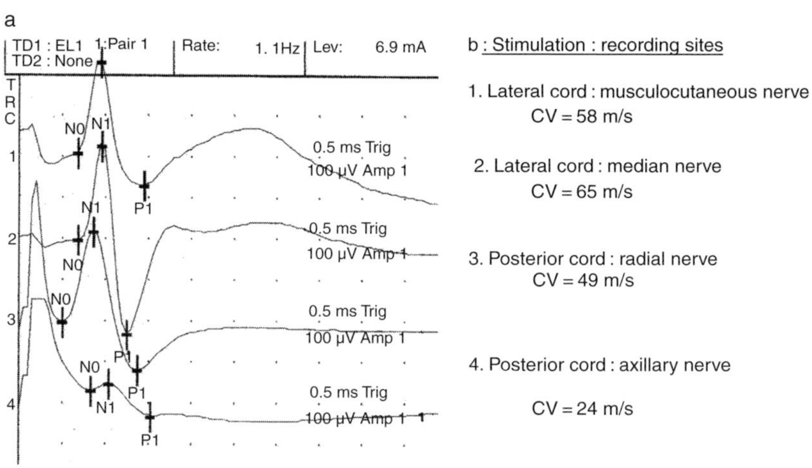

Figure 12.14 (a) Recordings of CNAPs after supramaximal electrical stimulation; (b) Stimulation and recording sites with calculated conduction velocities (CVs) for, each segment tested.

Direct suprascapular nerve stimulation with recordings from scalp and cervical spine (Figure 12.15a) showed peripheral-to-central continuity of the posterior column–medial lemniscal somatosensory pathways from the suprascapular nerve, while direct suprascapular nerve stimulation with recordings from the supraspinatus muscle (Figure 12.15b) demonstrated nerve to muscle continuity.

Case example 12.4

A 17-year-old high-school student was injured while playing football two months prior to surgery. He sustained a complete tear of his left knee anterior cruciate ligament and a left peroneal nerve injury. Outpatient EMG failed to show conduction along the peroneal nerve. The patient was taken to the operating room for peroneal nerve exploration.

Initial dissection identified the common peroneal nerve. However, direct nerve electrical stimulation failed to elicit a CNAP, suggesting that the nerve segment tested was non-functional. The surgeon was advised to extend the dissection and search for a nerve segment that could be electrically activated. Several nerve segments were tested at different levels. No responses were obtained and the dissection was extended to the level of the tibial nerve (Figure 12.16a).

Direct tibial nerve electrical stimulation using a tripolar stimulator and a bipolar hook electrode yielded a well-formed CNAP with a conduction velocity of 53 m/s (Figure 12.16b). This assured the IOM team and the surgeon that the IOM equipment was functioning properly.

Finally, direct electrical stimulation of the proximal peroneal and distal sciatic nerve stump revealed a well-

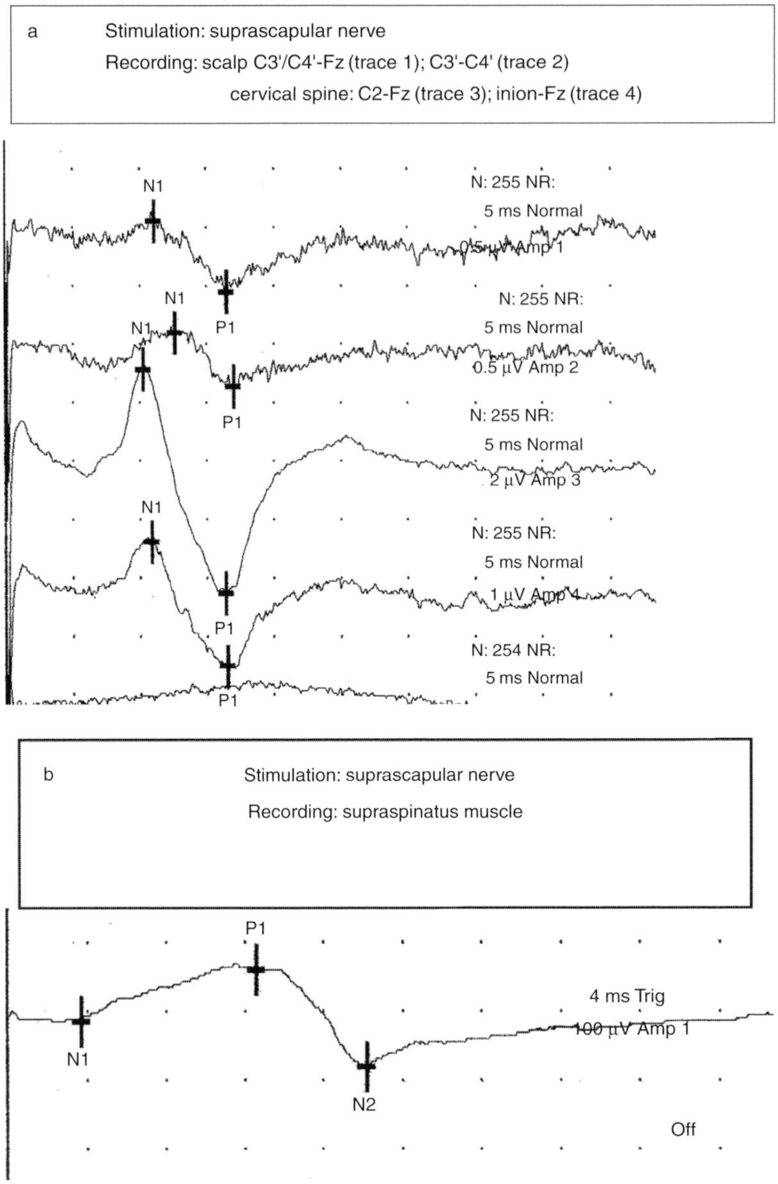

Fgure 12.15 Direct suprascapular nerve stimulation with recordings from scalp and cervical spine (a) and the supraspinatus muscle (b).

defined CNAP with a moderate degree of diminished conduction velocity (33 m/s) in comparison with that of the tibial nerve (Figure 12.16c). This site, which contains functional nerve axons, as demonstrated by the presence of the CNAP, was suitable for receiving the graft.

Case example 12.5

Figure 12.17 shows an example of multimodality IOM of continuous EMG activity and posterior tibial cortical SEPs in a 34-year-old male with a history of Tarlov cyst undergoing surgical resection. High frequency, grouped neurotonic discharges can be seen in trace 3 and slower frequency and less complex EMG activity in trace 6, corresponding to the left and right anal sphincters, respectively. In general, the presence of neurotonic discharges raises the possibility of nerve injury, more so than the type of MUPs seen in trace 6.

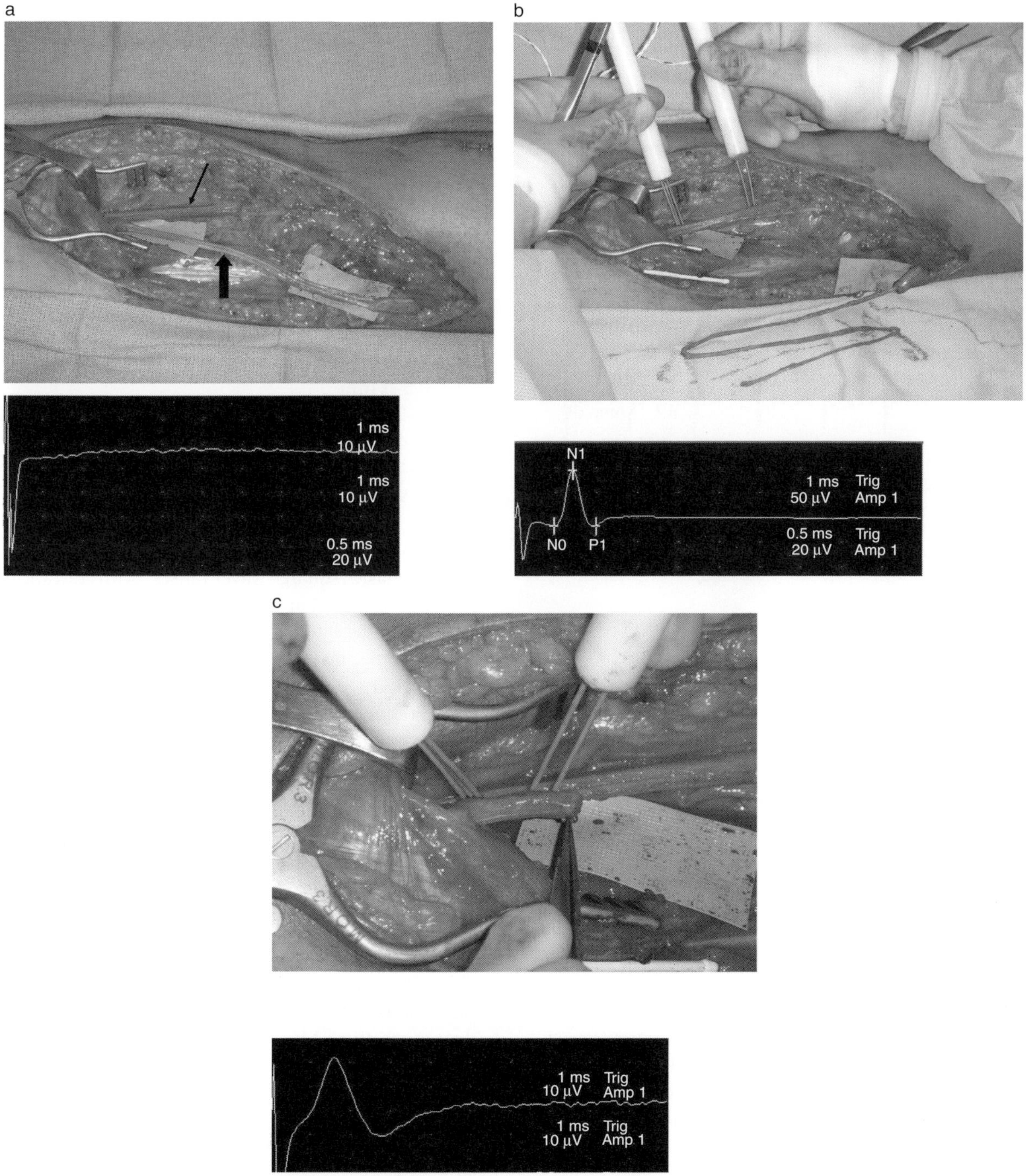

Figure 12.16 (a) A functioning peroneal nerve segment could not be found. Note the locations of the abnormal-appearing peroneal (large arrow) and normal tibial nerves (small arrow). (b) Direct tibial nerve electrical stimulation yielded a well-formed CNAP and a conduction velocity of 53 m/s. The sural nerve graft at the distal stump of the peroneal nerve is also visible. (c) Direct electrical stimulation of the proximal peroneal and distal sciatic nerve stump revealed a well-defined CNAP.

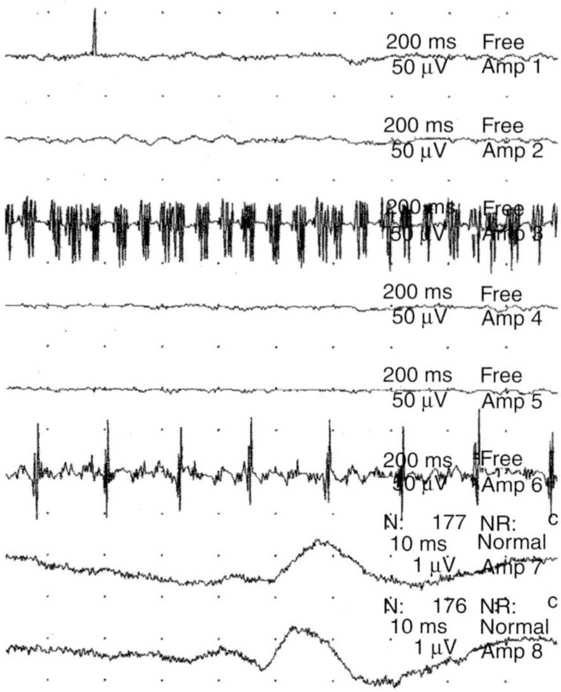

Traces 1–3: all left sided muscles: anterior tibialis; medial gastrocnemius; anal sphincter

Traces 4–6: all right sided muscles: anterior tibialis; medial gastrocnemius; anal sphincter

Trace 7: cortical SSEP after left median nerve stimulation

Trace 8: cortical SSEP after right median nerve stimulation

Figure 12.17 An example of multimodality IOM of continuous EMG activity and posterior tibial cortical SEPs: see text.

Questions

1. When using the Sunderland peripheral nerve injury classification, which degree of injury is consistent with preserved continuity of the endoneurial sheath with Wallerian degeneration occurring below the lesion and axonal regeneration occurring within the original endoneurial tube, eventually restoring the original pattern of innervation?
   (a) First-degree injury
   (b) Second-degree injury
   (c) Third-degree injury
   (d) Fourth-degree injury
   (e) Fifth-degree injury
2. When performing a CNAP, stimulation using a bipolar electrode is best described as:
   (a) Placement of the cathode directly on the nerve and the anode a short distance away but not in contact with the nerve
   (b) Placement of a cathode between two anodes directly on the nerve
   (c) Placement of the cathode and the anode directly on the nerve
   (d) Placement of the anode on the nerve and the cathode a short distance away but not in contact with the nerve
   (e) None of the above
3. When recording CNAPs the stimulator interelectrode distance for larger neural structures such as brachial plexus and sciatic nerve should be:
   (a) 0.5–1 mm
   (b) 1–2 mm
   (c) 3–4 mm
   (d) 5–7 mm
   (e) 8–9 mm
4. Which of the following is a typical parameter for intraoperative nerve stimulation?
   (a) Biphasic square wave pulses with duration between 0.05 and 0.1 ms
   (b) Stimulus intensity of 150 V
   (c) Stimulation rates of 15 Hz when recording CMAPs
   (d) Frequency rates of 1 Hz when using signal averaging
   (e) Stimulus intensity of 50.0 mA
5. According to primate studies, the presence of a recordable non-averaged CNAP indicates that there are how many functional axons in that nerve segment?
   (a) 500–1000
   (b) 1000–2000
   (c) 2000–3000
   (d) 3000–4000
   (e) 4000–5000
6. Intraoperative EMG activity can still be adequately monitored with what degree of pharmacologic neuromuscular blockade?
   (a) 25%
   (b) 45%
   (c) 55%
   (d) 65%
   (e) 75%
7. Intraoperative direct nerve stimulation will generate which of the following recordable responses from the innervated muscle?
   (a) CMAPs

(b) CNAPs
(c) SEPs
(d) Neurogenic TcMEPs
(e) All of the above

8. Neurotonic discharges have a typical frequency discharge rate of which of the following?
(a) 10–25 Hz
(b) 25–100 Hz
(c) 50–200 Hz
(d) 200–300 Hz
(e) Greater than 300 Hz

9. Monitoring which of the following is considered the most useful and specific in determining ventral nerve root avulsion?
(a) SEPs
(b) Neurogenic TcEPs
(c) CNAPs
(d) CMAPs
(e) EMG

10. What is the maximum level of pharmacologic neuromuscular blockade where EMG activity can still be adequate monitored?
(a) 15%
(b) 25%
(c) 50%
(d) 75%
(e) 95%

References

Brown, W. F. and Veitch, J. (1994). AAEM Minimonograph #42: Intraoperative monitoring of peripheral and cranial nerves. *Muscle & Nerve*, **17**, 371–377.

Burkholder, L. M., Houlden, D. A., Midha, R., Weiss, E., and Vennettilli, R. (2003). Neurogenic motor-evoked potentials: role in brachial plexus surgery. Case report. *Journal of Neurosurgery*, **98**, 607–610.

Campbell, W. W. (2008). Evaluation and management of peripheral nerve injury. *Clinical Neurophysiology*, **119**, 1951–1965.

Carvalho, G. A., Nikkhah, G., Mattehies, C., Penkert, G., and Samil, M. (1997). Diagnosis of root avulsions in traumatic brachial plexus injuries: value of computerized tomography myelography and magnetic resonance imaging. *Journal of Neurosurgery*, **86**, 69–76.

Crum, B. A. and Strommen, J. A. (2007). Peripheral nerve stimulation and monitoring during operative procedures. *Muscle & Nerve*, **35**, 159–170.

Daube, J. R. and Harper, C. M. (1989). Surgical monitoring of cranial and peripheral nerves. In *Neuromonitoring in Surgery*, ed. J. E. Desmedt. Amsterdam: Elsevier, pp. 115–138.

Gunnarsson, T., Krassioukov, A. V., Sarjeant, R., and Fehlings, M. G. (2004). Real-time continuous intraoperative electromyographic and somatosensory evoked potential recordings in spinal surgery: correlation of clinical and electrophysiologic findings in a prospective, consecutive series of 213 cases. *Spine*, **29**, 677–684.

Harper, C. M. (1998). Cranial nerve monitoring. In *1998 American Association of Electrodiagnostic Medicine Annual Meeting, Course F: Update in Intraoperative Monitoring Syllabus*, pp. 23–31.

Harper, C. M. (2005). Preoperative and intraoperative electrophysiologic assessment of brachial plexus injuries. *Hand Clinics*, **21**, 39–46.

Hashimoto, T., Mitomo, M., Hirabuki, N., *et al.* (1991). Nerve root avulsion of birth palsy: comparison of myelography with CT myelography and somatosensory evoked potential. *Radiology*, **178**, 841–845.

Holland, N. R. (1998). Intraoperative electromyography during thoracolumbar spinal surgery. *Spine*, **23**, 1915–1922.

Holland, N. R. (2002). Intraoperative electromyography. *Journal of Clinical Neurophysiology*, **19**, 444–453.

Kam, P. C. A., Kavanaugh, R., and Yoong, F. F. Y. (2001). The arterial tourniquet: pathophysiological consequences and anaesthetic considerations. *Anaesthesia*, **56**, 534–545.

Kim, D. H., Cho, Y. I., Tiel, R. L., and Kline, D. G. (2003a). Outcomes of surgery in 1019 brachial plexus lesions treated at Louisiana State University Health Sciences Center. *Journal of Neurosurgery*, **98**, 1005–1016.

Kim, D. H., Han, K., Tiel, R. L., Murovic, J. A., and Kline, D. G. (2003b). Surgical outcomes of 654 ulnar nerve lesions. *Journal of Neurosurgery*, **98**, 993–1004.

Kim, D. H., Murovic, J. A., Kim, Y.-Y., and Kline, D. G. (2006a). Surgical treatment and outcomes of 45 patients with posterior interosseous nerve entrapments and injuries. *Journal of Neurosurgery*, **104**, 766–777.

Kim, D. H., Murovic, J. A., Kim, Y.-Y., and Kline, D. G. (2006b). Surgical treatment and outcomes of 15 patients with anterior interosseous nerve entrapments and injuries. *Journal of Neurosurgery*, **104**, 757–765.

Kizilay, A., Aladag, I., Cokkeser, Y., *et al.* (2003). Effects of partial neuromuscular blockade on facial nerve monitorization in otologic surgery. *Acta Oto-laryngologica*, **123**, 321–324.

Kline, D. G. and DeJonge, B. R. (1968). Evoked potentials to evaluate peripheral nerve injuries. *Surgery, Gynecology and Obstetrics*, **127**, 1239–1248.

Kline, D. G. and Happel, L. T. (1993). Penfeld lecture: a quarter century's experience with intraoperative nerve action potential recording. *The Canadian Journal of Neurological Sciences*, **20**, 3–10.

Kline, D. G., Hackett, E. R., and May, P. R (1969). Evaluation of nerve injuries by evoked potentials and electromyography. *Journal of Neurosurgery*, **31**, 128–136.

Kline, D. G., Kim, D. H., Midha, R., Harsh, C., and Tiel, R. L. (1998). Management and results of sciatic nerve injuries: a 24-year experience. *Journal of Neurosurgery*, **89**, 13–23.

Kliot, M. and Slimp, J. (1994). Techniques for assessment of peripheral nerve function at surgery. In *Intraoperative Monitoring Techniques in Neurosurgery*, ed. C. M. Loftus and V. C. Traynelis. New York: McGraw-Hill, pp. 275–285.

Krause, F. (1912). *Surgery of the Brain and Spinal Cord Based on Personal Experience*. Vol. II. New York: Rebman Company, pp. 740.

Lennon, R. L., Hosking, M. P., Daube, J. R., and Welna, J. O. (1992). Effect of partial neuromuscular blockade on intraoperative electromyography in patients undergoing resection of acoustic neuromas. *Anesthesia and Analgesia*, **75**, 729–733.

Maggi, S. P., Lowe, J. B., and Mackinnon, S. E. (2003). Pathophysiology of nerve injury. *Clinical and Plastic Surgery*, **30**, 109–126.

Møller, A. R. (1994). Monitoring techniques in cavernous sinus surgery. In *Intraoperative Monitoring Techniques in Neurosurgery*, ed. C. M. Loftus and V. C. Traynelis. New York: McGraw-Hill, pp. 141–155.

Oberle, J., Antoniadis, G., Kast, E., and Richter, H.-P. (2002). Evaluation of traumatic cervical nerve root injuries by intraoperative evoked potentials. *Neurosurgery*, **51**, 1182–1190.

Seddon, H. J. (1943). Three types of nerve injury. *Brain*, **66**, 237–288.

Spinner, R. J. and Kline, D. G. (2000). Surgery for peripheral nerve and brachial plexus injuries or other nerve lesions. *Muscle & Nerve*, **23**, 680–695.

Sunderland, S. (1990). The anatomy and physiology of nerve injury. *Muscle & Nerve*, **13**, 771–784.

Tiel, R. L., Happel, L. T., and Kline, D. G. (1996). Nerve action potential recording method and equipment. *Neurosurgery*, **39**, 103–109.

Turkof, E., Millesi, H., Turkof, R., Pfundner, P., and Mayr, N. (1997). Intraoperative electroneurodiagnostics (transcranial electrical motor-evoked potentials) to evaluate the functional status of anterior spinal roots and spinal nerves during brachial plexus surgery. *Plastic and Reconstructive Surgery*, **99**, 1632–1641.

Veale, J. L., Mark, R. F., and Rees, S. (1973). Differential sensitivity of motor and sensory fibres in human ulnar nerve. *Journal of Neurology, Neurosurgery and Psychiatry*, **36**, 75–86.

Williams, H. G. and Terzis, J. K. (1976). Single fascicular recordings: an intraoperative diagnostic tool for the management of peripheral nerve lesions. *Plastic and Reconstructive Surgery*, **57**, 562–569.

Yates, S. K., Hurst, L. N., and Brown, W. F. (1981). The pathogenesis of pneumatic tourniquet paralysis in man. *Journal of Neurology and Neurosurgery*, **44**, 759–767.

Yingling, C. D. and Ashram, Y. A. (2005). Intraoperative monitoring of cranial nerves in neurotologic surgery. In *Cummings: Otolaryngology: Head and Neck Surgery*, 4th edn. Philadelphia: Mosby, pp. 2542–2556.

13 Special considerations in pediatric surgical monitoring

Khaled M. Zamel

Over the past few decades, neurophysiologic intraoperative monitoring (NIOM), aided by a wide variety of rapidly advancing electrophysiologic techniques and better anesthesia, has become broadly available. Cumulative experience gained by neurophysiologists, anesthesiologists, and surgeons helped in demonstrating the usefulness of NIOM in preventing postoperative neurologic deficits.

In fact, the value of NIOM in preventing postoperative deficits from neural injury has been determined for several different types of surgery. For example, in the case of spinal surgeries, a large multicenter study conducted in 1995 established that NIOM using somatosensory evoked potentials (SEPs) reduced the risk of postoperative paraplegia by 60% (Nuwer *et al.*, 1995). Subsequently, more recent studies documented that recording both motor-evoked potentials (MEPs) and SEPs following transcranial electrical stimulation of the brain provided a more comprehensive assessment of spinal-cord function and hence further improved the ability to prevent neural injury (Nagle *et al.*, 1996; Pelosi *et al.*, 2002; Schwartz *et al.*, 2007; Sutter *et al.*, 2007c; Weinzierl *et al.*, 2007).

In addition to monitoring the spinal cord, the Therapeutics and Technology subcommittee of the American Academy of Neurology recommended monitoring in two other types of surgeries as useful and non-investigational. The first type of surgery involves the use of electroencephalography (EEG), compressed spectral array, and SEPs in monitoring carotid endarterectomy (CEA) and other brain surgeries that potentially compromise cerebral blood flow. The use of brainstem auditory evoked potentials (BAEPs) and cranial nerve monitoring in surgeries performed in the region of the brainstem or inner ear (Lopez, 2004) is also recommended.

Nevertheless, despite these recommendations and the widespread use of NIOM, there is, in general, a lack of universally accepted guidelines on the utility and interpretation of NIOM in many other conditions. Many investigators, however, strongly believe that sufficient evidence already exists to mandate the use of certain NIOM modalities.

Much of the published literature on NIOM has focused on the adult population. However, children are similarly at risk for neurological complications during various surgical interventions (Harper and Nelson, 1992). Neurophysiological testing modalities used in adults can also be applied to the pediatric population but to be successful, certain modifications of the parameters and methods typically used in adults need to be applied in children (Sala *et al.*, 2002).

In children, there are age-related differences in neurophysiologic parameters for the neurophysiologic testing modalities available. A large body of literature supports the idea that different neural structures in children function and respond to testing differently than in adults, Maturational evolution is complex, as it is not linear over time and has age-specific patterns that are seen at discrete developmental periods. One might argue that these differences are irrelevant for the purpose of NIOM as the patient acts as his or her own control and that in many NIOM techniques, the methodology used in adults can be extrapolated to children. However, it is important for the clinical neurophysiologist to understand these differences which impact not only the interpretation but may also require technical modifications critical to accomplishing a successful NIOM procedure.

In general, neurophysiologic studies in children carry unique challenges and are best accomplished by experienced teams. Extra time is frequently needed, especially for extraoperative studies. These may take longer for explanation of the procedure to the parents and patient as well as to administer sedation. To understand, interpret and react to NIOM data obtained in children appropriately, it is important to have sufficient knowledge of the different factors that could contribute to these unique maturational differences in the pediatric population.

Skeletal factors

Younger children have greater surface area to volume ratios, shorter extremities, and thicker subcutaneous fat than in older children. A greater surface area makes a child more susceptible to temperature changes. In the operation room, this can lead to significant changes, especially a reduction in peripheral conduction, as well as neuromuscular transmission, with cooling of the body and extremities.

There is also a greater chance of shock artifact, owing to spread of the electrical current due to the short distance between electrodes placed on a small head or a short extremity. The shorter distances make it more likely for small errors in measurement to have a larger effect on the results. This may be more critical in analyzing preoperative studies.

Finally, thicker subcutaneous fat, especially in newborns and young infants (baby fat) can obscure important bony landmarks and, in some cases, make it more difficult for the electrical stimulus to reach deep seated nerves. This is also important to consider when needle electromyography (EMG) is planned as part of the NIOM session due to the thick subcutaneous layer, longer EMG needles may be required to reach the desired muscle.

In children, continuous intraoperative recording of EMG activity in cranial nerve-innervated muscles can be very helpful in minimizing morbidity while monitoring surgeries in the posterior fossa (Grabb *et al.*, 1997). A clear understanding of cranial muscle anatomy is essential for selection of the appropriate muscle and, in children, several muscles are often monitored as a back-up. These include the extraocular muscles, medial rectus (oculomotor nerve), superior oblique (trochlear nerve) and the lateral rectus (abducens nerve). The masseter and temporalis muscles are used to monitor the trigeminal nerve; the frontalis, orbicularis oris and oculi for the facial nerve; the vocalis and cricothyroid muscles for the vagus nerve; the trapezius and sternomastoid for the spinal accessory nerve, and the genioglossus muscle for the hypoglossal nerve (Harper and Nelson, 1992).

It is important to mention that some precautions should be taken in consideration in using TES in young children (Sala *et al.*, 2002). The use of needle or screw-like electrodes is contraindicated in a child under the age of 18 months, if the anterior fontanel has not closed. The location of such electrodes should also be modified if the child has a ventriculoperitoneal shunt which can be damaged while inserting these electrodes.

Central nervous system maturational factors

Compared with adults, the motor and sensory pathways of the brain, brainstem, and spinal cord in children are immature, due to developing myelination. The effects of maturation on the sensory pathways have been studied extensively. Age-specific differences have been identified for sensory evoked responses. Differences are seen in latencies and amplitudes, as well as morphology of the recorded waveforms and are in general linear in relation to age. Brainstem auditory evoked potentials (BAEPs) have been recorded at all ages, starting even in fetal life (Staley *et al.*, 1990; Soundararajan and Cunliffe, 2007).

The myelination of the auditory nerve begins at the 24th week post-conception (Eggermont, 1985; 1995). This myelination progresses in a caudal to rostral direction that correlates with the latency and amplitude changes seen in BAEP responses. Eggermont described an exponential growth model with equal maturation rate for each station, with the maturation of evoked potentials from auditory brainstem to auditory cortex representing equal percentage changes in equal time periods (Eggermont, 1992). At full term, most of the peripheral structures are well myelinated while the myelination of the central components (the cochlear nucleus, superior olivary complex, and inferior colliculi) continues through the first few months of infancy (Rotteveel *et al.*, 1987a; 1987b). Subsequently, the latency of BAEP waveform I, which reflects conduction in the auditory nerve, reaches adult values within the first 2–8 weeks of life (Zimmerman *et al.*, 1987; Picton *et al.*, 1992; Levy, 1997) while waveform V, which reflects conduction through the brainstem and brain, does not reach adult values before the age of 18–36 months.

The poorly myelinated motor axons in young children usually lead to a higher voltage threshold. With stimulation of the corticospinal motor pathways by transcranial electric stimulation (TES) to trigger motor-evoked potentials (MEPs), the responses are more difficult to obtain (Lieberman *et al.*, 2006). Lieberman and his group found that younger age was associated with an increase in the threshold voltage needed to elicit a sufficient MEP response. Lieberman also reported that younger age was associated with

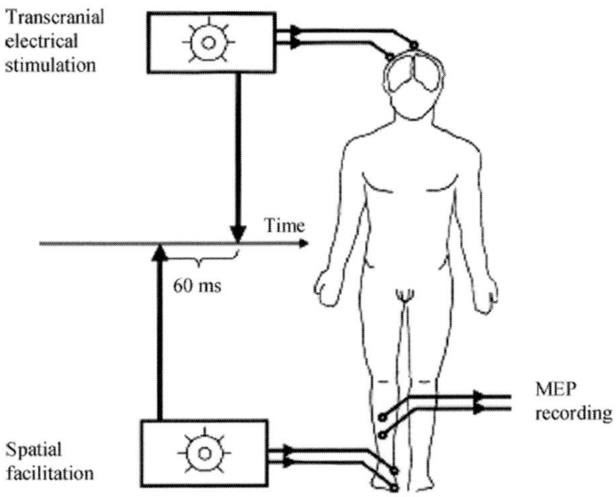

Figure 13.1 Schematic drawing of intraoperative methodology for eliciting and recording motor evoked potentials from the spinal cord and limb muscles. In this set-up, spatial facilitation is achieved by applying an electrical stimulus to the medial border of the plantar arch of one side with needle electrodes 60 ms before the transcranial electrical stimulus while recording MEPs from both tibialis muscles (train-of-ten pulses at 500 Hz; pulse duration 0.5 ms; strength 20–60 mA (Frei *et al.*, 2007, with permission from Lippincott, Wolters Kluwer Health).

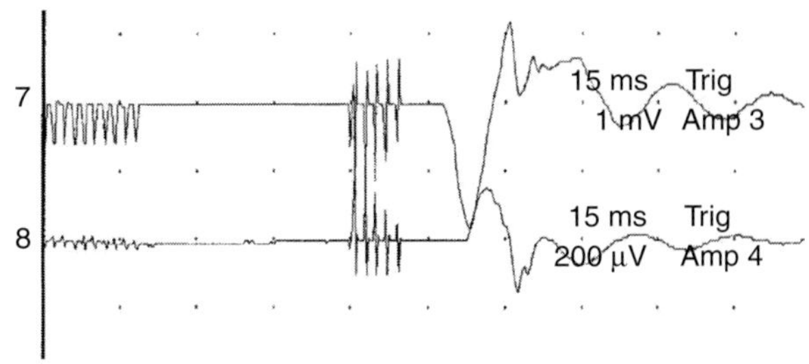

Figure 13.2 Example of motor-evoked potentials in the left (7) and right (8) tibialis anterior muscle. The recording starts with ten pulses applied to the plantar arch of the left foot: 60 ms later, a train-of-five pulse is delivered transcranially. A much larger MEP signal of 3 mV after facilitation of the left foot is demonstrated in comparison with the response of 320 µV on the right side (Frei *et al.*, 2007, with permission from Lippincott, Wolters Kluwer Health).

longer stimulating pulse trains and a greater need to adjust scalp electrodes.

There was also a similar, but weaker relationship between threshold voltage and increasing body surface area, weight, and height. This was noted earlier by Morota and his group (Morota *et al.*, 1997), who recognized that MEPs are more difficult to monitor in children and suggested that incomplete myelination of the corticospinal pathways and increased vulnerability of immature neuronal structures to compression could be contributing factors in the difficulty in monitoring of MEPs in children. In the same study, Morota demonstrated that the ability to elicit MEP responses before myelotomy was associated with a significantly better surgical outcome in adults, while the same association was inapplicable in their small population of eight children.

The MEP responses are also more likely to be depressed at varying anesthetic levels, even with the use of temporal facilitation. Therefore, the recording of reliable MEPs in young children using standard techniques is frequently unsuccessful. The addition of a conditioning stimulation using spatial facilitation combined with ketamine as the anesthetic agent was found to be particularly useful for recording MEPs in young children (Frei *et al.*, 2007). In Frei's study, the standard TES-MEP was obtained using temporal facilitation with a train of five pulses. Spatial facilitation was then added by stimulating the medial border of the plantar arch of one side with needle electrodes 60 ms before TES was started, while recording MEPs from both tibialis muscles (Figures 13.1 and 13.2). With the use of spatial facilitation, reliable signals were successfully monitored in an additional 19% of the patients (24 of 134). The efficacy of spatial facilitation was greater in children younger than 6 years of age (33%, 7/21) and in children 6 to 12 years of age (33%, 9/27) than in children between 12 and 18 years of age (9%, 8/86).

Studies using transcranial magnetic stimulation (TMS) to produce MEPs have also found an association between younger age and an increase in the threshold voltage needed to trigger MEPs. Earlier work by Eyre and his group showed that this threshold voltage does not decrease to adult levels until the age of 16 years. With background muscle activation, MEP responses using TMS can be elicited even from infants but, with the muscles at rest, these responses are not reliable till the age of six years (Koh and Eyre, 1988).

Resection during epilepsy surgery, electrical stimulation of the cerebral cortex for functional brain mapping, and identification of eloquent cortex might be unsuccessful in the pediatric population if conventional adult stimulation parameters are used. Electrical stimulation with higher intensity is frequently needed to produce the desired response (Signorelli *et al.*, 2004) with the youngest children requiring the highest intensities. Special stimulation paradigms that rely on increments in both stimulus intensity and pulse duration were designed for use in young children in order to minimize the energy required to elicit a response.

Anesthesia

The basic principles of applying anesthesia during NIOM in children are similar to those in adults: these are essentially to stabilize the depth of anesthesia and use steady continuous infusions instead of bolus administration as much as possible (see Chapter 5). However, the effects of anesthetic drugs on the nervous system of children can be greater than in adults.

Of particular concern is the attenuation of cortical potentials in response to inhalational anesthetics. The effect of anesthesia on the evoked potentials has been extensively evaluated in adults (Banoub *et al.*, 2003). Several studies were more focused on the changes seen in children. A reduction in the amplitude of somatosensory evoked potentials (SEPs) due to anesthesia is more likely to occur in children (Harper and Nelson, 1992). In their study, Harper and Nelson found that in 98 consecutive patients, the scalp potential disappeared in 14 patients between 5 and 185 minutes after halogenated anesthetics were initiated. The mean age of those who lost the scalp SEPs (13.5 ± 5.2 years) was significantly lower than those who retained a recordable SEP throughout surgery (22.2 ± 12 years; $p < 0.001$).

A dose-dependent amplitude reduction and latency prolongation of SEPs has been demonstrated with the use of inhalational anesthetics, such as sevoflurane, with further suppression after the addition of nitrous oxide (da Costa *et al.*, 2001). Large interindividual variation of anesthesia-induced SEP amplitude reduction has been noted to be more likely in children younger than five years old than in older children (Helmers and Hall, 1994; Westeren-Punnonen *et al.*, 2008). This may be related to maturational changes that take place in the somatosensory system in the first decade of life. These changes are thought to be a combination of asynchronous myelination and synaptogenesis.

As with the effect of inhalational anesthetics on SEPs, in their study using TES to elicit MEPs, Lieberman and his group reported that younger children received significantly lower levels of isoflurane than older patients but comparable doses of propofol (Lieberman *et al.*, 2006). In fact, the increasing use of MEPs in addition to SEPs to monitor scoliosis surgeries is additionally challenging to the anesthetist. The use of TIVA (total intravenous anesthesia) with propofol and minimizing neuromuscular blockade provides additional benefits in assuring successful monitoring in children.

As part of the anesthesiologists' routine preoperative evaluation of children planned for spine surgery, effort should be made for a careful examination and knowledge of associated medical conditions. (Soundararajan and Cunliffe, 2007). Problems leading to hypoxia or hypercarbia, in addition to vulnerability to hypothermia, increase difficulty in recording the cortical SEPs. When facing these problems, it would be helpful to plan for SEP recordings directly from the spinal cord using epidural electrodes. Responses from the spinal cord are much less sensitive than those from the cortex to these above-mentioned factors as well as to the effect of anesthetic agents.

Age-related differences are also seen in the effect of anesthetics on motor-evoked potentials (MEPs) triggered by transcranial electric stimulation as discussed in the previous section.

In addition to changes in neural conduction, there are age-related differences in cerebrovascular physiology which are also important. Children younger than 2 may have lower autoregulatory reserves and may be at increased risk of cerebral ischemia. A larger percentage of cardiac output is directed to the brain in children, since a child's head accounts for 19% of the body surface as opposed to only 9% in adults. This can place infants at higher risk of significant hemodynamic instability during neurosurgical procedures (Soriano and Bozza, 2006). If a major craniotomy is anticipated, close monitoring of the hemodynamic status is needed to avoid cerebral hypoperfusion, which can also lead to unsuccessful monitoring.

Special indications for NIOM in children

Virtually all the NIOM modalities, as well as the indications for monitoring in the adult population, can be applied in children. However, a few conditions form the vast majority of monitored surgeries in the pediatric population, some of them unique to children. Most of the literature on NIOM in children is derived from data obtained during scoliosis surgeries. Other common indications for NIOM in children include cauda equina surgeries, typically for release of tethered cord, and selective dorsal rhizotomy for management of cerebral palsy.

Scoliosis

It is estimated that 1–3% of adolescents (children aged 10–16 years) are at risk of having some degree of idiopathic scoliosis, defined as a Cobb angle of at least 10° (lateral spinal curvature). An additional rotational deformity is common (kyphosis). Only about 10% of adolescents with scoliosis require an intervention. Surgery is usually indicated when the Cobb angle exceeds 40°. Scoliosis may be associated with a variety of neurological disorders, including cerebral palsy, congenital spinal anomalies, Chiari malformation, spinal-cord tumors, and many neuromuscular disorders, such as hypotonia and spinal muscular atrophy.

The common surgical intervention to correct scoliosis is posterior spinal fusion, which is sometimes preceded by an anterior spinal release in severe cases of scoliosis. A variety of instruments is used during scoliosis surgery. This includes placement of Harrington rods to straighten the vertebral column. These rods are typically anchored to the vertebral processes using pedicle screws and sublaminar hooks or wires. The mechanical straightening of the spine (known as distraction) is a critical stage during the surgery, during which spinal-cord injury or ischemia could occur.

Surgical correction of scoliosis is the most common procedure involving NIOM in children. Modern NIOM has substituted the use of the "wake-up" test, which involved intermittently and temporarily lightening the patient's anesthesia level during the surgery to evaluate the patient's motor function (Vauzelle *et al.*, 1973). It is obviously challenging to use this crude method, especially in uncooperative children. In addition, it carries the risk of injury due to the patient's movement and has the potential for psychologic disturbance due to post-operative memory recall (Padberg *et al.*, 1998).

For nearly two decades, monitoring spinal-cord functions during scoliosis surgery has traditionally and most commonly been achieved using SEPs even though it is limited to testing only the ascending sensory pathways. Nevertheless, monitoring SEPs usually acts as a good surrogate for motor pathways as most spinal-cord lesions would involve both pathways. The posterior tibial nerve is typically selected while the peroneal nerve can be used as an alternative. Sensory pathways from the arms with median nerve SEPs may also be monitored. A detailed description of the methods and technique of SEPs is included in Chapter 6. As mentioned earlier, a large multicenter study in 1995 established the value of SEPs in reducing the risk of postoperative paraplegia (Nuwer *et al.*, 1995). However, despite its value and for the reasons mentioned, several cases have been reported where the motor pathways were affected without a change in the SEPs (Ginsburg *et al.*, 1985; Lesser *et al.*, 1986; Zornow *et al.*, 1990; Minahan *et al.*, 2001; Jones *et al.*, 2003; Hilibrand *et al.*, 2004). Subsequently, more recent studies noted that the addition of motor-evoked potentials (MEPs) to the SEPs provides a more comprehensive assessment of the spinal-cord function and hence further improves the ability to prevent neural injury (Nagle *et al.*, 1996; Pelosi *et al.*, 2002; Schwartz *et al.*, 2007; Sutter *et al.*, 2007c; Weinzierl *et al.*, 2007).

Recording myogenic MEPs using transcranial electrical stimulation is now considered the gold standard for monitoring motor pathways during spinal-cord surgery, although SEPs continue to have a role in providing continuous assessment of the integrity of the dorsal column (Deletis and Sala, 2008). Further detailed discussion on MEPs is available in Chapter 7.

Another modality that has gained popularity is the use of triggered EMG during spinal intraosseous pedicle screw placement. Pedicle screws offer several advantages over traditional instruments, such as rods and hooks. Pedicle screws avoid the use of instrumentation inside the spinal canal and allow for better correction and control over the spine. However, there is a risk of causing significant neurological injury to the nerve roots or even the spinal cord. In addition to the use of intraoperative imaging, pedicle screw triggered EMG stimulation can be very helpful in avoiding such injury (Figures 13.3 and 13.4). A recent study on the use of lumbar pedicle screws by Raynor *et al.* (2007) demonstrated that the probability of detecting a breach caused by pedicle screws using triggered EMG stimulation increases with decreasing EMG thresholds with a 0.31% probability if an EMG response is noted with stimulation >8.0 mA, 17.4% for 4.0 to 8.0 mA, 54.2% for <4.0 mA, and 100% for <2.8%.

As with scoliosis surgery, multimodal NIOM using combined SEPs and MEPs, as well as continuous EMG monitoring, has become routine and is now commonly used for spinal-cord surgeries, such as correction of congenital spine anomalies, resection of spinal-cord tumors, and anterior or posterior decompressions in spinal stenosis (Kothbauer *et al.*, 1997; Sala *et al.*, 2002; Sutter *et al.*, 2007a; 2007b; Hsu, 2009).

Selective dorsal rhizotomy

Electrophysiological methods to manage spasticity in children with cerebral palsy have been developed using selective ablation of the abnormal sensory nerve rootlets. This was first attempted with some success by

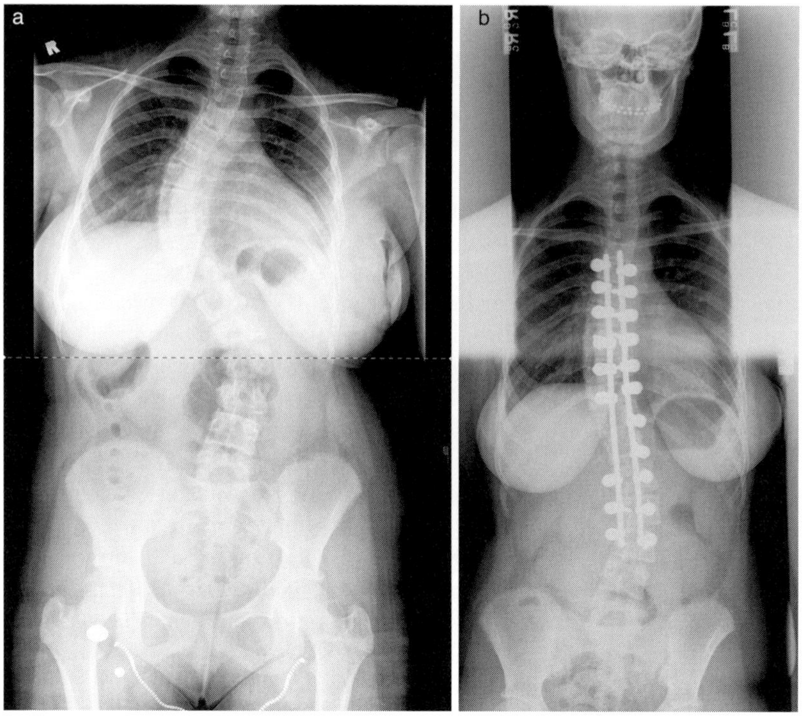

Figure 13.3 Radiographs of a teenager with progressive S-shaped scoliosis treated by posterior instrumentation with rods and screws. (a) Preoperative standing posterior–anterior (PA) shows dextroscoliosis with a thoracic curvature from T5 to T12 measuring 61 degrees and levoscoliosis with a lumbar curvature from L1 to L4 measuring 56 degrees. (b) Postoperative standing PA radiograph following posterior fusion that resulted in an excellent correction with the thoracic dextroscoliosis measuring 30° and the lumbar levoscoliosis measuring 31°. Note the pedicle screws at all levels of the spinal rods except on the right at T11 and T12 levels that were removed when triggered EMG stimulation (see Figure 13.4) suggested a breach of the pedicle wall at these two levels.

Foerster (1913). It took more than 60 years before the more refined selective dorsal rhizotomy using NIOM was introduced (Fasano, 1979). This technique involves selectively sectioning the sensory rootlets, to achieve a balance between reduction of spasticity and preservation of motor function.

The use of neurophysiologic monitoring during selective dorsal rhizotomy (SDR) is now considered both safe and effective, and is achieving significant functional benefits. A detailed description of the methodology and technique of SDR can be found in Chapter 15.

Tethered cord syndrome

Tethered cord syndrome (TCS) is a congenital condition, usually present in young children, in which the caudal end of the spinal cord is anchored to fixed bony or inelastic tissues (Khealani and Husain, 2009). Lower spinal-cord dysfunction usually follows, owing to impaired oxidative metabolism resulting from the over stretched tissues. Consequently, children with TCS can develop a combination of motor and sensory deficits in the legs as well as sphincter dysfunction.

Early surgery to release the "tethering" of the spinal cord can prevent further deficits and could possibly lead to reversal of some or all of the deficits. The surgery involves careful dissection of the lumbosacral nerve roots, therefore NIOM can potentially help in both mapping and preventing unwanted injury to neural structures. Ideally, this should be performed with measures to monitor sensory, motor, and sphincter functions. Routine SEPs can be obtained with stimulation of the posterior tibial nerves. Continuous electromyography (EMG) from the lower extremities can be obtained from selected muscle groups. This can be used to record spontaneous EMG activity that would indicate irritation of the nerve roots or to monitor triggered responses by stimulating the structures before dissection is attempted to determine if they contain neural elements. Similarly, EMG activity can be monitored from the anal sphincter, typically by inserting needle electrodes in the external anal sphincter to record both spontaneous and triggered EMG activity (Khealani and Husain, 2009).

Questions

1. During NIOM, children are in general more susceptible to temperature changes than adults.
   (a) True
   (b) False

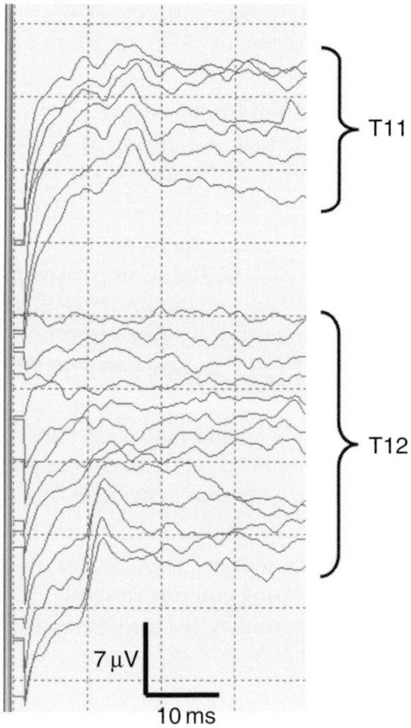

Figure 13.4 Stimulus evoked EMG recorded from a pair of bipolar needle electrodes placed over the right lower abdominal muscles in response to lower thoracic pedicle screw stimulation. Compound muscle action potential (CMAP) was evoked with stimulation of the right sided T11 pedicle screw at higher than 2 mA and of the right T12 at higher than 5 mA. This indicated current flow to the neural tissues due to reduced impedance, probably caused by a break of the pedicle wall. Palpation of the medial wall of both pedicles revealed a breach and both screws were removed by the surgeon.

2. All of the following statements are true with regard to the myelination of the auditory pathways in children, except:
   (a) Changes in the amplitudes and latencies of the recorded BAEPs are in general linear in relation to age
   (b) The myelination of the auditory nerve has been determined to start as early as the 34th week post-conception
   (c) The myelination of the auditory pathways progresses in a caudal to rostral direction that, in general, correlates with the latency and amplitude changes seen on BAEP responses
   (d) The central components of the auditory pathways continue to mature through the first few months of infancy
3. Which of the following is *true* with regard to the maturational changes of the auditory pathways in children?
   (a) At full term, most of the peripheral structures are still poorly myelinated
   (b) The latency waveform V of the BAEP reaches adult values within the first two to eight weeks of life
   (c) The latency waveform I of the BAEP reaches adult values by the age of 18 months
   (d) An exponential growth model with equal maturation rate for each station has been proposed to describe the development of the central auditory pathways
4. Monitoring the corticospinal motor pathways by transcranial electric stimulation (TES) to trigger motor-evoked potentials (MEPs) in children:
   (a) Usually require less voltage than is needed for adults to elicit a sufficient MEP
   (b) The MEP responses are less vulnerable to anesthesia than in adults
   (c) When added to temporal facilitation, spatial facilitation can improve the monitoring of reliable MEPs during spinal surgeries in children,
   (d) None of the above
5. In preparing for resective epilepsy surgery in children:
   (a) Electrical stimulation of the cerebral cortex usually requires higher intensities than in adults
   (b) The energy required to elicit a response during functional brain mapping can be minimized by using paradigms that rely on increments in both stimulus intensity and pulse duration
   (c) Both statements are false
   (d) Both statements are true
6. Which of the following statements is true with regard to the effects of anesthesia during NIOM in children?
   (a) Interindividual large variation of anesthesia-induced SEP amplitude reduction has been noted to be more likely to occur in children younger than five years than in older children
   (b) The use of inhalational anesthetics can lead to a dose-dependent amplitude reduction and latency prolongation of SEPs
   (c) Total intravenous anesthesia (TIVA) can be very helpful in enhancing the MEP responses
   (d) All of the above
7. The following is true about the NIOM of scoliosis surgeries:
   (a) Monitoring SEPs is useful as it is very sensitive in detecting injury of all parts of the spinal cord

(b) Higher thresholds for triggered EMG activity through pedicle screw stimulation indicate a high probability of a pedicle wall breach
(c) Monitoring MEPs can detect spinal cord injuries that could otherwise be missed by monitoring SEPs alone
(d) None of the above

References

Banoub, M., Tetzlaff, J. E., and Schubert, A. (2003). Pharmacologic and physiologic influences affecting sensory evoked potentials – Implications for perioperative monitoring. *Anesthesiology*, **99**, 716–737.

da Costa, V. V., Saraiva, R. A., de Almeida, A. C., *et al.* (2001). The effect of nitrous oxide on the inhibition of somatosensory evoked potentials by sevoflurane in children. *Anaesthesia*, **56**, 202–207.

Deletis, V. and Sala, F. (2008). Intraoperative neurophysiological monitoring of the spinal cord during spinal cord and spine surgery: A review focusing on the corticospinal tracts. *Clinical Neurophysiology*, **119**, 248–264.

Eggermont, J. J. (1985). Evoked-potentials as indicators of auditory maturation. *Acta Oto-Laryngologica*, **421**, 41–47.

Eggermont, J. J. (1992). Development of auditory evoked-potentials. *Acta Oto-Laryngologica*, **112**, 197–200.

Eggermont, J. J. (1995). Evoked-potentials as indicators of auditory development. *International Journal of Pediatric Otorhinolaryngology*, **32**, S183–S186.

Fasano, V. A., Barolat-Romana, G., Zeme, S., and Squazzi, A. (1979). Electrophysiological assessment of spinal circuits in spasticity by direct dorsal root stimulation. *Neurosurgery*, **4**, 146–151.

Foerster, O. (1913). On the indications and results of the excision of posterior spinal nerve roots in men. *Surgery, Gynecology and Obstetrics*, **16**, 463–475.

Frei, F. J., Ryhult, S. E., Duitmann, E., *et al.* (2007). Intraoperative monitoring of motor evoked potentials in children undergoing spinal surgery. *Spine*, **32**, 911–917.

Ginsburg, H. H., Shetter, A. G., and Raudzens, P. A. (1985). Postoperative paraplegia with preserved intraoperative somatosensory evoked-potentials. *Journal of Neurosurgery*, **63**, 296–300.

Grabb, P. A., Albright, A. L., Sclabassi, R. J., and Pollack, I. F. (1997). Continuous intraoperative electromyographic monitoring of cranial nerves during resection of fourth ventricular tumors in children. *Journal of Neurosurgery*, **86**, 1–4.

Harper, C. M. and Nelson, K. R. (1992). Intraoperative electrophysiological monitoring in children. *Journal of Clinical Neurophysiology*, **9**, 342–356.

Helmers, S. L. and Hall, J. E. (1994). Intraoperative somatosensory-evoked potential monitoring in pediatrics. *Journal of Pediatric Orthopaedics*, **14**, 592–598.

Hilibrand, A. S., Schwartz, D. M., Sethuraman, V., Vaccaro, A. R., and Albert, T. J. (2004). Comparison of transcranial electric motor and somatosensory evoked potential monitoring during cervical spine surgery. *Journal of Bone and Joint Surgery – American Volume*, **86A**, 1248–1253.

Hsu W, Pradilla., G., Constantini, S., and Jallo, G. I. (2009). Surgical considerations of spinal ependymomas in the pediatric population. *Child's Nervous System*, **25**, 1253–1259.

Jones, S. J., Buonamassa, S., and Crockard, H. A. (2003). Two cases of quadriparesis following anterior cervical discectomy, with normal perioperative somatosensory evoked potentials. *Journal of Neurology, Neurosurgery and Psychiatry*, **74**, 273–276.

Khealani, B. and Husain, A. M. (2009). Neurophysiologic intraoperative monitoring during surgery for tethered cord syndrome. *Journal of Clinical Neurophysiology*, **26**, 76–81.

Koh, T. H. and Eyre, J. A. (1988). Maturation of corticospinal tracts assessed by electromagnetic stimulation of the motor cortex. *Archives of Disease in Childhood*, **63**, 1347–1352.

Kothbauer, K., Deletis, V., and Epstein, F. J. (1997). Intraoperative spinal cord monitoring for intramedullary surgery: an essential adjunct. *Pediatric Neurosurgery*, **26**, 247–254.

Lesser, R. P., Raudzens, P., Luders, H., *et al.* (1986). Postoperative neurological deficits may occur despite unchanged intraoperative somatosensory evoked-potentials. *Annals of Neurology*, **19**, 22–25.

Levy, S. (1997). Brain stem auditory evoked potentials in pediatrics. In *Evoked Potentials in Clinical Medicine*, ed. K. Chiappa. Philadelphia: Lippincot-Raven. pp. 269–282.

Lieberman, J. A., Lyon, R., Feiner, J., Diab, M., and Gregory, G. A. (2006). The effect of age on motor-evoked potentials in children under propofol/isoflurane anesthesia. *Anesthesia and Analgesia*, **103**, 316–321.

Lopez, J. R. (2004). The use of evoked potentials in intraoperative neurophysiologic monitoring. *Physical Medicine and Rehabilitation Clinics of North America*, **15**, 63–84.

Minahan, R. E., Sepkuty, J. P., Lesser, R. P., Sponseller, P. D., and Kostuik, J. P. (2001). Anterior spinal cord injury with preserved neurogenic 'motor' evoked potentials. *Clinical Neurophysiology*, **112**, 1442–1450.

Morota, N., Deletis, V., Constantini, S., *et al.* (1997). The role of motor-evoked potentials during surgery for intramedullary spinal cord tumors. *Neurosurgery*, **41**, 1327–1336.

Nagle, K. J., Emerson, R. G., Adams, D. C., *et al.* (1996). Intraoperative monitoring of motor-evoked potentials: a review of 116 cases. *Neurology*, **47**, 999–1004.

Nuwer, M. R., Dawson, E. G., Carlson, L. G., Kanim, L. E. A., and Sherman, J. E. (1995). Somatosensory-evoked potential spinal-cord monitoring reduces neurologic deficits after scoliosis surgery – results of a large multicenter survey. *Electroencephalography and Clinical Neurophysiology*, **96**, 6–11.

Padberg, A. M., Wilson-Holden, T. J., Lenke, L. G., and Bridwell, K. H. (1998). Somatosensory and motor evoked potential monitoring without a wake-up test during idiopathic scoliosis surgery – an accepted standard of care. *Spine*, **23**, 1392–1400.

Pelosi, L., Lamb, J., Grevitt, M., *et al.* (2002). Combined monitoring of motor and somatosensory evoked potentials in orthopaedic spinal surgery. *Clinical Neurophysiology*, **113**, 1082–1091.

Picton, T., Taylor, M. J., and Pelosi, L. (1992). Brainstem auditory evoked potentials in pediatrics. In *Electrodiagnosis in Clinical Neurology*, ed. M. Aminoff. London: Churchill Livingstone, pp. 537–569.

Raynor, B. L., Lenke, L. G., Bridwell, K. H., Taylor, B. A., and Padberg, A. M. (2007). Correlation between low triggered electromyographic thresholds and lumbar pedicle screw malposition. *Spine*, **32**, 2673–2678.

Rotteveel, J. J., Colon, E. J., Stegeman, D. F., and Visco, Y. M. (1987a). The maturation of the central auditory conduction in preterm infants until 3 months post term. 1. Composite group averages of brainstem (ABR) and middle latency (MLR) auditory evoked-responses. *Hearing Research*, **26**, 11–20.

Rotteveel, J. J., DeGraaf, R., Colon, E. J., Stegeman, D. F., and Visco, Y. M. (1987b). The maturation of the central auditory conduction in preterm infants until 3 months post term. 2. The auditory brainstem responses (ABRS). *Hearing Research*, **26**, 21–35.

Sala, F., Krzan, M. J., and Deletis, V. (2002). Intraoperative neurophysiological monitoring in pediatric neurosurgery: why, when, how? *Child's Nervous System*, **18**, 264–287.

Schwartz, D. M., Auerbach, J. D., Dormans, J. P., *et al.* (2007). Neurophysiological detection of impending spinal cord injury during scoliosis surgery. *Journal of Bone and Joint Surgery – American Volume*, **89A**, 2440–2449.

Signorelli, F., Guyotat, J., Mottolese, C., *et al.* (2004). Intraoperative electrical stimulation mapping as an aid for surgery of intracranial lesions involving motor areas in children. *Child's Nervous System*, **20**, 420–426.

Soriano, S. G. and Bozza, P. (2006). Anesthesia for epilepsy surgery in children. *Child's Nervous System*, **22**, 834–843.

Soundararajan, N. and Cunliffe, M. (2007). Anaesthesia for spinal surgery in children. *British Journal of Anaesthesia*, **99**, 86–94.

Staley, K., Iragui, V., and Spitz, M. (1990). The human-fetal auditory evoked-potential. *Electroencephalography and Clinical Neurophysiology*, **77**, 1–5.

Sutter, M., Deletis, V., Dvorak, J., *et al.* (2007a). Current opinions and recommendations on multimodal intraoperative monitoring during spine surgeries. *European Spine Journal*, **16**, S232–S237.

Sutter, M., Eggspuehler, A., Grob, D., *et al.* (2007b). The diagnostic value of multimodal intraoperative monitoring (MIOM) during spine surgery: a prospective study of 1017 patients. *European Spine Journal*, **16**, S162–S170.

Sutter, M., Eggspuehler, A., Grob, D., *et al.* (2007c). The validity of multimodal intraoperative monitoring (MIOM) in surgery of 109 spine and spinal cord tumors. *European Spine Journal*, **16**, S197–S208.

Vauzelle, C., Stagnara, P., and Jouvinro, P. (1973). Functional monitoring of spinal-cord activity during spinal surgery. *Clinical Orthopaedics and Related Research*, 173–178.

Weinzierl, M. R., Reinacher, P., Gilsbach, J. M., and Rohde V. (2007). Combined motor and somatosensory evoked potentials for intraoperative monitoring: intra- and postoperative data in a series of 69 operations. *Neurosurgical Reviews*, **30**, 109–116.

Westeren-Punnonen, S., Ypparila-Wolters, H., Partanen, J., *et al.* (2008). Somatosensory evoked potentials by median nerve stimulation in children during thiopental/sevoflurane anesthesia and the additive effects of ketoprofen and fentanyl. *Anesthesia and Analgesia*, **107**, 799–805.

Zimmerman, M. C., Morgan, D. E., and Dubno, J. R. (1987). Auditory brainstem evoked-response characteristics in developing infants. *Annals of Otology, Rhinology, and Laryngology*, **96**, 291–299.

Zornow, M. H., Grafe, M. R., Tybor, C., and Swenson, M. R. (1990). Preservation of evoked-potentials in a case of anterior spinal artery syndrome. *Electroencephalography and Clinical Neurophysiology*, **77**, 137–139.

14 Intraoperative neurophysiologic monitoring of vascular disorders

Jaime R. López

Introduction

The use of neurophysiologic intraoperative monitoring (IOM) in the surgical and endovascular treatment of vascular disorders has become more widespread and, in some centers, an indispensable adjunct in the management of these cases. In addition to hemorrhage, cerebral ischemia is a possible and feared complication in the surgical and endovascular treatment of these disorders. Although IOM may not be specific in identifying cerebral hemorrhage, it can be of great utility in identifying cerebral ischemia. Thus, different techniques have been used over the past several years in an attempt to identify and potentially reverse cerebral ischemia.

Several different types of electrophysiologic techniques have been shown to be useful in identifying intraoperative cerebral ischemia and altering the intraoperative management of these cases to potentially reverse cerebral ischemia and improve patient neurologic outcome. The goals of this chapter are to describe the different IOM techniques that have been shown to be of value in the intraoperative management of these disorders, review the effects of cerebral ischemia on the different electrophysiologic techniques currently being used, and provide an understanding of the vascular disorders, with a primary emphasis on cerebrovascular disorders, where IOM may be a useful adjunct.

Regional cerebral blood flow and electrophysiologic studies

Owing to the potential risk of cerebral infarction during the operative treatment of cerebrovascular disorders, different techniques have been studied to determine their utility in identifying and potentially reversing cerebral ischemia prior to this producing permanent cerebral injury. Electroencephalography (EEG) and somatosensory evoked potentials (SEPs) have been most extensively studied and are the most commonly used techniques. The utility of these techniques as indirect markers of cerebral ischemia dates back several decades; several previous animal and human studies have demonstrated an excellent correlation between these techniques and regional cerebral blood flow (rCBF). Obviously, these techniques would be clinically more useful if they could be used intraoperatively to identify ischemic levels prior to this, so that irreversible cerebral injury can be prevented.

Several investigators (Sharbrough *et al.*, 1973; Sundt *et al.*, 1974; 1981) showed that in patients undergoing carotid endarterectomies (CEA) major EEG changes occurred with rCBF < 10 ml/100 g/min, less severe EEG changes were noted at levels at 10–18 ml/100 g/min, and a critical level was defined as 15 ml/100 g/min.

Consistent with human clinical studies, animal experiments have also shown changes in electrophysiologic studies that correlate well with rCBF indicative of cerebral ischemia. Primate studies have demonstrated that cortical SEPs are present at levels of rCBF ≥ 16 ml/100 g/min and absent at levels <12 ml/100 g/min, with a sharp decline in amplitude at levels between 14 and 16 ml/100 g/min and a 50% amplitude reduction corresponding to a rCBF of 16 ml/100 g/min (Branston *et al.*, 1974, 1977; Symon, 1980, 1984).

Ischemia appears to prolong the central conduction time (CCT) with a corresponding threshold of rCBF 15 ml/100 g/min which is similar to thresholds previously reported on the basis of amplitude reduction (Hargadine *et al.*, 1980; Lesnick *et al.*, 1984). The threshold relationship between cerebral blood flow and alteration of electrophysiologic studies appears to be a result of cellular dysfunction. There is strong experimental evidence correlating cellular membrane failure with a threshold level of rCBF (Symon, 1980; Astrup, 1982). Branston *et al.* (1977) reported that at rCBF levels of 12–16 ml/100 g/min the SEP was abolished, and that this was associated with small self-limiting increases in extracellular potassium. They were also able to demonstrate that a massive irreversible rise in extracellular potassium, associated with structural changes of infarction, occurred at an rCBF of 7.6–11.4 ml/100 g/min.

In middle cerebral artery (MCA) occlusion baboon chronic stroke models, areas of infarction corresponded to blood flow levels of 10 (ml/100 g)/min or below (Symon, 1980; Symon *et al.*, 1975). In primate models of acute stroke, cerebral infarcts occurred only in areas where the rCBF measured ≤ 12 ml/100 g/min (Morawetz *et al.*, 1978; Jones *et al.*, 1981). There is also experimental evidence suggesting a differential susceptibility of neural tissue to local ischemia as one descends the neuraxis. This is shown by preserved electrophysiologic function of the brainstem and spinal cord at systemic hypotension levels that affect cortical SEPs (Branston *et al.*, 1984).

Therefore, the following are reasonable conclusions from these studies:

1. Major EEG changes occur with an rCBF of <10 ml/100 g/min, with a critical level defined as 15 ml/100 g/min.
2. A 50% reduction in amplitude of the cortical SEP, which corresponds experimentally to an rCBF of 14–16 ml/100 g/min, is indicative of cerebral ischemia.
3. This "critical" rCBF is probably higher than that associated with cerebral infarction (rCBF ≤12 ml/100 g/min).
4. The changes caused by cerebral ischemia are potentially reversible if the rCBF is increased to a level >16 ml/100 g/min.

More recently, motor-evoked potentials after transcranial and direct cortical electrical stimulation have also been used as a method to assess cerebral ischemia in aneurysm surgery. Horiuchi *et al.* (2005) reported that blood flow insufficiencies in both the lenticulostriate and middle cerebral artery cortical branches perfusing the corticospinal tracts could be detected by intraoperative motor-evoked potential monitoring, while finding that SEPs proved to be unreliable.

Of course, it is important to keep in mind the functional pathways being monitored by the electrophysiologic modalities. One would not expect SEPs to be affected in corticospinal dysfunction since their neural pathway lies in the posterior column–medial lemniscal system. In addition, Sasaki *et al.* (2007) indicate that motor-evoked potential monitoring is currently the most reliable technique for detecting blood flow disturbances in the internal carotid artery (ICA) and MCA regions. However, a review of the literature fails to show similar experimental or clinical evidence outlining specific rCBF levels and corresponding changes in motor-evoked potentials to those established for EEG and SEPs.

There appear to be no clearly established or demonstrated cerebral ischemia threshold rCBF levels and their corresponding motor-evoked potential changes in the literature at this time. This is not to say that these threshold levels do not exist but seemingly similar studies to those previously performed using SEPs in animal models of cerebral ischemia have yet to be performed. Obviously, further studies outlining the specific rCBF levels corresponding to motor-evoked potential changes would be a welcome addition to our understanding of the physiologic changes occurring during cerebral ischemia.

Operative procedures where IOM is commonly used

It is important to have an appropriate understanding of the operative strategies used in the treatment of cerebrovascular and vascular disorders. Treatment of these disorders is typically performed surgically or through endovascular interventional procedures. In addition, the vascular territories at risk will either be intracranial or extracranial. The most common vascular disorders requiring operative or endovascular intervention include:

1. High-grade carotid artery stenosis,
2. Cerebral aneurysms,
3. Cerebral arteriovenous malformations,
4. Cerebral cavernous malformations,
5. Vascular occlusive disorders, such as moyamoya disease,
6. Thoracic and thoracoabdominal aortic aneurysm (TAAA).

High-grade carotid artery stenosis

Carotid endarterectomy (CEA) was first introduced successfully as a treatment for cerebral ischemia associated with internal carotid artery (ICA) stenosis in 1953 (DeBakey, 1975). Even though no controlled trials on the utility of CEA had been performed, the use of CEA increased dramatically through the mid-1980s. However, the North American Symptomatic

Carotid Endarterectomy Trial (NASCET) in 1991 unequivocally showed the benefit of CEA over medical management in symptomatic patients with high-grade stenosis. This trial was followed in 1995 by a finding of a modest benefit of CEA (versus medical therapy) in asymptomatic patients with high-grade carotid artery stenosis (Executive Committee for the Asymptomatic Carotid Atherosclerosis Study, 1995). The NASCET trial also emphasized the importance of maintaining a low perioperative complication rate because complication rates >5.8% (strokes and deaths) significantly diminished the benefit of CEA.

Currently, high-grade carotid artery stenosis is also commonly treated with endovascular techniques, such as carotid angioplasty and placement of stents.

IOM of carotid artery surgical and endovascular procedures

Special attention should be paid to the possibility of cerebral ischemia as a direct consequence of surgical and endovascular procedures involving the carotid artery, since in most instances transient interruption of carotid artery blood flow is unavoidable in these procedures. The common, internal, and external carotid arteries are routinely cross-clamped in CEA. Ipsilateral hemispheric cerebral ischemia can result if intracranial collateral cerebral circulation is inadequate. Similarly, angioplasty or carotid stent placement also transiently interrupts the carotid artery blood flow and carries the risk of transient cerebral ischemia.

In the case of CEA, a variety of techniques have been used in an attempt to identify cerebral ischemia. These can be classified into two main categories: techniques that assess cerebrovascular integrity, such as stump pressure measurements, radionuclide cerebral blood flow measurements, and transcranial Doppler; and electrophysiologic assessment of cerebral function that indirectly measures cerebral ischemia, including EEG, SEPs, and brainstem auditory evoked potentials (BAEPs). More recently, near infrared spectroscopy (NIRS) has also been used to identify cerebral ischemia but has demonstrated mixed results. This will be discussed in more detail later in this chapter.

If cerebral ischemia is identified, then such interventions as induced systemic hypertension or intraluminal shunting can be employed as a means of reversing the ischemia. An argument could be made that, to ensure adequate cerebral perfusion, an indwelling carotid shunt should always be used. However, the routine use of shunts in CEA is controversial because it carries the risk of embolization, possible stroke (Fode *et al.*, 1986; Halsey, 1992), and arterial damage (Loftus *et al.*, 1986). In addition, shunt malfunctions have also been reported (Whittemore, 1980; Lindsey, 1984). As a result, there is no clear consensus regarding the necessity for an indwelling arterial shunt during CEA. Surgeons appear to follow one of three strategies in regard to the use of shunts: always shunt, never shunt, or shunt selectively, based on the results of IOM (Fisher *et al.*, 1995; Loftus and Quest, 1995).

In the case where IOM does not identify cerebral ischemia, adequate cerebral perfusion is assumed to be present, presumably because there is sufficient collateral circulation. In such a case, a shunt would not be used and there would be no possibility of shunt-related embolic events. Although IOM is likely to identify hemispheric and regional (ICA, MCA and anterior cerebral artery (ACA)) cerebral ischemia accurately, it may not identify relatively small focal areas of ischemia or infarct.

Continuous EEG monitoring

Monitoring of EEG activity in operative procedures involving the carotid artery has been in use for over 40 years: EEG was first used in 1965, in CEA surgery (Perez-Borja and Meyer, 1965). An advantage of EEG is that it is produced by spontaneously generated electroencephalographic brain wave activity, is easily recorded, and requires no stimulus input. Over the years, the use of EEG as a technique for intraoperative monitoring of cerebral ischemia has become more popular for several reasons: readily available equipment, ease of set-up, familiarity with the technique, and reliability in determining carotid artery cross clamp-dependent ischemia. Several different studies have demonstrated the ability of EEG to detect ischemia and predict postoperative neurological deficits reliably (Sharbrough *et al.*, 1973; Sundt *et al.*, 1981; Wassmann *et al.*, 1984; Cho, 1986). These studies were able to show that patients with EEG changes consistent with ischemia during CEA were found to have a higher incidence of stroke as opposed to patients who did not develop similar changes. Redekop and Ferguson (1992) reported that in their series of 293 CEAs the postoperative neurological deficit risk was much higher in a subgroup of patients with major EEG changes (18%) than in those without changes (2%). Zampella *et al.* (1991) reported that in their study, 43 of 59 patients developed EEG changes and, of these, 17% (10/59) developed postoperative complications. In their study of 1145 consecutive CEAs, Sundt *et al.* (1981), stated that they "never had a patient emerge from anesthesia with a new deficit that was not predicted by the EEG." The literature also suggests that EEG monitoring during CEAs has influenced intraoperative management of these cases, most notably with a reduction in the frequency of carotid shunts and, ultimately improving neurological outcome (Wassmann *et al.*, 1984; Cho *et al.*, 1986).

Interpreting intraoperative EEG changes

Not all intraoperative EEG changes are caused by cerebral ischemia. It is important to recognize the pattern of EEG changes and the cerebral localization of these changes. The major causes of intraoperative EEG changes can be divided into three main categories:

Ischemia

Changes in EEG consistent with ischemia are the most important changes that the interpreter needs to be able to recognize consistently. These should also be differentiated from EEG changes that are not a result of cerebral ischemia. These major changes can be characterized as follows:

(a) Generalized slowing and decreased amplitude in the ischemic hemisphere is typically observed during ischemia (Perez-Borja and Meyer, 1965; Trojaborg and Boyson, 1973).
(b) Attenuation of the anesthetic-induced fast rhythms is also commonly seen (Chiappa *et al.*, 1979).
(c) Ipsilateral hemispheric EEG changes are the most common findings during ischemia after unilateral carotid cross-clamping. However, one should also keep in mind that bilateral EEG changes can be seen in patients with compromised collateral cerebral circulation (Chiappa *et al.*, 1979). Furthermore, patients with abnormal preoperative EEG have also been noted to have a higher frequency of intraoperative EEG changes (Sundt *et al.*, 1981).
(d) In some cases, bilateral, unilateral, or contralateral EEG changes consistent with ischemia can be seen when the mean arterial pressure (MAP) drops below a certain critical level that is dependent on the patient's baseline MAP (personal observation). Some of these unusual and more complex EEG pattern changes can result from patients with tandem intracranial vascular lesions, which may be bilateral, ipsilateral, or contralateral to the diseased carotid artery.

Anesthesia

Anesthetic agents can have a profound effect on the EEG. These can range from mild non-specific EEG changes to an isoelectric pattern. During induction, the EEG can display rhythmic beta activity, in the 12–18 Hz (beta) range, which typically appears over the anterior hemispheres. Later during the procedure, with lighter levels of steady-state anesthesia, the amplitude of the EEG increases, becomes widespread and slows to 8–14 Hz. Intermittent delta activity may also be seen anteriorly. In addition, burst suppression patterns can be seen at high doses of many inhaled and intravenous anesthetic agents and barbiturates. Generalized slowing of the EEG can be seen with bolus administration of narcotic agents, such as remifentanyl (personal observation). It is important to keep in mind that EEG changes due to anesthetic agents are typically bilateral and symmetrical.

Technical factors

Poor electrode placement, faulty recording electrodes, high electrode impedances, and contamination of the EEG signal by ambient electrical noise can lead to poor quality recordings. Attention to proper electrode placement and recording techniques and the identification and elimination of operating room equipment producing excessive electrical noise is of utmost importance, to obtain adequate, accurate, and useful EEG recordings.

Critical intraoperative EEG changes

Ipsilateral EEG changes after carotid artery cross-clamping are most probably a result of reduced cerebral perfusion to that particular hemisphere. To reverse cerebral ischemia, arterial flow to that hemisphere needs to be restored. Surgeons who use IOM will usually shunt once the EEG shows major changes. There is clinical evidence to suggest that shunting in such cases leads to improved overall neurological outcome (Sundt *et al.*, 1981; Sundt, 1983; Messick *et al.*, 1984; Cho *et al.*, 1986). In an interesting study by Cho *et al.* (1986), comparing neurological outcome in two groups of CEAs performed either with or without EEG IOM, the use of IOM and selective shunting was found to reduce the frequency of carotid shunts from 49% to 12% and to decrease major neurological morbidity and mortality from 2.3% to 1.1%.

Computer-processed EEG

In addition to the monitoring of conventional EEG, different forms of computer-processed EEG have also been used in an attempt to facilitate interpretation (Myers *et al.*, 1977; Chiappa *et al.*, 1979). Fourier analysis of EEG data produces a power versus frequency display that can be used by individuals less skilled in reading conventional EEG. The computer-processed EEG can be displayed in different formats, including compressed spectral array (CSA), or by areas of density (intensity), termed density-modulated spectral array (DSA) (Levy *et al.*, 1980). These methods simplify the interpretation of the EEG and show ischemic changes as loss of amplitude in the power spectrum for all frequencies and a shift of the EEG spectrum to the lower frequency range.

Several studies have found these techniques useful in identifying cerebral ischemia and predicting neurological outcome following CEA (Chiappa *et al.*, 1979; Rampil *et al.*, 1983; Ahn *et al.*, 1988). Some authors have claimed that computer-processed EEG is more useful than analog EEG (Chiappa *et al.*, 1979; Ahn *et al.*, 1988). However, Kearse *et al.* (1993) demonstrated, in a prospective study of 103 patients undergoing CEA monitored with analog EEG and DSA, that DSA was not sufficiently sensitive in detecting mild cerebral ischemia, which was identified by characteristic analog EEG changes. It is quite likely that the experience of the individual interpreting conventional EEG plays a critical role in proper EEG analysis. Conversely, persons with limited EEG experience might be advised to use computer-processed EEG techniques, while understanding that these techniques may lack sensitivity in detecting mild EEG patterns consistent with cerebral ischemia.

Somatosensory evoked potentials

As a result of the strong correlation between SEPs and cerebral ischemia, SEPs have gained popularity as useful intraoperative markers of cerebral ischemia. Their use has become more widespread and their application extends into a variety of cerebrovascular disorders. The use of SEPs in CEAs is based on the fact that the cerebral territory at risk is typically the ipsilateral hemisphere, specifically the ICA and MCA vascular territories. Therefore, whatever IOM protocol is used to monitor CEAs should use techniques that can assess ischemia in these territories.

Cortical evoked potentials can identify regional vascular ischemic changes but these are determined by the particular neural generator associated with that specific evoked potential (EP). In the case of CEA surgery, the MCA vascular territory at risk encompasses the area of the primary somatosensory cortex in the parietal lobe. Median nerve SEPs are generated by the primary somatosensory cortex and reflect the cerebral functional status of the cortical region representing the hand, which is supplied by the MCA and, thus, at risk during carotid cross-clamping (Figure 14.1). During IOM, SEPs have several advantages over EEG, including relative resistance to general anesthesia, the use of fewer electrode sites, easier reporting and interpretation, and simpler serial comparison (Markand, 1994).

Several studies have looked at the use of SEPs in CEA. Moorthy *et al.* (1982) were first to report the use of median nerve SEPs during CEA and to correlate SEP changes with neurological deterioration. A subsequent study of 38 CEAs (10 done under general and 28 with local anesthesia) monitored with SEPs by Markand *et al.* (1984) reported that one patient had a sudden loss of consciousness and loss of cortical SEPs following carotid cross-clamping. The SEP changes were reversed and the patient regained consciousness after placement of a shunt.

In three other cases, SEP changes were also reversed by shunting. The majority of clinical studies support the reliability and accuracy of SEPs in CEAs (Gigli *et al.*, 1987; Schweiger *et al.*, 1991; Tiberio *et al.*, 1991; Amantini *et al.*, 1992; Haupt and Horsch, 1992; Kearse *et al.*, 1992). Pedrini *et al.* (1998), in a study of 196 patients, reported a reliability of 86.7%, a sensitivity of 100%, and a specificity of 86.5%. As with any clinical assessment tool, the occurrence of false-positive and false-negative results is well recognized in the literature. Fortunately, these rates tend to be relatively low (Lam *et al.*, 1991; Amantini *et al.*, 1992; Wöber *et al.*, 1998). A review of the literature by Manninen *et al.* (2001) found a false-negative rate that ranged from 0% to 3.5%.

However, at least two studies have been published that concluded that SEPs are not sensitive enough markers of cerebral ischemia and are unreliable predictors of neurologic outcome (Lam *et al.*, 1991;

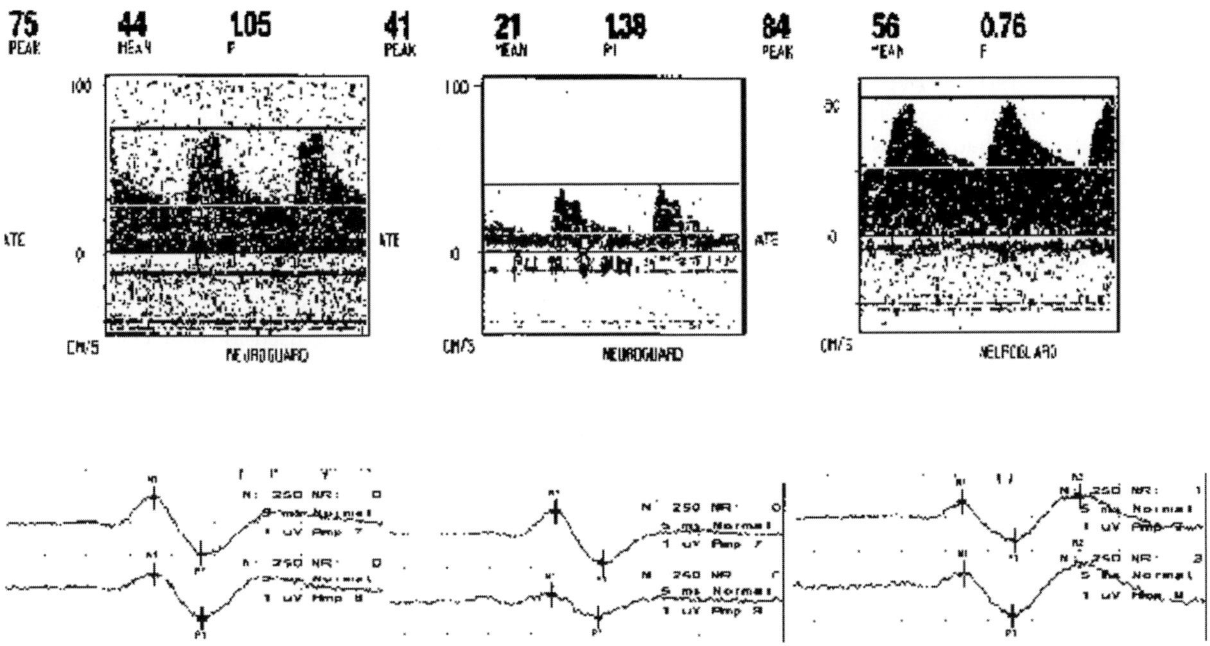

Figure 14.1 Patient undergoing left CEA. Left upper panel: baseline TCD trace and cortical SEPs after left and right median nerve stimulation (traces 1 and 2, respectively). Middle panel: reduced TCD trace and >50% amplitude reduction of the left cortical (right median nerve stimulation, trace 2). These findings are consistent with left cerebral ischemia. Note the recovery of the SEP and improved TCD velocity on the right panel, which occurred after placement of a left carotid shunt.

Wöber *et al.*, 1998). One study concluded that SEPs are not sufficiently sensitive to detect the onset of cerebral ischemia, since neither a 50% amplitude reduction nor an increase in CCT has been established as a physiological marker of intraoperative ischemia (Lam *et al.*, 1991). In this study, instead of using neurologic outcome as the endpoint, the SEP criteria were measured against the EEG gold standard. The results indicated that 10/23 patients with EEG evidence of ischemia after cross-clamping had prolongation of the CCT but only one had an amplitude reduction of 50% or more.

Wöber *et al.* (1998), in a study of 32 patients, reported that the intraoperative cortical SEP latencies and amplitudes did not differ between shunted and non-shunted patients. They concluded that a decreased cortical SEP and delayed cortical SEP were unreliable predictors of neurologic outcome in CEA patients and unsuitable as criteria for selective shunting. Furthermore, they performed a meta-analysis of 15 studies and reported that neurologic outcome was not improved by selective shunting based on SEP monitoring. However, their comparison of different studies from different centers is complicated by the lack of uniform criteria for shunt placement and non-uniform definition of what constitutes significant SEP changes. The authors themselves also recognized other shortcomings of their meta-analysis, such as a possible underestimation of the number of transient neurologic deficits in some studies and a lack of reporting in other studies, inclusion of multiple serial studies published by the same group of authors, and the possible discrepancies in data that appeared to occur in some of these studies.

In contrast, Amantini *et al.* (1992) concluded that there is a higher incidence of false-positives with EEG and, subsequently, the use of shunts, since there is less agreement in identifying EEG evidence of the ischemic risk when compared with SEPs, where there is more agreement regarding the criteria for what constitutes cerebral ischemia. Even though different investigators have used different criteria, such as amplitude reduction, latency delay, prolongation of CCT, or a combination of the three to signify cerebral ischemia, most studies have found that a 50% amplitude reduction of the N19–P24 or a 1 ms delay of the CCT correlates best with postoperative neurological deficits (Brinkman *et al.*, 1984; Gigli *et al.*, 1987; De Vleeschauwer *et al.*, 1988; Schweiger *et al.*, 1991; Haupt and Horsch, 1992; Kearse *et al.*, 1992; 1993).

Fisher *et al.* (1995) reviewed seven studies comprising 3028 patients, and reported that the cumulative results revealed 5.6% (170) of patients had a significant decline in intraoperative SEPs (owing to surgical manipulation) and that, of these, 20% (34) developed significant postoperative deficits. In addition, Lam *et al.* (1991) reviewed the reported sensitivity and specificity in six studies (including theirs) between 1985 and 1991. They found that the sensitivities ranged from 83% to 100%, and the specificities ranged from 83% to 99%. They concluded that conventional EEG and SEP monitoring during CEAs had similar sensitivity and specificity.

In a separate study, Brinkman *et al.* (1984) found that intraoperative SEP amplitude decrements of 50% or more were associated with worse neuropsychologic performance, determined by comparing preoperative and postoperative neuropsychologic testing.

The overall conclusion from these clinical studies is that SEP IOM for CEA is reliable in identifying intraoperative ischemia and useful in improving neurological outcome during CEA.

Transcranial Doppler (TCD)

In the past, monitoring strategies have focused on the detection of cerebral hypoperfusion and possible ischemia with the use of EEG and SEPs. Although these techniques have proven to be useful, embolic events and subcortical infarcts have not always been reliably identified. Transcranial Doppler offers the advantage of being able to detect embolic events as well as to measure blood flow directly. The disadvantages are that one is limited to a single vessel and that reduced flow does not always correlate with cerebral function. In some instances, combined use of SEPs and TCD can provide useful information regarding cerebral ischemia (Figure 14.1).

Transcranial Doppler, which is based on the Doppler effect, measures blood flow velocity and direction in large intracranial arteries, such as the ICA, MCA, ACA, basilar, and vertebral arteries. The TCD probe produces ultrasound waves, which are transmitted to the intracranial vessel and then become reflected off moving red blood cells. The frequency of the reflected signal will change in direct proportion to the velocity of the moving red blood cells. This can be clinically useful and may provide important information about specific intracranial vascular abnormalities. For example, increased blood flow velocity can be seen in vessel stenosis. Furthermore, it is believed that changes in flow velocity reflect changes in cerebral blood flow as long as fluctuations in arterial blood pressure and arterial CO_2 content remain small. Hemodynamic compromise is thought to occur when there is a reduction in mean flow velocities or when there is slow flow acceleration (Sloan, 2006). Transcranial Doppler can also detect microscopic emboli; this may be useful in certain clinical situations.

With respect to the use of TCD for IOM, it has been and continues to be used in some centers but is not as commonly used as EEG or SEP and its use and application tends to be limited to CEA. Some of the factors limiting its more widespread use are technical, such as: the requirement for personnel specifically trained in performing the study; probe dislocation, given the relative proximity of the probe to the operative field; and the lack of an adequate acoustic window in a significant number of patients.

However, there exists a body of literature in the use of TCD in CEA. In a multicenter, retrospective trial involving 1495 patients undergoing CEA, Halsey (1992) found that severe cerebral ischemia developed during ICA cross-clamping if, within the first minute, the mean velocity (MV) dropped to 0–15% of baseline, mild ischemia developed if velocity was 16–40%, and ischemia was absent if >40%. The mean velocity baseline was established immediately prior to cross-clamping. This study also showed that insertion of a shunt was associated with a significant risk of stroke in those patients experiencing mild or no ischemia, as detected by TCD, and that the best results were obtained using selective shunting (Table 14.1).

Moreover, in 50 consecutive CEAs monitored with TCD under local anesthesia, Bornstein *et al.* (1996) found that the velocity changes were similar in those who developed complications and those who did not. In their study, eight patients developed intraoperative neurological symptoms and three developed minor postoperative strokes. They concluded that their results did not support the view that transcranial Doppler monitoring is helpful in deciding whether to use a shunt during CEA.

Moritz *et al.* (2007) compared the accuracy of four cerebral monitoring techniques in detecting ischemia during CEA using regional anesthesia in 48 patients. They compared TCD, stump pressure measurements (SP), near infrared spectroscopy (NIRS), and SEPs, and concluded that all modalities provided similar accuracy in detecting cerebral ischemia but that the

Table 14.1

Degree of ischemia	Incidence of stroke	Probability (p)
Severe (MV = 0%–15%)		
Shunt	0/74 (0%)	< 0.0001
No shunt	6/13 (46%)	
Mild (MV = 16%–40%)		
Shunt	3/77 (3.9%)	$0.1 > p > 0.05$
No shunt	1/159 (0.6%)	
None (MV > 40%)		
Shunt	6/136 (4.4%)	< 0.001
No shunt	7/1016 (0.7%)	

MV = mean velocity
Adapted from Halsey, 1992.

lowest accuracy was observed with SEP monitoring. They also reported a high rate of technical difficulties in the TCD group, which prevented TCD monitoring in 10 (21%) patients. Their overall conclusion was that SP and NIRS might be techniques to use in practice. However, their study has some limitations, specifically the relatively low number of patients studied, the high number of patients who could not be monitored with TCD, and the relatively low stimulus intensity (10 mA) used to obtain the median nerve SEP. Obviously, they used relatively low stimulus intensities, to avoid discomfort in awake patients.

In contrast, Rowed *et al.* (2004) compared the use of EEG and SEPs in 156 patients and TCD in 91 during CEA using general anesthesia. Unfortunately, only 57/91 patients were monitored throughout the procedure with TCD. They concluded that the SEPs were the most reliable IOM method to identify intraoperative ischemia. The SEPs accurately identified intraoperative strokes without false-positive or false-negative results noted. Overall, TCD has some utility but has not been shown to be superior to EEG or SEPs, as an intraoperative marker of cerebral ischemia during CEA and the technical difficulties make its use much more limited.

Near infrared spectroscopy (NIRS)

Recently, there has been growing interest in the use of NIRS as a non-invasive means of assessing cerebral oxygenation. This technology uses light in the near-infrared range to evaluate brain oxygen saturation (Sloan, 2006). Cerebral oximetry is based on the Lambert–Beer law, which states that light transmission through a solution is a function of the density or concentration of absorbing molecules in the solution and the pathlength (Wahr *et al.*, 1996), and the fact that oxygenated and deoxygenated hemoglobin have characteristic and distinct absorbance spectra (Davies and Janelle, 2006). Since skin, skull, and other biological tissues are relatively transparent to the near-infrared wavelength of light, deeper tissues, such as the brain, can be analyzed.

Currently only one cerebral oximetry monitor is approved for clinical use in the United States, the INVOS series (Somanetics Corp, Troy, MI). This device is known as a "saturation monitor." It measures the ratio of hemoglobin and oxyhemoglobin by using a single light-emitting diode to transmit infrared light at two wavelengths (724 and 810 nm), with two silicon photodiodes as light detectors (Davies and Janelle, 2006). A more detailed review of this technology is beyond the scope of this chapter but the reader is referred to the excellent reviews by Wahr *et al.* (1996) and Davies and Janelle (2006). Suffice it to say that the possibility of obtaining cerebral oximetry non-invasively and in real time is quite attractive.

There have been several reports in the literature on the use of NIRS in different cerebrovascular surgical procedures, most commonly in CEAs. As previously noted, Moritz *et al.* (2007) compared the accuracy of different cerebral monitoring systems in detecting cerebral ischemia in 48 patients undergoing carotid surgery during regional anesthesia and ultimately concluded that stump pressure measurements and NIRS might be superior in clinical practice.

However, in an earlier study, Beese *et al.* (1998) compared the use of NIRS and SEPs for the detection of cerebral ischemia in 317 patients undergoing CEA. The results revealed that all 287 patients with preserved cortical SEPs remained neurologically intact. The regional cerebral saturation (rSo_2) obtained by using the cerebral oximeter showed a statistically significant decrease in patients without as well as with consecutive loss of the cortical SEPs. The authors also reported that the substantial interindividual variability of rSo_2 and derived change of rSo_2 did not allow for a definition of a threshold value indicating the need of shunt placement. Their overall conclusions were a reaffirmation of the reliability of SEPs for the detection of clamp-related cerebral hypoperfusion and that therapeutic interventions based on monitoring with cerebral oximetry are not justified as long as rSo_2 threshold values indicating critical cerebral ischemia are not defined.

In a separate prospective study, Rigamonti *et al.* (2003) compared continuous non-invasive monitoring of rSo_2 and the occurrence of clinical and EEG signs of cerebral ischemia during carotid cross-clamping in 50 patients undergoing CEA surgery using a cervical plexus block. In five patients (10%), carotid cross-clamping was associated with clinical and EEG signs of cerebral ischemia. All of these patients were treated with the placement of a shunt, which completely resolved their symptoms. No patient experienced a permanent neurological injury. In four of these five patients, EEG signs of cerebral ischemia were noted, while the percentage rSo_2 reduction from baseline during cross-clamping was 17% in patients requiring shunt and only 8% in those who did not require shunt. An attempt was made to establish a threshold that could identify cerebral ischemia but such a threshold revealed a 44% sensitivity, 82% specificity, and a 94% negative predictive value. The authors concluded that they were not able to identify an rSo_2 threshold that could be used alone to predict the need for shunt placement because of the low sensitivity and specificity.

Overall, reviews of the literature regarding the clinical application of NIRS for identification of cerebral ischemia suggest persistent problems in both sensitivity and specificity, with a number of authors questioning not only the clinical application but the accuracy of the technology in its current state (Schwarz *et al.*, 1996; Davies and Janelle, 2006).

Thoracic and thoracoabdominal aortic aneurysm (TAAA) surgical repair

Spinal cord ischemia and infarct with resultant paraparesis or paraplegia are potential complications of (TAAA) surgical repair, since the surgical procedure involves cross-clamping the aorta. Aortic clamping can cause an interruption of the blood supply to the spinal cord from the intercostal and lumbar arteries. Unfortunately, the thoracic spinal cord is highly dependent on the aortic vascular supply and an interruption in supply can lead to spinal ischemia and possible infarct. Clinically, this can cause paraparesis or paraplegia. Svensson *et al.* (1993), in a study of 1509 patients, stated the overall risk of paraplegia after TAAA repair to be 16%. Fortunately, this risk has dropped to as low as 3.3% as a result of the use of surgical adjuncts such as distal aortic perfusion, moderate hypothermia, cerebral spinal fluid drainage, and systemic reimplantation of the intercostal arteries (Achouh *et al.*, 2007).

Previously, a variety of different surgical adjuncts were studied in an attempt to preserve spinal cord function. Unfortunately, most of the strategies failed to reduce postoperative spinal cord injury significantly (Livesay *et al.*, 1985). However, Verdant (1992) reported no occurrence of paraplegia or other spinal cord ischemic deficits in a series of 267 consecutive patients undergoing surgery for descending thoracic aortic aneurysm. The authors used a 9 mm Gott shunt for distal perfusion in all cases to provide optimal protection of the surrounding organs during aortic occlusion.

In addition to using intraoperative strategies to prevent spinal cord ischemia, intraoperative spinal cord monitoring using SEPs was introduced as a method for assessing spinal cord function in TAAA repair in the late 1980s. This allowed for the intraoperative detection of spinal cord ischemia. More recently, transcranial electrical motor-evoked potentials (TcMEPs), usually in conjunction with SEPs, have also been used to monitor spinal cord function.

One of the utilities of IOM is that it can be used to guide surgical strategy in the repair of TAAA, and potentially modify the course of spinal cord impairment before permanent injury has occurred (Schepens *et al.*, 1994; Etz *et al.*, 2006). Schepens *et al.* (1994) reported the use of SEPs in 43 patients undergoing TAAA repair. They found the incidence of paraplegia and paraparesis was 7% and 5%, respectively. Of the 18 patients (42%) without SEP changes, one awoke paraplegic immediately after the surgery and two others developed delayed onset paraplegia. Overall, these results indicate a higher incidence of immediate postoperative paraplegia or paraparesis in the groups with transient or permanent SEP changes 4/24 (16%) than in the group without changes 1/18 (5.5%). The authors concluded that the use of SEPs enhanced safety and offered an improvement in the surgical strategy.

The clinical utility of SEPs was again demonstrated in a prospective study of 444 consecutive patients undergoing surgical treatment of descending TAAA (Achouh *et al.*, 2007). Patients with SEP changes were classified into three groups: 1: absence of signs of spinal cord dysfunction throughout the case; 2: transient signs of spinal cord dysfunction that normalized

Table 14.2 Prospective study of 444 patients undergoing surgical treatment of descending TAAA. The table shows the summary of patient outcomes in each group

Neurological deficit	Group 1	Group 2	Group 3
Immediate	3/357 (0.8%)	2/65 (3.1%)	3/22 (13.6)
Delayed	11/357 (3.0%)	2/65 (3.1%)	1/22 (4.5%)
30-day mortality	41/357 (11.5%)	18/65 (27.7%)	11/22 (50.0%)

Modified from Achouh *et al.*, 2007

at the end of the procedure; and 3: persistent signs of spinal cord dysfunction at the end of the procedure. Their results revealed that 87/444 (19.6%) patients showed SEP changes (groups 2 and 3). Immediate corrective measures were instituted and SEP changes normalized in 65/87 (75%; group 2) patients. In 22/444 (5%) of patients the SEP changes were permanent (group 3), and 149/444 (34%) of patients had SEP changes consistent with peripheral nerve ischemia (Table 14.2).

A standardized intervention protocol aimed at reversing spinal cord ischemia was used when SEP changes indicated spinal cord dysfunction. This included reimplantation of intercostal arteries, lowering cerebrospinal fluid pressure, increase of hemoglobin via transfusion, and increasing distal aortic perfusion pressure. The sensitivity and specificity of IOM of SEPs in predicting combined immediate and delayed neurologic deficits was 30.0% and 80.9%, respectively, with a negative predictive value of 96.1%.

As previously mentioned, TcMEPs have recently been used to monitor spinal cord function in TAAA. In a retrospective study of 100 patients undergoing TAAA repair, TcMEPs were used to monitor spinal cord function without segmental spinal cord artery reimplantation. Immediate postoperative paraplegia occurred in one patient (1/100), who demonstrated persistent SEP and TcMEP changes. The authors concluded that with monitoring of SEPs and TcMEPs the surgery can be safely performed with sacrificing and without reimplantation of as many as 18 intercostal and lumbar arteries (Etz *et al.* 2006).

In conclusion, the published data suggest that IOM using SEPs or a multimodality approach consisting of SEPs and TcMEPs is a useful adjunct in the surgical management of TAAA, and in many instances alters the surgical strategy.

Monitoring intracranial cerebrovascular disorders

The rationale for the use of IOM in the operative treatment of intracranial cerebrovascular disorders is similar to that mentioned for CEA. Patients undergoing surgical or endovascular treatment of cerebral aneurysms, cerebral arteriovenous malformations (AVMs), cerebral cavernous malformations (CAVMALs), and intracranial vascular occlusive disorders, as well as those undergoing deliberate cerebral vessel occlusion, may be good candidates for IOM.

Cerebral aneurysm surgery

The initial use of SEPs and BAEPs for IOM of cerebral aneurysm and AVM surgical procedures was based on the strong experimental correlation between SEP changes and cerebral ischemia. In addition, SEPs have the added benefit that they can identify regional cortical ischemic changes, which correspond to different cerebral vascular territories. Whereas EEG provides information about cortical function, SEPs have the advantage that they can provide functional information regarding the cortical and subcortical somatosensory pathways. As previously mentioned, the median nerve SEP cortical neural generator is located in the primary somatosensory parietal cortex representing the hand. This cortical region is supplied by the internal carotid and middle cerebral arteries. In addition, the median nerve SEPs provide information on thalamic activity supplied by the posterior cerebral artery (Buchtal and Belopavlovic, 1992).

In contrast, the cortical generator of the posterior tibial nerve SEP is located in the parasagittal somatosensory parietal cortex representing the leg, which is supplied by the ACA. Posterior circulation and brainstem ischemia can be identified using a combination of SEPs, BAEPs, and motor-evoked potentials. Table 14.3 lists the surgical procedures of common cerebral aneurysm and AVMs, the probable corresponding vascular territories at risk, and the recommended IOM modality.

These electrophysiologic techniques can be very useful in identifying focal or regional cerebral ischemia in the surgical or endovascular treatment of

Table 14.3 Suggested IOM techniques to use in the following cerebrovascular surgical procedures

Procedure and cerebrovascular territory at risk	Recommended IOM technique					
	EEG	Median nerve SEP	Posterior tibial nerve SEP	BAEP	TcEP	EMG
Aneurysm and AVM surgery						
ICA	×	×	×		×	
MCA	×	×	×		×	
ACA	×	×	×		×	
V-B		×	×	×	×	Possibly
Moyamoya disease (MCA branch)	×	×	×			

cerebral aneurysms, which can arise from different intracranial arteries. As a result, SEPs and BAEPs have been used to assess cerebral function in intracranial aneurysm and AVM surgical and endovascular procedures (Chang *et al.*, 1999; López *et al.*, 1999; Paulsen *et al.*, 1999; Liu *et al.*, 2003). The utility of these electrophysiologic techniques is enhanced further if they are used in combination, providing multimodality recordings.

More recently, TcMEPs have been used in conjunction with SEPs and BAEPs as a comprehensive IOM protocol in cerebral aneurysm surgery (Quiñones-Hinojosa *et al.*, 2004). Direct cortical motor-evoked potentials, combined with SEPs, have also been used by other investigators to monitor cortical and subcortical corticospinal tracts (Horiuchi *et al.*, 2005; Sasaki *et al.*, 2007). Although EEG can also reliably detect cortical ischemia, it has had limited use in intracranial non-epilepsy surgery because the craniotomy and brain relaxation produce airspaces between the dura and arachnoid that interfere with EEG recording from the underlying cortex (Emerson and Turner, 1993).

A review of the literature suggests that IOM of cerebral aneurysm surgery is useful and that it can alter the surgical management. Palatinsky *et al.* (1994) reported the clinical outcomes of 10 studies involving MCA aneurysm surgery and found the reliability of median nerve SEPs in predicting outcome ranged between 78% and 100%. Friedman *et al.* (1987) found that a decrease or disappearance of the cortical SEP amplitude accurately predicted postoperative deficits.

In a separate study of 53 MCA aneurysm procedures, Friedman *et al.* (1991) found new postoperative deficits in four of four patients who experienced a significant change in SEP with incomplete return to baseline or loss of the SEP. In addition, a new postoperative deficit was seen in one of five patients with significant SEP changes that recovered back to baseline levels. Only 1 of 37 patients without SEP changes developed a new postoperative deficit.

In a study of 134 aneurysms (39 ICA and 40 MCA), Schramm *et al.* (1990) reported loss or alteration of SEPs in 12 cases of temporary occlusion, two cases of accidental clipping, one case of permanent occlusion, and one case each of retraction of the MCA and cerebellar retraction. In 15 of 17 instances of SEP changes, the surgical course was altered. Revised surgical management included repositioning of vascular clips and repositioning of retractors.

Similar findings were found in a more recent study of 70 surgically treated cerebral aneurysms in 49 consecutive patients (López *et al.*, 1999). The authors reported that the intraoperative benefit of electrophysiological monitoring was seen in 24% of patients: 13 patients (26%) had transient SEP changes and two patients (4%) had transient BAEP changes. In nine cases, placement of aneurysm clips resulted in SEP or BAEP changes, usually within several minutes. Rapid removal or adjustment of clip position resolved the SEP and BAEP changes in all these cases. In four cases, transient hypotension caused SEP and BAEP changes. Additionally, adjustment of retractors resolved changes in one patient. Of the two patients with permanent IOM changes, one developed a new neurologic deficit and one died.

In addition, in rare instances, cortical SEP changes may identify a peripheral ischemic process, such as a malfunctioning blood pressure cuff causing arm ischemia (Friedman *et al.*, 1991). López *et al.* (1999) also demonstrated that IOM using SEPs and BAEPs was accurate in identifying ischemia in both the anterior and posterior cerebral vasculature and that postoperative neurological deficits were predicted equally in both groups, based on EP changes.

Previously, Manninen *et al.* (1994) reported that dual-modality monitoring using SEPs and BAEPs was useful during posterior fossa aneurysm surgery. They reported that 10 of 67 patients (70 vertebrobasilar aneurysms) showed BAEP changes, with 6 patients developing a postoperative neurologic deficit. There were SEP changes in 14 patients, with 8 developing new postoperative deficits. In addition, all of the patients who had permanent EP changes developed new postoperative neurological deficits. They also found the incidence of false-negative and false-positive results for both modalities to be 20% and 30%, respectively.

Quiñones-Hinojosa *et al.* (2004) used TcMEPs to monitor basilar artery aneurysm surgery. In their study, they monitored 30 consecutive patients using TcMEPs in addition to SEPs and EEG. Changes were seen in 10 patients. Isolated TcMEP and SEP changes were seen in one and five patients, respectively. In patients where both TcMEP and SEP changes occurred, the TcMEP abnormalities were reported to be more "robust" and to occur earlier than SEP abnormalities. The authors concluded that TcMEP monitoring can be safely and easily performed and that TcMEPs may be more sensitive than SEPs in identifying basilar and perforating brainstem artery ischemia.

The concept of multimodality monitoring was further expanded by Neuloh and Schramm (2004), who used a combination of TcMEPs, SEPs, and microvascular Doppler ultrasonography in cerebral aneurysm surgery. They concluded that TcMEPs were superior to the other modalities in detecting motor impairment due to subcortical ischemia and correlated with similar findings from other studies (Horiuchi *et al.*, 2005; Sasaki *et al.*, 2007).

Overall, it is reasonable to conclude that the combined results and findings of these studies suggest that cerebral and brainstem ischemia can be identified using a multimodality IOM approach.

Cerebral AVMs

Although the number of published studies investigating the use of IOM in cerebral AVM surgery is less than that reported for aneurysm surgery, there still appears to be support for the use of IOM in these cases. Chang *et al.* (1999) reported the use of SEPs and BAEPs in the surgical treatment of cerebral AVMs in 53 patients (54 procedures). Five of the 54 procedures (9%) had SEP changes (four transient, one permanent). In the four cases of transient SEP changes, three patients developed a new, but transient, neurologic postoperative deficit and one exhibited no clinical deficit. The sole patient with permanent SEP loss developed a prolonged transient hemiparesis and hemisensory loss lasting three weeks. All transient SEP changes resolved with adjustment or removal of clips on feeding vessels or with elevating the mean arterial pressure. One patient had a permanent change in the BAEP and experienced new permanent neurologic deficits. In addition, one patient without IOM changes developed a new neurologic deficit. The authors reported the sensitivity of SEP and BAEP for predicting new postoperative transient or permanent deficits to be 86% and the specificity to be 98%.

Moyamoya disease

Moyamoya disease is a progressive occlusive disease of the terminal ICA, which can involve the proximal MCA, ACA, and posterior cerebral artery (PCA). As the disease progresses, it can lead to hemodynamic cerebral ischemia. The etiology is unknown but appears not to be atherosclerotic or inflammatory. The disorder usually involves the brain bilaterally, but can also be unilateral. There is compensatory enlargement of perforating arteries, also referred to as moyamoya ("puff of smoke") vessels, in the basal ganglia.

The surgical treatment of this disorder involves direct and indirect revascularization techniques to increase cerebral blood flow. The more commonly performed direct revascularization procedure requires grafting the superior temporal artery (STA) to a branch of the MCA. In doing this, a portion of the MCA supply is compromised during the period that the STA–MCA anastomosis is being performed, raising the risk of cerebral ischemia. Even though there are no currently published series in the use of IOM for these cases, cerebral ischemia is still a possible complication and IOM using a combination of EEG and SEPs may prove to be useful in the intraoperative management of these surgical procedures.

Interventional neuroradiologic procedures

Interventional neuroradiologic procedures have become much more common and are performed routinely in many institutions. The potential for cerebral ischemia, analogous to that seen in cerebrovascular surgical cases, exists and can be a cause for an adverse outcome. In many instances, these procedures are not monitored for the possibility of cerebral ischemia; this is shown by the few reports of IOM monitoring in interventional neuroradiologic procedures. Nonetheless, there are some reports in the literature. Electroencephalography and evoked potentials (EPs) have been used successfully to monitor selected interventional procedures. The most common neuroradiologic procedures monitored are carotid artery angioplasty, deliberate arterial occlusion of a variety of extracranial and intracranial arteries, coiling of intracerebral aneurysms, and glue embolization of cerebral AVMs. The modalities chosen to monitor such cases depend on the cerebrovascular territory at risk and whether the patient will be awake or under general anesthesia during the procedure.

The recordings are performed in a similar fashion to those described previously. One important exception is when the patient is kept awake. In this situation, concomitant neurologic examination in association with "superselective" intra-arterial amobarbital sodium (amytal sodium) testing can augment IOM. This technique will be described later in this chapter.

Paulsen *et al.* (1999) reported the use of SEPs in 23 embolization sessions (17 patients) undergoing treatment of rolandic AVMs. In their study, two patients were unable to undergo embolization because of positive results of the amobarbital sodium testing. One case showed a >50% amplitude reduction of the cortical SEP, and physical examination showed a transient new neurologic deficit. The second case, under general anesthesia, revealed a significant amplitude reduction of the cortical SEP. The authors concluded that the combined use of SEPs and neurologic physical examination was a valuable tool in the endovascular treatment of rolandic AVMs.

In a separate study, Paiva *et al.* (1995) used EEG in monitoring 19 patients during cyanocrylate AVM embolization and found that focal or diffuse low-frequency EEG abnormalities were sometimes followed by poor clinical outcomes.

Dietz *et al.* (1993) studied 17 patients undergoing balloon occlusion test of the ICA owing to tumor infiltrate, using a combination of clinical, electrophysiologic (median nerve SEPs and transcortical motor EPs), and Doppler sonographic monitoring to detect severe cerebral complications. They concluded that neurophysiologic monitoring played an important role in predicting cerebral complications after permanent occlusion of the ICA. In other instances, EEG alone has been used to assess cerebral ischemia. Cloughesy *et al.* (1993) used continuous polygraph and quantitative EEG monitoring, along with repeated detailed clinical examinations, in 17 carotid balloon test occlusions in 16 patients. Four of 17 consecutive cases showed changes in the patient's EEG or examination results during carotid occlusion; two-thirds of patients with EEG changes also had clinical changes; one-third with EEG changes had no associated clinical changes; and one patient had clinical changes but no concomitant EEG changes. The investigators concluded that continuous EEG monitoring and repeated clinical examinations provide useful ways of evaluating cerebral perfusion during carotid test occlusions.

Overall, there is strong clinical evidence that IOM is a useful adjunct in the surgical and endovascular management of cerebral aneurysms and AVMs.

At our institution, endovascular treatment of cerebral AVMs is typically performed using superselective intra-arterial amobarbital sodium (amytal sodium) testing of the feeding artery prior to embolization. Superselective intra-arterial amobarbital sodium testing consists of injecting a small dose of amobarbital sodium, typically 50 mg, directly into the pedicle that is to be embolized or the vessel that is to be occluded. The rationale is that if there is a significant change in the EPs, EEG, or the patient's physical examination after the amobarbital sodium injection, then that particular vascular segment is probably supplying normal brain tissue and should not be embolized.

If no changes are detected in the baseline electrophysiologic studies or the patient's baseline physical examination after the amobarbital sodium test injection, then embolization is performed in the tested vessel. If changes are detected after the test, then that particular vessel is not embolized and a different region of the same artery, usually more proximal to the AVM, or a different feeding artery is tested anew. Table 14.4 lists the recommended IOM modalities, based on the probable cerebrovascular territories at

Table 14.4 Suggested IOM techniques to use in the endovascular treatment of the following cerebrovascular disorders

Procedure and cerebrovascular territory at risk	Recommended IOM technique					
	EEG	Median nerve SEP	Posterior tibial nerve SEP	BAEP	TcMEP (if procedure is done under general anesthesia)	Modified neurologic physical examination
1. Aneurysm coiling, 2. AVM, 3. Deliberate artery occlusion.						
ICA	×	×	×	×	×	Yes, if done with patient awake
MCA	×	×	×		×	Yes, if done with patient awake
ACA	×	×	×		×	Yes, if done with patient awake
V-B		×	×	×	×	Yes, if done with patient awake

risk in the endovascular treatment of common cerebrovascular disorders.

Definition of critical changes

The following criteria were used to define critical EP changes: 50% or more reduction in the amplitude of cortical component of the SEP, an increase in the central conduction time (CCT) of 1 ms or more, or loss of waves III or V of the BAEP, or any combination. In addition, any alteration found in a neurologic physical examination after amobarbital sodium testing, and not due to anesthetic or sedating medication effects, from the patient's pre-testing baseline physical examination results during endovascular treatment was also considered a critical change.

Possible interventions in response to critical changes

In most instances the surgical or endovascular management of cases is altered when IOM changes are detected. Below are strategies that can be employed in some commonly performed procedures.

Endovascular cases

With respect to endovascular cases, revised management in response to IOM changes typically consists of the following:

AVMs

(a) Repositioning the embolizing microcatheter within the same vessel and retesting with amobarbital sodium and serial clinical examinations,
(b) Testing a different vessel altogether with amobarbital sodium and clinical examination, prior to embolization.

Deliberate transient balloon test occlusion of extracranial or intracranial cerebral arteries

(a) Retesting at a different segment of the vessel using electrophysiologic studies and clinical examinations,
(b) Halting the procedure if the changes persist, as this indicates that permanent occlusion will likely result in a neurologic deficit.

Cerebrovascular disorders

In the case of the surgical treatment of cerebrovascular disorders, the following are common strategies used in an attempt to reverse IOM changes.

Aneurysm

(a) Reposition or remove temporary aneurysm clips,
(b) Reposition the permanent clip,
(c) Increase the MAP to improve overall cerebral perfusion,
(d) Reduce cerebral retraction, if possible.

AVMs

(a) Reduce brain retraction, if possible,
(b) Increase the MAP to improve overall cerebral perfusion,
(c) If placed, reposition or replace temporary clips.

CEAs

(a) Increase the MAP to improve overall cerebral perfusion,
(b) Place an indwelling carotid shunt.

Extracranial-to-intracranial bypass procedures (moyamoya)

(a) Increase the MAP to improve overall cerebral perfusion.

Tables 14.3 and 14.4 summarize the IOM techniques that we typically use to monitor cerebral AVM surgery and endovascular interventional cases in our institution. As previously mentioned, the suggested techniques are based on regional cerebral vascular territories at risk during surgery or endovascular treatment and the corresponding neural generator of the evoked potential. It is also important to note that all surgical procedures at our institution are performed using induced mild hypothermia to provide neuroprotection, with the target temperature being 33 °C. When appropriate, we will also administer barbiturates to induce a burst-suppression pattern on the EEG in an attempt to provide a greater degree of neuroprotection.

Stimulating and recording techniques

Many different stimulation and recording protocols can be used to obtain adequate evoked potentials. The following is not meant to be the definitive treatise on the matter. Instead, the objective is to provide a description of the methods and techniques that we have previously partially described and have found useful in obtaining consistent and reproducible peripheral, spinal, and cerebral SEPs and BAEPs (López *et al.*, 1999).

Recordings should be obtained as early as possible during the endovascular or surgical procedure. Initial baseline recordings should be obtained after anesthesia induction and prior to the first incision, and at regular intervals during the surgical procedure, even if the risk of neural injury is low. Simultaneous and serial documentation of the anesthetic levels, mean arterial pressure, temperature, and surgical events should be performed. Monitoring should be done continuously during the period when the nervous system is at risk and the EP morphology, latency, and amplitude compared serially. Furthermore, monitoring should continue until the patient is awake or as close as possible to being awake, since ischemia can occur during the period of wound closure. Changes in SEPs that are considered critical are as follows: 50% or more reduction in the amplitude of cortical (N19-P24/P40-N45) component of the SEP or an increase in the central conduction time of 1 ms or more, or loss of waves III or V of the BAEP.

Stimulation

1. The SEPs are typically recorded after bilateral independent median nerve stimulation using standard surface stimulating electrodes at the wrist. Stimulation is at a rate of 2 to 5 Hz with pulse duration of 0.1 to 0.3 ms using a minimum level of constant-current stimulation intensity approximately 50% above the threshold levels sufficient to produce a thumb twitch. The ground electrode is placed on the arm proximal to the stimulating electrode. In most instances, 150–250 stimulations are averaged and produce adequate and reproducible cortical SEPs.
2. The BAEPs are obtained after bilateral independent ear stimulation, using ear inserts. The preferred stimuli to use are clicks with an alternating or rarefaction polarity. Stimulation rates between 9.7 and 11.7 Hz and an average of 500–1000 stimulations usually produce a full complement of auditory brainstem wave components.
3. The TcMEPs are performed by stimulating the motor cortex. If possible, each hemisphere should be activated independently and done sequentially. This can be done using several different methods with standard surface EEG, subdermal needle, or corkscrew electrodes placed on the scalp. The commonly used C3–C4 or C1–C2 montages will usually generate myogenic MEPs from bilateral upper and lower extremities if sufficiently high stimulation intensities are used. However, lowest-level threshold stimulation levels required to activate the hemisphere of interest should be used. Deep subcortical motor pathways may be activated if stimulation levels are too high, which can then bypass higher cortical ischemia. When excessively high transcranial electrical stimulation levels are

used, it is possible to generate myogenic MEPs from the contralateral limbs while bypassing cortical ischemia.

(a) To generate TcMEPs, anodal stimulation is required. Thus, the electrode corresponding to the cerebral hemisphere of interest should be connected to the stimulator anode contact. Occasionally, we will attempt to isolate the hemisphere of interest using a C3/4–Cz or C3/4–Fz montage, to obtain myogenic MEPs from the contralateral limbs. For example, if the cerebral hemisphere of interest is on the right, then we may primarily stimulate that hemisphere using a C4–Cz or C4–Fz montage and expect to obtain myogenic MEPs from the left upper and possibly the left lower extremity.
(b) Specific stimulation parameters can vary, and depend on several factors. However, reproducible myogenic TcMEPs can usually be obtained using the following settings: pulse duration of 50 μs, stimulus train of 3–6, inter-stimulus interval of 1.5–4.5 ms, and a maximum intensity of 500 V. Proper stimulation settings need to be established for each case and can vary widely between patients.
(c) Direct cortical stimulation can also be used to obtain myogenic TcMEPs. However, this will not be reviewed, since a detailed description of this technique is beyond the scope of this chapter.

Recording

Recordings can be made using a variety of electrodes. Standard surface EEG or subdermal needle electrodes are the most commonly employed. However, if surface electrodes are used, they should be attached with collodion to reduce the risk of accidental detachment. The electrode impedance should be maintained at less than 5000 Ω throughout the procedure. For SEPs, electrodes are placed on the scalp at C3′, C4′, Cz, Cz′, and a reference electrode is placed at the midfrontal (Fz) location. Using the 10–20 international electrode placement system, C3′, C4′, and Cz′ electrode sites are 2 cm behind their corresponding non-prime electrode positions. Subcortical and brachial plexus (Erb's point) evoked potentials can be monitored using other available channels. The low and high filters are set at 30 Hz and 3 kHz. For BAEPs, A1 and A2 (left and right ear-lobe or mastoid) are also placed.

Suggested montages

Median nerve somatosensory evoked potentials

(a) C3′/ C4′–Fz (contralateral cortex–midfrontal reference).
(b) Contralateral cortex–ipsilateral cortex. These montages may also provide recordings of near-field cortical SEP components (N19, P24, P40, and N45).
(c) Cervical spine, typically C5 or C7, referenced to Fz or contralateral shoulder Fz. This set-up may provide identification of subcortical far-field potentials (P14, N18) and permit monitoring of the CCT (time interval between P14 and N19). Central conduction time reflects the intracranial conduction time between foramen magnum and somatosensory cortex.
(d) Ipsilateral Erb's point–contralateral Erb's point allows for recording of a peripheral (brachial plexus) evoked potential.

Posterior tibial somatosensory evoked potentials

(a) CZ–Fz,
(b) Cz′–Fz,
(c) C5/C7–Fz,
(d) Ipsilateral popliteal fossa–knee reference.

Brainstem auditory evoked potentials

Scalp electrodes for EEG are placed in a similar manner to that listed previously. Electrodes are placed in the vertex region at Cz and Cz′ and on both ears (A1 and A2).

Electroencephalography

A minimum of eight channels are recommended, with four electrodes placed on each side of the head. For example, the following sample montage is useful for monitoring of moyamoya surgery: F3/F4, C3/C4, P3/P4, T3/T4, referenced to Fz. At our institution, we have also found it useful to add occipital electrodes (O1 and O2) to this protocol.

TcMEPs

Myogenic recording of compound muscle action potentials can be easily done using surface or subdermal needle electrodes. We prefer to use needle electrodes mainly because of convenience. They are easy to place, are relatively secure, and require no electrolyte contact gel. In situations where the muscles are

Table 14.5 TcMEP stimulation and recording parameters

	Stimulation parameters			
	Intensity (V)	**Pulse duration (ms)**	**Stimulus train**	**ISI (ms)**
	100–500	0.05	3–6	1.5–4.0
		Recording parameters		
	Muscles	Filter settings	Amp gain	Sweep
Upper extremity	Abductor pollicus brevis (thenar)	5–2000 Hz	50–1000 μV	10–20 ms/div
	Forearm flexors or extensors	5–2000 Hz	50–1000 μV	10–20 ms/div
Lower extremity	Tibialis anterior	5–2000 Hz	50–1000 μV	10–20 ms/div
	Abductor hallucis	5–2000 Hz	50–1000 μV	10–20 ms/div

ISI = inter-stimulus interval

superficial, the needle electrodes will penetrate muscle and then effectively become intramuscular recording electrodes. Table 14.5 shows examples of recording and stimulation parameters and lists commonly used muscles for recording myogenic TcMEPs.

Conclusion

Intraoperative neurophysiologic monitoring is a dynamic, evolving field in clinical neurophysiology, which continues to expand into new areas, such as IOM in the interventional radiology suite. The application of these techniques in the surgical and endovascular treatment of cerebrovascular and vascular disorders requires a clear and detailed understanding of neurology, neurosurgery, neuroanatomy, and neurophysiology, as well as many aspects of clinical neurophysiology, including EEG, SEPs, BAEPs, and TcMEPs. In addition, knowledge of the effects of anesthetic agents on electrophysiologic studies is imperative in the proper application of IOM.

This review demonstrates that mastery of these concepts allows for proper application of IOM techniques in vascular disorders, which leads to altered surgical and endovascular management of these cases and, ultimately, the promise of improved patient outcomes. So far, the predominance of clinical data supports the utility of IOM in vascular disorders. However, the field is still lacking in prospective clinical trials and rigorous outcome studies, which indicates that further study is needed. This is to be expected in a relatively young and developing area of medicine.

Case examples for Chapter 14

Case example 1

Figure 14.2a shows a cerebral angiogram of a patient with a basilar tip aneurysm prior to treatment with endovascular coils. Baseline electrophysiologic studies were performed in the endovascular suite prior to placement of aneurysm coils, and the results are shown in Figure 14.2b. During treatment, a sudden loss of the right upper and lower SEPs was noted and the neuroradiologist was notified (Figure 14.2c). Cerebral angiography revealed occlusion of the left PCA (Figure 14.2d). Blood flow was restored after the prompt administration of tPA (tissue plasminogen activator), and Figure 14.2e shows recovery of the left cortical SEPs after right upper and lower extremity stimulation. A final cerebral angiogram (Figure 14.2f) after coiling revealed no arterial occlusion and the patient displayed no new neurologic deficits.

Case example 2

A 42-year-old female with a right frontal arteriovenous malformation (Figure 14.3a, b) was scheduled for surgical resection. Somatosensory evoked potentials were monitored during surgery (Figure 14.3c, d). A >50% amplitude reduction of the right cortical SEPs was associated with a drop of the mean arterial pressure below 60 mmHg (Figure 14.3d), and a rapid recovery occurred after the MAP was raised above 70 mmHg (Figure 14.3e).

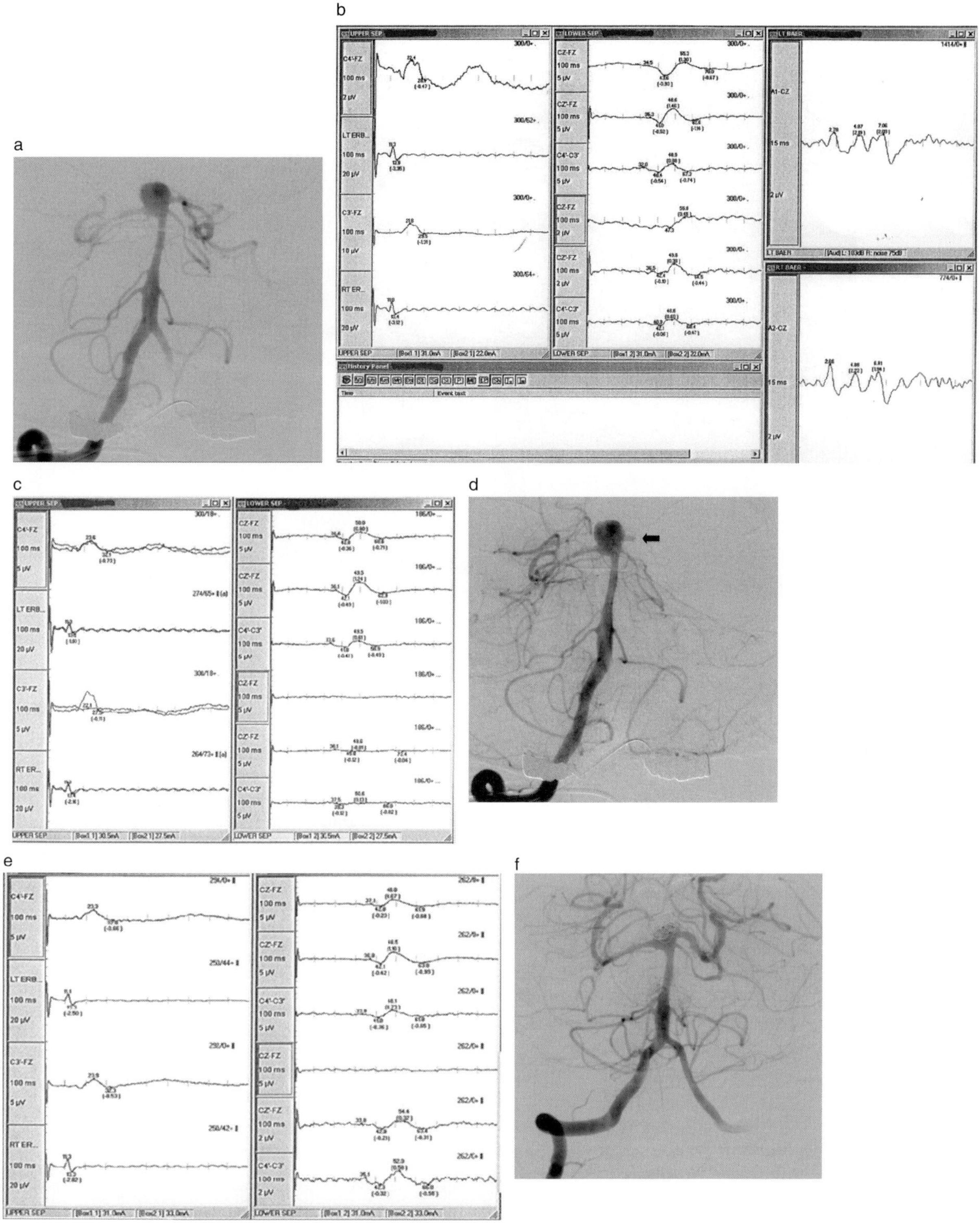

Figure 14.2 (a) Cerebral angiogram of a patient with a basilar tip aneurysm prior to treatment with endovascular coils. (b) Results of baseline electrophysiologic studies. The left panel shows the cortical (trace 1) and Erb's point (trace 2) SEPs after left median nerve stimulation and the corresponding EPs (traces 3 and 4) after right median nerve stimulation. The middle panel shows the cortical SEPs after left posterior tibial nerve (traces 1–3) and right posterior tibial nerve stimulation, respectively. The right upper panel shows the BAEP after left ear stimulation and the right lower panel shows the BAEP after right ear stimulation. (c) A sudden loss of the right upper and lower SEPs. (d) Cerebral angiogram, showing occlusion of the left PCA (arrow). (e) Recovery of the left cortical SEPs after right upper and lower extremity stimulation. (f) Final cerebral angiogram after coiling, revealing no arterial occlusion.

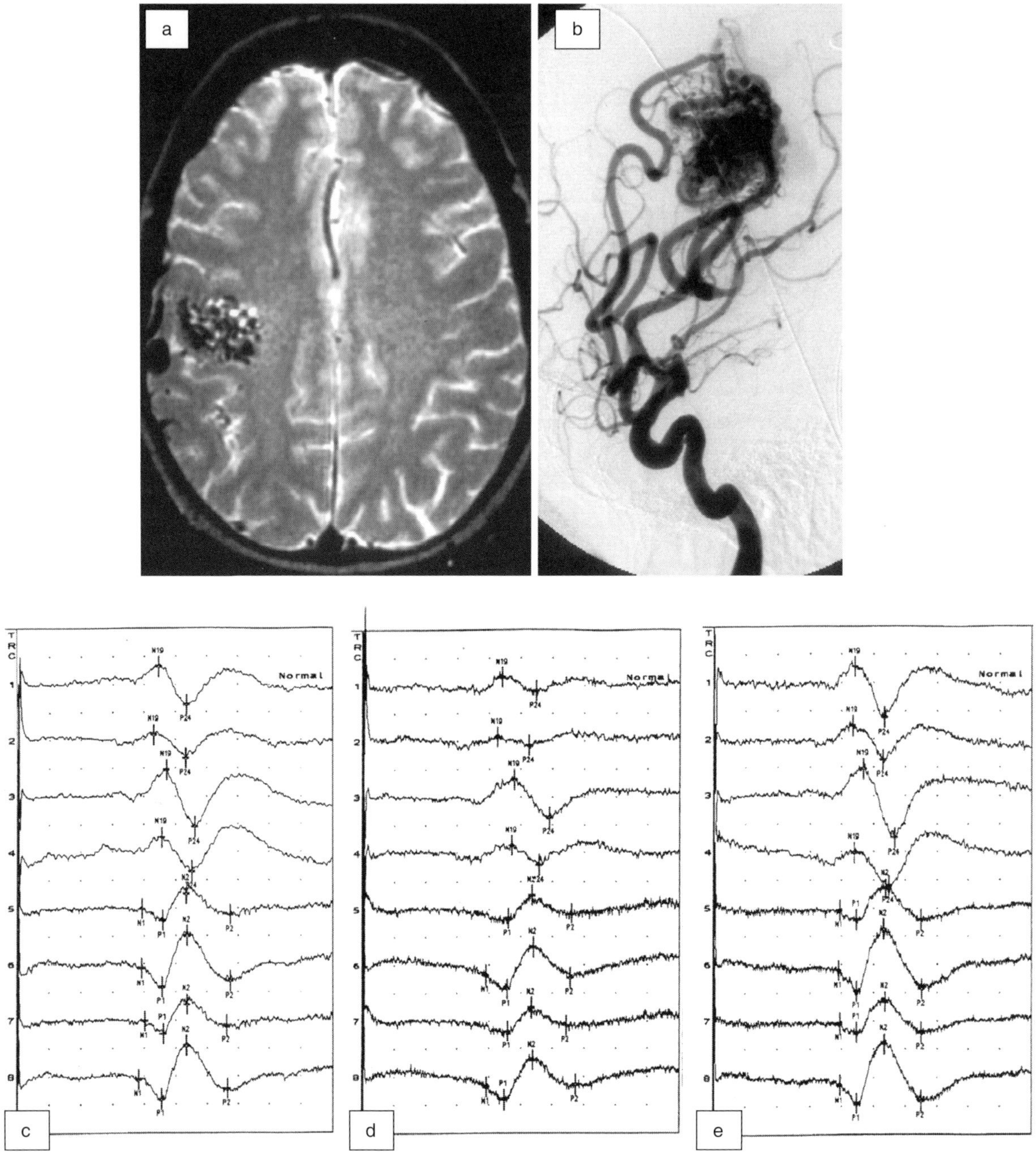

Figure 14.3 (a) MRI scan, showing right frontal AVM. (b) Cerebral angiogram, showing the right frontal AVM. (c) Baseline scalp recorded cortical SEPs after stimulation of the following nerves: left median (traces 1, 2); right median (traces 3, 4); left posterior tibial (traces 5, 6); and right posterior tibial (traces 7, 8). (d) A >50% amplitude reduction of the right cortical SEPs (traces 1 and 2) associated with a drop of the mean arterial pressure below 60 mmHg. (e) Rapid recovery of the right cortical SEPs (traces 1, 2) after the mean arterial pressure is raised above 70 mmHg.

Questions

1. Which of the following is an intervention that can be used in an attempt to reverse IOM changes consistent with cerebral ischemia in cerebral aneurysm and AVM surgery?
   (a) Placement of an indwelling shunt
   (b) Reducing the anesthetic levels
   (c) Increasing the mean arterial pressure
   (d) Increasing the TcMEP stimulus intensity
   (e) Increasing core body temperature
2. The use of IOM of TAAA is primarily to identify and prevent:
   (a) Cerebral ischemia
   (b) Cerebral emboli
   (c) Cervical spinal cord ischemia
   (d) Thoracic spinal cord ischemia
   (e) Lower limb ischemia
3. Which of the following is (or are) commonly monitored in CEA surgery?
   (a) TcMEPs
   (b) Radionuclide cerebral blood flow measurements
   (c) SEPs
   (d) BAEPs
   (e) All of the above
4. Generalized bilateral slowing of the EEG is noted during right carotid cross-clamping during a right CEA. What is the most likely cause of the EEG slowing?
   (a) Right cerebral hemisphere ischemia
   (b) Left cerebral hemisphere ischemia
   (c) Bolus administration of propofol
   (d) Low recording electrode impedances
   (e) Lack of a right carotid indwelling shunt
5. What are the common neuroradiologic procedures that are monitored?
   (a) Carotid artery angioplasty
   (b) Deliberate intracranial arterial occlusion
   (c) Endovascular coiling of intracerebral aneurysms
   (d) Embolization of cerebral AVMs
   (e) All of the above
6. A fundamental difference in monitoring some endovascular versus surgical procedures is that the patient may be awake: how does this affect the IOM protocol?
   (a) It does not; IOM should be exactly the same
   (b) Clinical examination should be included
   (c) SEPs cannot be monitored in awake patients
   (d) BAEPs cannot be monitored in awake patients
   (e) EEG is the only reliable method in this setting
7. Aphasia that develops five minutes after a deliberate balloon test occlusion of the left ICA but without cortical SEP or EEG changes is indicative of:
   (a) Left hemisphere ischemia
   (b) Technically inadequate SEPs and EEG
   (c) Sedative medication side-effects
   (d) Faulty clinical examination
   (e) Poor balloon placement
8. Experimental animal studies show that a 50% amplitude reduction of the cortical SEP corresponds to an rCBF of which of the following?
   (a) 23–25 ml/100 g/min
   (b) 20–22 ml/100 g/min
   (c) 17–19 ml/100 g/min
   (d) 14–16 ml/100 g/min
   (e) 11–13 ml/100 g/min
9. Which of the following techniques would be most useful in IOM of a left ACA aneurysm?
   (a) Cortical SEPs after left median nerve stimulation
   (b) Cortical SEPs after right median nerve stimulation
   (c) Cortical SEPs after left posterior tibial nerve stimulation
   (d) Cortical SEPs after right posterior tibial nerve stimulation
   (e) EEG
10. Intraoperative EEG rhythmic beta activity in the 12–18 Hz range is commonly due to which of the following?
   (a) Anesthetic effects
   (b) Cerebral ischemia
   (c) Poor recording electrode placement
   (d) Electrical artifact
   (e) Hypothermia

References

Achouh, P. E., Estrera, A. L., Miller, C. C., *et al.* (2007). Role of somatosensory evoked potentials in predicting outcome during thoracoabdominal aortic repair. *Annals of Thoracic Surgery*, **84**, 782–788.

Ahn, S., Jordan, S. E., Nuwer, M. R., Marcus, D. R., and Moore, W. S. (1988). Computed electroencephalographic topographic brain mapping. A new and accurate monitor of cerebral circulation and function for patients having carotid endarterectomy. *Journal of Vascular Surgery*, **8**, 247–254.

Amantini, A., Bartelli, M., de Scisciolo, *et al.* (1992). Monitoring of somatosensory evoked potentials during carotid endarterectomy. *Journal of Neurology*, **239**, 241–247.

Astrup, J. (1982). Energy-requiring cell functions in the ischemic brain: their critical supply and possible inhibition in protective therapy. *Journal of Neurosurgery*, **56**, 482–497.

Beese, U., Langer, H., Lang, W., and Dinkel, M. (1998). Comparison of near-infrared spectroscopy and somatosensory evoked potentials for the detection of cerebral ischemia during carotid endarterectomy. *Stroke*, **29**, 2032–2037.

Bornstein, N. M., Rossi, G. B., Treves, T. A., and Shifrin, E. G. (1996). Is transcranial Doppler effective in avoiding the hazards of carotid surgery? *Cardiovascular Surgery*, **4**, 335–337.

Branston, N. M., Symon, L., Crockard, H. A., and Pasztor, E. (1974). Relationship between the cortical evoked potential and local cortical blood flow following acute middle cerebral artery occlusion in the baboon. *Experimental Neurology*, **45**, 195–208.

Branston, N. M., Strong, A. J., and Symon, L. (1977). Extracellular potassium activity, evoked potential and tissue blood flow: relationships during progressive ischaemia in baboon cerebral cortex. *Journal of the Neurological Sciences*, **32**, 305–321.

Branston, N. M., Ladds, A., Symon, L., and Wang, A. D. (1984). Comparison of the effects of ischaemia on early components of the somatosensory evoked potential in brainstem, thalamus, and cerebral cortex. *Journal of Cerebral Blood Flow and Metabolism*, **4**, 68–81.

Brinkman, S. D., Braun, P., Ganji, S., Morrell, R. M., and Jacobs, L. A. (1984). Neuropsychological performance one week after carotid endarterectomy reflects intraoperative ischemia. *Stroke*, **15**, 497–503.

Buchtal, A. and Belopavlovic, M. (1992). Somatosensory evoked potentials in cerebral aneurysm surgery. *European Journal of Anaesthesiology*, **9**, 493–497.

Chang, S. D., López, J. R., and Steinberg, G. K. (1999). The usefulness of electrophysiological monitoring during resection of central nervous system vascular malformations. *Journal of Stroke and Cerebrovascular Diseases*, **8**, 412–422.

Chiappa, K. H., Burke, S. R., and Young, R. R. (1979). Results of electroencephalogram monitoring during 367 carotid endarterectomies: Use of a dedicated minicomputer. *Stroke*, **10**, 381–388.

Cho, I., Smullens, S. N., Streletz, L. J., and Fariello, R. G. (1986). The value of intraoperative EEG monitoring during carotid endarterectomy. *Annals of Neurology*, **20**, 508–551.

Cloughesy, T. F., Nuwer, M. R., Hoch, D., *et al.* (1993). Monitoring carotid test occlusions with continuous EEG and clinical examination. *Journal of Clinical Neurophysiology*, **10**, 363–369.

Davies, L. K. and Janelle, G. M. (2006). Conclusion: all cardiac surgical patients should not have intraoperative cerebral oxygenation monitoring. *Journal of Cardiothoracic and Vascular Anesthesia*, **20**, 450–455.

DeBakey, M. E. (1975). Successful carotid endarterectomy for cerebrovascular insufficiency: 19-year follow-up. *Journal of the American Medical Association*, **233**, 1083–1085.

De Vleeschauwer, P., Horsch, S., and Matamoros, R. (1988). Monitoring of somatosensory evoked potentials in carotid surgery: results, usefulness and limitations of the method. *Annals of Vascular Surgery*, **2**, 63–68.

Dietz, A., von Kummer, R., Adams, H. P., *et al.* (1993). Balloon occlusion test of the internal carotid artery for evaluating resectability of blood vessel infiltrating cervical metastasis of advanced head and neck cancers – Heidelberg experience. *Laryngorhinootologie*, **72**, 558–567.

Emerson, R. G. and Turner, C. A. (1993). Monitoring during supratentorial surgery. *Journal of Clinical Neurophysiology*, **10**, 404–411.

Etz, C. D., Halstead, J. C., Spielvogel, D., *et al.* (2006). Thoracic and thoracoabdominal aneurysm repair: is reimplantation of spinal cord arteries a waste of time? *Annals of Thoracic Surgery*, **82**, 1670–1678.

Executive Committee for the Asymptomatic Carotid Atherosclerosis Study (1995). Endarterectomy for asymptomatic carotid artery stenosis. *Journal of the American Medical Association*, **273**, 1421–1428.

Fisher, R. S., Raudzens, P., and Nunemacher, M. (1995). Efficacy of intraoperative neurophysiological monitoring. *Journal of Clinical Neurophysiology*, **12**, 97–109.

Fode, N. C., Sundt, T. M., Robertson, J. T., Peerless, S. J., and Shields, C. B. (1986). Multicenter retrospective review of results and complications of carotid endarterectomy in 1981. *Stroke*, **17**, 370–376.

Friedman, W. A., Kaplan, B. L., Day, A. L., Sypert, G. W., and Curran, M. T. (1987). Evoked potential monitoring during aneurysm operation: observations after 50 cases. *Neurosurgery*, **20**, 678–687.

Friedman, W. A., Chadwick, M. A., Verhoeven, F. J., Mahla, M., and Day, A. L. (1991). Monitoring of somatosensory evoked potentials during surgery for middle cerebral artery aneurysms. *Neurosurgery*, **29**, 83–88.

Gigli, G. L., Caramia, M., Marciani, M. G., *et al.* (1987). Monitoring of subcortical and cortical somatosensory evoked potentials during carotid endarterectomy: comparison with stump pressure levels. *Electroencephalography and Clinical Neurophysiology*, **68**, 424–432.

Halsey, J. H. (1992). Risks and benefits of shunting in carotid endarterectomy. *Stroke*, **23**, 1583–1587.

Hargadine, J. R., Branston, N. M., and Symon, L. (1980). Central conduction time in primate brain ischemia – a study in baboons. *Stroke*, **11**, 637–642.

Haupt, W. F. and Horsch, S. (1992). Evoked potential monitoring in carotid surgery: a review of 994 cases. *Neurology*, **42**, 835–838.

Horiuchi, K., Suzuki, K., Sasaki, T., *et al.* (2005). Intraoperative monitoring of blood flow insufficiency during surgery of middle cerebral artery aneurysms. *Journal of Neurosurgery*, **103**, 275–283.

Jones, T. H., Morawetz, R. B., Crowell, R. M., *et al.* (1981). Thresholds of focal ischemia in awake monkeys. *Journal of Neurosurgery*, **54**, 773–782.

Kearse, L. A., Brown, E. N., and McPeck, K. (1992). Somatosensory evoked potential sensitivity relative to electroencephalography for cerebral ischemia during carotid endarterectomy. *Stroke*, **23**, 498–505.

Kearse, L. A., Martin, D., McPeck, K., and Lopez-Breshnahan, M. (1993). Computer-derived density spectral array in detection of mild analog electroencephalographic ischemic pattern changes during carotid endarterectomy. *Journal of Neurosurgery*, **78**, 884–890.

Lam, A. M., Manninen, P. H., Ferguson, G. G., and Nantua, W. (1991). Monitoring electrophysiologic function during carotid endarterectomy: a comparison of somatosensory evoked potentials and conventional electroencephalogram. *Anesthesiology*, **75**, 15–21.

Lesnick, J. E., Michele, J. J., Simeone, F. A., DeFeo, S., and Welsh, F. A. (1984). Alteration of somatosensory evoked potentials in response to global ischemia. *Journal of Neurosurgery*, **60**, 490–494.

Levy, W. J., Shapiro, H. M., Maruchak, G., and Meathe, E. (1980). Automated EEG processing for intraoperative monitoring: A comparison of techniques. *Anesthesiology*, **53**, 223–236.

Lindsey, R. L. (1984). A simple solution for determining shunt flow during carotid endarterectomy. *Anesthesiology*, **61**, 215–216.

Liu, A. Y., López, J. R., Do, H. M., *et al.* (2003). Neurophysiological monitoring in the endovascular therapy of aneurysms. *American Journal of Neuroradiology*, **24**, 1520–1527.

Livesay, J. J., Cooley, D. A., Ventemiglia, R. A., *et al.* (1985). Surgical experience in descending thoracic aneurysmectomy with and without adjuncts to avoid ischemia. *Annals of Thoracic Surgery*, **39**, 3–46.

Loftus, C. M. and Quest, D. O. (1995). Technical issues in carotid artery surgery 1995. *Neurosurgery*, **36**, 629–647.

Loftus, C. M., Dyste, G. N., Reinarz, S. J., and Hingtgen, W. L. (1986). Distal cervical carotid dissection after carotid endarterectomy: a complication of indwelling shunt? *Neurosurgery*, **19**, 441–445.

López, J. R., Chang, S. D., and Steinberg, G. K. (1999). The use of electrophysiological monitoring in the interoperative management of intracranial aneurysms. *Journal of Neurology, Neurosurgery and Psychiatry*, **66**, 189–196.

Manninen, P. H., Patterson, S., Lam, A. M., Gelb, A. W., and Nantua, W. E. (1994). Evoked potential monitoring during posterior fossa aneurysm surgery: a comparison of two modalities. *Canadian Journal of Anaesthesia*, **41**, 92–97.

Manninen, P., Tan, T., and Sarjeant, R. M. (2001). Somatosensory evoked potential monitoring during carotid endarterectomy in patients with a stroke. *Anesthesia and Analgesia*, **93**, 39–44.

Markand, O. N. (1994). Continuous assessment of cerebral function with EEG and somatosensory evoked potential techniques during extracranial vascular reconstruction. In *Intraoperative Monitoring Techniques in Neurosurgery*, ed. C. M. Loftus and V. C. Traynelis. New York: McGraw-Hill, pp. 19–32.

Markand, O. N., Dilley, R. S., Moorthy, S., and Warren, C. (1984). Monitoring of somatosensory evoked responses during carotid endarterectomy. *Archives of Neurology*, **41**, 375–378.

Messick, J. M., Sharbrough, F., and Sundt, T. (1984). Selective shunting on the basis of EEG and regional CBF monitoring during carotid endarterectomy. *International Anesthesiology Clinics*, **22**, 137–145.

Moorthy, S. S., Markand, O. N., Dilley, R. S., McCammon, R., and Warren, C. H. (1982). Somatosensory-evoked responses during carotid endarterectomy. *Anesthesia and Analgesia*, **61**, 879–883.

Morawetz, R. B., DeGiorlami, U., Ojemann, R. G., Marcoux, F. W., and Crowell, R. M. (1978). Cerebral blood flow determined by hydrogen clearance during middle cerebral artery occlusion in unanesthetized monkeys. *Stroke*, **9**, 143–149.

Moritz, S., Kasprzak, P., Arlt, M., Taeger, K., and Metz, C. (2007). Accuracy of cerebral monitoring in detecting cerebral ischemia during carotid endarterectomy. A comparison of transcranial Doppler sonography, near-infrared spectroscopy, stump pressure, and somatosensory evoked potentials. *Anesthesiology*, **107**, 563–569.

Myers, R. R., Stockard, J. J., and Saidman, L. J. (1977). Monitoring of cerebral perfusion during anesthesia by time-compressed Fourier analysis of the electroencephalogram. *Stroke*, **8**, 331–337.

Neuloh, G. and Schramm, J. (2004). Monitoring of motor-evoked potentials compared with somatosensory evoked potentials and microvascular Doppler ultrasonography in cerebral aneurysm surgery. *Journal of Neurosurgery*, **100**, 389–399.

North American Symptomatic Carotid Endarterectomy Trial Collaborators (1991). Beneficial effect of carotid endarterectomy in symptomatic patients with high grade stenosis. *New England Journal of Medicine*, **325**, 445–453.

Paiva, T., Campos, J., Baeta, E., *et al.* (1995). EEG monitoring during endovascular embolization of cerebral arteriovenous malformations. *Electroencephalography and Clinical Neurophysiology*, **95**, 3–13.

Palatinsky, E., DiScenna, A., McDonald, H., Whittingham, T., and Selman, W. (1994). SSEP and BAEP monitoring of temporary clip application and induced hypotension during cerebrovascular surgery. In *Intraoperative Monitoring Techniques in Neurosurgery*, ed. C. M. Loftus and V. C. Traynelis. New York: McGraw-Hill, pp. 61–71.

Paulsen, R. D., Steinberg, G. K., Norbash, A. M., López, J. R., and Marks, M. (1999). Embolization of rolandic cortex arteriovenous malformations. *Neurosurgery*, **44**, 479–484.

Pedrini, L., Tarantini, S., Cirelli, M. R., *et al.* (1998). Intraoperative assessment of cerebral ischaemia during carotid surgery. *International Angiology*, **17**, 10–14.

Perez-Borja, C. and Meyer, J. S. (1965). Electroencephalographic monitoring during reconstructive surgery of the neck vessels. *Electroencephalography and Clinical Neurophysiology*, **18**, 162–169.

Quiñones-Hinojosa, A., Alam, M., Lyon, R., Yingling, C. D., and Lawton, M. T. (2004). Transcranial motor-evoked potentials during basilar artery aneurysm surgery: technique application for 30 consecutive patients. *Neurosurgery*, **54**, 16–24.

Rampil, I. J., Holzer, J. A., Quest, D. O., Rosenbaum, S. H., and Correll, J. W. (1983). Prognostic value of computerized EEG analysis during carotid endarterectomy. *Anesthesia and Analgesia*, **62**, 186–192.

Redekop, G. and Ferguson, G. (1992). Correlation of contralateral stenosis and intraoperative electroencephalogram changes with risk of stroke during carotid endarterectomy. *Neurosurgery*, **34**, 191–194.

Rigamonti, A., Scandroglio, M., Minicucci, F., *et al.* (2003). A clinical evaluation of near-infrared cerebral oximetry in the awake patient to monitor cerebral perfusion during carotid endarterectomy. *Journal of Clinical Anesthesia*, **17**, 426–430.

Rowed, D. W., Houlden, D. A., Burkholder, L. M., and Taylor, A. B. (2004). Comparison of monitoring techniques for intraoperative cerebral ischemia. *The Canadian Journal of Neurological Sciences*, **31**, 347–356.

Sasaki, T., Kodama, N., Matsumoto, M., *et al.* (2007). Blood flow disturbance in perforating arteries attributable to aneurysm surgery. *Journal of Neurosurgery*, **107**, 60–67.

Schepens, M. A., Boezeman, E. H., Hamerlijnck, R. P., TerBeek, H., and Vermeulen, F. E. (1994). Somatosensory evoked potentials during exclusion and reperfusion of critical aortic segments in thoracoabdominal aortic aneurysm surgery. *Journal of Cardiac Surgery*, **9**, 692–702.

Schramm, J., Koht, A., Schmidt, G., *et al.* (1990). Surgical and electrophysiological observations during clipping of 134 aneurysms with evoked potential monitoring. *Neurosurgery*, **26**, 61–70.

Schwarz, G., Litscher, G., Kleinert, R., and Jobstmann, R. (1996). Cerebral oximetry in dead subjects. *Journal of Neurosurgery and Anesthesiology*, **8**, 189–193.

Schweiger, H., Kamp, H. D. and Dinkel, M. (1991). Somatosensory-evoked potentials during carotid artery surgery: experience in 400 operations. *Surgery*, **109**, 602–609.

Sharbrough, F., Messick, J., and Sundt, T. M. (1973). Correlation of continuous electroencephalograms with cerebral blood flow measurements during carotid endarterectomy. *Stroke*, **4**, 674–683.

Sloan, M. A. (2006). Prevention of ischemic neurologic injury with intraoperative monitoring of selected cardiovascular and cerebrovascular procedures: roles of electroencephalography, somatosensory evoked potentials, transcranial Doppler, and near-infrared spectroscopy. *Neurologic Clinics*, **24**, 631–645.

Sundt, T. M. (1983). The ischemic tolerance of neural tissue and the need for monitoring and selective shunting during carotid endarterectomy. *Stroke*, **14**, 93–98.

Sundt, T. M., Sharbrough, F. W., Anderson, R. E., and Michenfelder, J. D. (1974). Cerebral blood flow measurements and electroencephalograms during carotid endarterectomy. *Journal of Neurosurgery*, **41**, 310–320.

Sundt, T. M., Sharbrough, F., Piepgras, D. G., *et al.* (1981). Correlation of cerebral blood flow and electroencephalographic changes during carotid endarterectomy with results of surgery and hemodynamics of cerebral ischemia. *Mayo Clinic Proceedings*, **56**, 533–543.

Svensson, L. G., Crawford, E. S., Hess, K. R., Coselli, J. S., and Safi, H. J. (1993). Experience with 1509 patients undergoing thoracoabdominal aortic operations. *Journal of Vascular Surgery*, **17**, 357–370.

Symon, L. (1980). The relationship between CBF, evoked potentials and the clinical features in cerebral ischaemia. *Proceedings of the 23rd Scandinavian Neurological Congress. Acta Neurologica Scandinavica*, **62** (suppl.), 175–190.

Symon, L., Crockard, H. A., Dorsch, N. W. C., Branston, N. M., and Juhasz, J. (1975). Local cerebral blood flow and vascular reactivity in a chronic stable stroke in baboons. *Stroke*, **6**, 482–492.

Tiberio, G., Floriani, M., Giulini, S. M., *et al.* (1991). Monitoring of somatosensory evoked potentials during carotid endarterectomy: Relationship with different haemodynamic parameters and clinical outcome. *European Journal of Vascular Surgery*, **5**, 647–653.

Trojaborg, W. and Boyson, G. (1973). Relationship between EEG, regional cerebral blood flow and internal carotid artery pressure during carotid endarterectomy. *Electroencephalography and Clinical Neurophysiology*, **34**, 61–69.

Verdant, A. (1992). Descending thoracic aortic aneurysms: Surgical treatment with the Gott shunt. *Canadian Journal of Surgery*, **35**, 493–496.

Wahr, J., Tremper, K. K., Samra, S., and Delpy, D. T. (1996). Near-infrared spectroscopy: theory and applications. *Journal of Cardiothoracic Vascular Anesthesia*, **10**, 406–418.

Wassmann, H., Fischdick, G., and Jain, K. K. (1984). Cerebral protection during carotid endarterectomy–EEG monitoring as a guide to the use of intraluminal shunts. *Acta Neurochirurgica*, **71**, 99–108.

Whittemore, A. D. (1980). Carotid endarterectomy and alternative approach. *Archives of Surgery*, **115**, 940–942.

Wöber, C., Zeitlhofer, J., Asenbaum, S., *et al.* (1998). Monitoring of median nerve somatosensory evoked potentials in carotid surgery. *Journal of Clinical Neurophysiology*, **15**, 429–438.

Zampella, E., Morawetz, R. B., McDowell, H. A., *et al.* (1991). The importance of cerebral ischemia during carotid endarterectomy. *Neurosurgery*, **29**, 727–731.

15 Selective dorsal rhizotomies: technique and protocol for monitoring

Gloria M. Galloway

Background, rationale, and functional assessments

Selective dorsal rhizotomy (SDR) is a surgical procedure used to treat spasticity, and has been in use since the early part of the twentieth century. It has become more frequently utilized for this purpose recently, partly because of complications associated with the use of the implantable pumps that deliver antispasmodic medication to treat the spasticity. The surgical procedure referred to as selective dorsal rhizotomy is primarily performed in the pediatric population, for spasticity occurring predominately as a consequence of cerebral palsy, although it can be used for spasticity from other causes as well. In most of these cases the cerebral palsy is idiopathic but it may be a result of an ischemic or vascular injury. Various forms of cerebral palsy exist, including the spastic, dystonic, and choreoatheotoid forms. The spastic form of cerebral palsy is the most common type and appears to be the most responsive to selective dorsal rhizotomy.

Reports of randomized clinical trials comparing outcomes using functional scores in patients with cerebral palsy undergoing selective dorsal rhizotomy and physiotherapy showed significant improvements in functional outcomes in those patients undergoing SDR compared with those patients undergoing physiotherapy alone (Phillips and Parks, 1989).

Several quantitative scales have been developed to measure function and range of motion of joints. One scale is called the *Gross Motor Functional Scale* (Gregson *et al.*, 1999) and is widely used in both clinical and research studies to follow patients with cerebral palsy. It is less precise over time in infants than in older children but still very useful (Ross and Engsberg, 2007). It measures function from Level 1 (best) to Level 5 (worst) and utilizes age ranges of gross functional abilities based on self-initiated movements of the child. Because it is difficult to be precise in regard to motor function in children younger than two years, this scale it is less useful in the younger age range.

Other functional scales in use include the *Ashworth scale*, the *Modified Ashworth Scale* (Bohannan and Smith, 1987), and the *Pediatric Evaluation of Disability scale* (Coster *et al.*, 2008). In addition, range-of-motion indicators can be used to assess function: one example uses the popliteal angle range (Thompson *et al.*, 2001) as compared with children without cerebral palsy (Katz *et al.*, 1992; Maathuis *et al.*, 2005).

Regardless of the etiology, treatment of this disabling spasticity is necessary to improve quality of life for both the patient and caregivers. Often, unsuccessful trials with medication, including both oral forms and repeated botulinum toxin injections, are used to treat the spasticity first. Botulinum toxin can be successfully injected on repeated occasions in cerebral palsy patients for focal muscle spasticity, particularly equine foot deformity, to aid in ambulation. Although very helpful in some patients, botulinum toxin has been shown initially to improve hip and knee range of motion significantly. The effectiveness of botulinum toxin lessens after repeated injections and there is less effect with botulinum toxin on muscle tone and spasticity than on joint range of motion. The muscles typically injected are the gastrocnemius, soleus, and the hamstrings. Because of significant spasticity, patients not only have difficulty with ambulation, but in those patients who are no longer ambulatory, it is often difficult to care for them, perform daily hygiene and transfer them in and out of bed or to a bath or chair.

When patients are being considered for selective dorsal rhizotomy, a thoughtful approach utilizing a team to include a neurologist, neurosurgeon, rehabilitative therapists, and even a psychologist. This multidisciplinary team approach to patient care and treatment involves many types of medical caregivers with different skills (van Shie *et al.*, 2005).

The indications for SDR in spastic cerebral palsy are well described by Abbott (1991). These indications are based on degree of ambulation or dependence and can be divided into six categories:

1. Patients independently ambulating without assistive devices: these patients may have early contracture or need tendon releases
2. Patients ambulating with assistive devices, most often due to trunk instability
3. Quadruped crawlers with impaired postural control of standing
4. Commando (belly) crawlers with severe postural impairments
5. Totally dependent, requiring assistance with all activities of daily living
6. Asymmetric spasticity

To assess the benefit of this procedure, quantifiable measures of functional ability should be used. These measures include the Gross Motor Function Measure, the Pediatric Evaluation of Disability Inventory (Haley *et al.*, 1993), the Edinburgh Visual Gait Score (Reed *et al.*, 1999; Kerr *et al.*, 2005), and Passive Range of Motion Score (Mittal *et al.*, 2005).

There has been significant debate as to whether there is greater potential benefit in using selective dorsal rhizotomy earlier in the course of disease rather than later (Hellelgard *et al.*, 2007). This is particularly of interest, given the greater recognition of complications associated with the use of implantable pumps to administer antispasmodic medications (Abbott *et al.*, 1993).

Preoperative assessment of the patient beforehand by the involved neurophysiologist allows for identification of the particular muscle groups most useful to monitor during the procedure, as each patient may present unique features and particular characteristics. Therefore a good working relationship with the surgeon involved allows for better coordination of care. This preoperative assessment may be done several days prior to the procedure or may need to be coordinated on the day of the actual surgery. Planning the procedure and which muscle groups should be monitored in a given patient is the key to a successful result (Turner, 2009).

In spastic cerebral palsy, there is a reduction in the spinal inhibitory influence on the alpha motor neuron pool. Spasticity occurs as a result of impairment of the normally inhibitory descending corticospinal tracts. Sensory information mediated through the posterior or dorsal roots is relayed to the anterior horn cells with reduced inhibition from the corticospinal tracts. This reduced inhibition results in an imbalance with increased tone with spasticity and clonus (Figure 15.1).

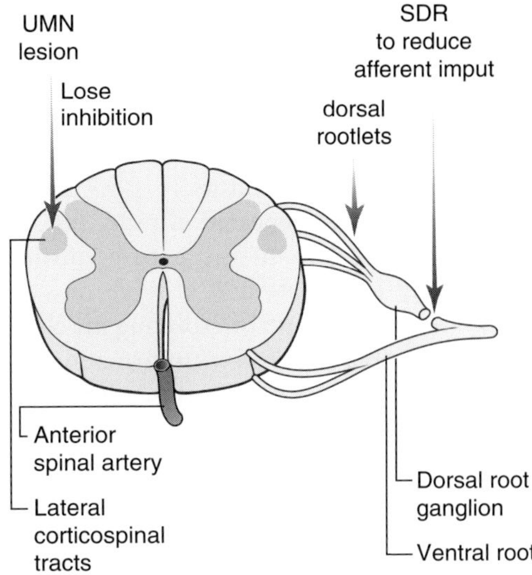

Figure 15.1 Influences on alpha motor neuron output in normal and abnormal muscle tone.

Selective dorsal rhizotomy does not remove all sensory input but is selective in partially removing sensory input so that the degree of motor tone is reduced but the patient is not left hypotonic (Peacock *et al.*, 1990). Patients who are still ambulating often depend on a degree of spasticity to walk. Even in those patients not still ambulatory, the goal of selective dorsal rhizotomy is to reduce the functionally limiting spasticity without abolishing the spasticity of any muscles needed for stability (Peacock *et al.*, 1987).

Clinical assessment and operating room set-up

In some ways, it is more important in these cases than in other surgeries to consider how the monitoring equipment should be set up in the operating room and where the various NIOM team members will be. It is necessary to know which set-up will work best for these cases in regard to equipment, personnel, and proximity to the operating field when creating a protocol for the NIOM team involved. As such, the clinical neurophysiologist will need to be in a position that is close to both the patient and the monitoring equipment. In this way, he or she can clinically palpate muscles as well as see the electromyographic signals at the time of stimulation, to interpret abnormality accurately (Fasano *et al.*, 1988). Typically, this means that the neurophysiologist will be close to the surgical staff and under draping to allow close patient contact.

In this same small space will be the monitoring equipment and a very skilled neurophysiology technician to handle the equipment. Then consideration will be made for all the other types of equipment needed in the surgical room, including the anesthesiology team and equipment and systemic monitors.

The procedure involves repeated clinical assessments of muscle contraction in the patient during the case. The ability to palpate and observe muscle contractions while reviewing and interpreting the electromyographic recording allows for confirmation of these muscle responses and allows observation of muscles involved in contraction that may not have been selected for monitoring, based on prior review. In this way assessment of clonus and hypertonicity is obtained, as well as of the elimination of clonus after sectioning occurs.

However, multitasking of some degree is required, depending on the set-up, in that both assessment of the patient's extremities under the sterile draping and interpretation of the electromyographic signals on the monitor will be needed as each rootlet is stimulated (Steinbok *et al.*, 1992a). In the case of failure of the monitoring equipment or inability to record muscle responses because of anesthesia or other factors, the clinical assessment provides a degree of safety in that there is often correlation of the findings. Therefore, in this procedure the neurophysiologist provides meaningful assessment and interpretation clinically, which may typically require being under draping, while reviewing the electrophysiological data simultaneously. It can be a very useful approach to complement and support the neurophysiologic assessment.

Methodology of the procedure

The procedure involves surgical lumbar laminectomy with an incision that exposes both the lumbar and sacral spinal nerves from the L2 to S2 levels. The surgeon identifies the S1 root, which is the largest nerve root in the lumbosacral area and serves as a useful landmark for spinal level identification below and above. At each spinal level, selective severing of several dorsal rootlets serves to decrease the facilitory afferent input.

The association of this procedure with neurophysiologic monitoring was first described by Fasano *et al.* (1980) and further modified by other investigators, particularly Peacock *et al.* (1987) and then Deletis and colleagues (1992). Fasano's technique involves stimulating sensory roots located in the cauda equina. The technique was modified by Peacock *et al.*, so that the rhizotomy was to occur at the level of the root exit foramina rather than at the level of the conus and the lower lumbar and sacral roots could be more accurately identified and stimulated, allowing sparing of the sphincter. Later techniques described by Deletis allowed for identification of nerve roots specifically involved in micturition with associated improvement in postoperative urological outcome.

The surgeon will identify the ventral and dorsal roots upon separation and, depending on style and experience, may proceed to dorsal root stimulation or may first have each ventral root stimulated and then marked before testing each dorsal root to assure proper identification. Once each dorsal root is stimulated, it is divided into its component rootlets. There are typically two to six rootlets per dorsal root. This division into component rootlets is done because each rootlet will need to be individually stimulated. It is very common to have an abnormal response from one rootlet in a given root and a normal response from the next tested rootlet in that same root.

The surgeon will opt to sever the most severely involved rootlets based on the clinical and electromyographic (EMG) response. Much greater stimulation intensity is needed for dorsal root stimulation than for ventral root stimulation. In a patient in whom the level of anesthesia is too great, stimulation of the dorsal root will not result in a muscle response. Although muscle relaxant is used during induction and opening it should be short lasting, so that it is not effectively present during the stimulation and electromyographic recording needed during SDR (Riegle *et al.*, 1992).

Recording needle electrodes are placed in several muscles bilaterally to cover the lumbar and upper sacral root levels. Surface electrodes could be used instead but are more likely than needle electrodes to be dislodged with patient positioning. The display gain on the computer monitor should be set to 100–200 μV/division and adjusted as needed; a sweep of 1000 ms or more is used (100–200 ms/division). Filter settings include a low frequency filter of 10 Hz and a high frequency filter of 2–5 kHz, although other combinations have been used, including 30 Hz to 3000 kHz, similar to that which might be used with diagnostic EMG recordings. Since the intensity of voltage needed to stimulate a rootlet is less than that needed to stimulate a root, adjustments in the stimulus

Table 15.1 Stimulation parameters for SDR

LFF	10 Hz
HFF	2–5 KHz
Display gain	100–200 ms/div
Stimulation rate	1 Hz
	30–50 Hz

intensity will be needed. See Table 15.1, which shows the stimulation parameters used at our institution and in common practice.

The muscles typically monitored in SDR are selected according to a patient's needs and desired outcome. Table 15.2 lists some recommended muscles that can be used in patients, depending on the particular clinical history and examination findings. In some of these muscles, it is desirable to spare the nerve root involved in innervation, while in other muscles it is desirable to sever the dorsal root involved in innervation to reduce exorbitant or inappropriate involvement that hampers patient function.

For the electrical testing paradigm of dorsal root stimulation, a testing stimulation of 0.1 ms duration constant-current stimulation at 1 Hz is given at low stimulation intensity and repeated to get a unilateral response. This is the *threshold* of that dorsal rootlet. This lowest voltage is the *threshold response* and testing of other nerve roots will probably be needed at or above this level. After threshold stimulation, then stimulation at 30–50 Hz (tetanic stimulation) for one second at each rootlet level is used. Tetanic stimulation will involve a significantly greater intensity of muscle response than threshold response and allow for interpretation of the range of abnormality and muscle involvement and the spread of response in that patient.

The stimulation required to generate muscle contraction with rootlet stimulation is significantly less than for stimulation of the entire root. If, however, a response is not obtained or high levels of stimulation are required, the level of neuromuscular blockade and anesthesia should be immediately reviewed and other factors searched for to determine why a response was not possible. The surgical field should be as dry as possible to avoid current spread and the surgeon should use a bipolar or hook stimulator to stimulate the nerve root or rootlet. Then responses are recorded from multiple muscles bilaterally on a computer screen, allowing for multiple channel analysis of

Table 15.2 Muscles typically used to record muscle activity during SDR

Muscle	Nerve for stimulation
Hip adductors (adductor magnus, longus, or brevis)	Obturator nerve L2–5
Hip abductor (gluteus medius)	Superior gluteal nerve L4, 5, S1
Hip extension (vastus lateralis, medialis)	Femoral nerve L2–4
Hip flexion (rectus femoris)	Femoral L2–4
Knee flexors (bicep femoris, long and short head)	Peroneal division, sciatic nerve L5, S1
Plantar flexion (gastrocnemius)	Tibial nerve S1,2
Ankle dorsiflexion (tibialis anterior)	Deep and common peroneal nerve: L4, 5
Rectal tone (external anal sphincter)	Pudendal nerve: S2–4

responses along several myotomes. At each level the surgeon indicates the level of the root or rootlets being stimulated to the interpreting neurophysiologist so that appropriate interpretation of muscle examination and electromyographic recording can take place. Figure 15.2 shows an algorithmic protocol for interpretation and assessment of IOM during SDR.

The external anal sphincter and the external urethral sphincter are both innervated by the pudendal nerve and supplied by S2–4 roots. The left and right sides of the external anal sphincter are evaluated with recording needle electrodes placed and referenced to each other. To monitor bladder function, voiding capability can be evaluated by monitoring vesical pressure or by monitoring muscle action potentials of the detrusor muscle.

Anesthetic protocol during SDR

Consideration of the anesthetic regimens utilized during SDR, as alluded to already, is important to the success of the procedure. This is not only true in regard to the use of neuromuscular blocking agents or paralytics but in the use of the anesthesia utilized to keep the patient unaware and to awaken without memory of the surgery. Premedication with midazolam is often used to relieve anxiety and such induction agents as IV propofol, vecuronium, fentanyl, and remifentanil are very commonly employed. After induction, vecuronium

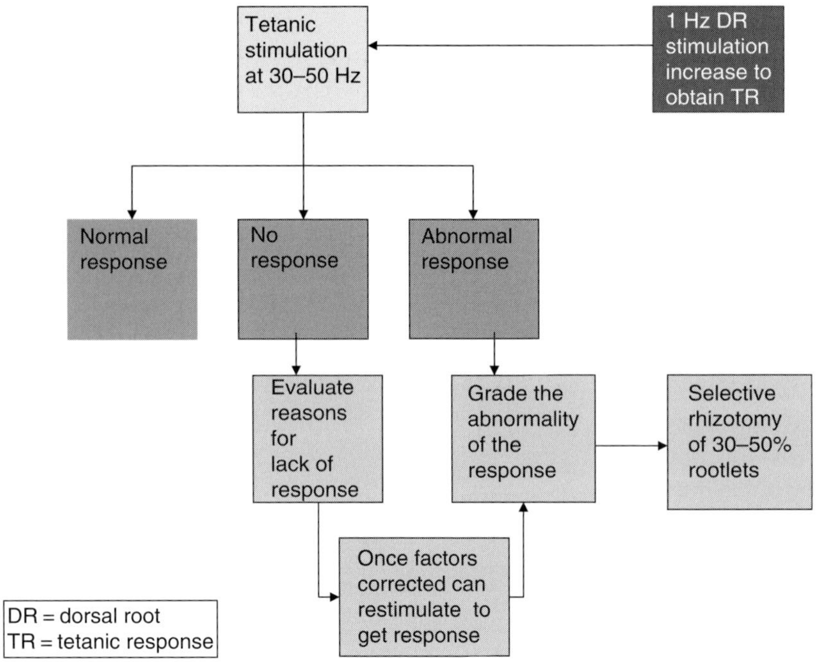

Figure 15.2 Protocol for interpretation and assessment of IOM during SDR.

should not be used because of its muscle relaxant effect. Infusions of remifentanil, sevoflurane, isoflurane, and nitrous oxide may be used. Propofol may be problematic in some children due to the development of intense muscle spasms with dorsal root stimulation when it is used (Riegle *et al.*, 1992).

One small study on 18 patients (Konya *et al.*, 2009) showed a beneficial effect of sevoflurane over isoflurane in preventing what is referred to as the brisk hyperactive response (BHR) after dorsal root stimulation. This can be seen even in normal dorsal rootlets. Using sevoflurane allows the interpretation of the abnormal muscle response at a given rootlet without BHR interference.

Frigon *et al.* (2008) studied the use of ketamine in 32 children undergoing SDR, who had received ketamine as part of a low dose regimen prior to dorsal root stimulation. In this study, although there was no significant difference in intraoperative electrophysiological responses in the group receiving the ketamine and those who did not, there was a lower maximal threshold for a muscle response with root stimulation. In addition the response to a train of stimulation lasted longer in the group receiving ketamine than in the group that did not. This could cause interpretation of a more intense response in the patient with dorsal root stimulation than would occur if the patient was not given ketamine. This finding of enhancement of muscle response to dorsal root stimulation is not surprising, since ketamine has been known to enhance cortical signals during SEP monitoring and is used with caution in patients with a strong history of seizures because of possible excitatory effects on cortical potentials (Zaarour *et al.*, 2007).

Interpretation

In normal patients, with a stimulation train of 30–50 Hz, there is normal inhibition in the response above 10 Hz owing to the interaction and modulating effects of spinal inhibitory interneurons. However, in spasticity or where there is lack of or reduction in upper motor neuron inhibition, several types of abnormal response can be seen. The degree of abnormality can be graded. The primary criteria for an abnormality after rootlet or root stimulation include clinical abnormalities in regard to the patient's muscle response to stimulation as well as abnormalities related to the electromyographic signal.

Both clinically and electromyographically, an abnormal response is one in which a muscle response is seen beyond the period of stimulation or in which there is a spread of the response to muscle groups not typically innervated by that sensory root level or

Table 15.3 Response grading system

Grade 0	Normal 1-second unsustained contraction with complete relaxation following stimulation
Grade 1	Sustained, beyond the period of stimulation, but limited to the myotome of stimulation
Grade 2	Sustained response and spread of muscle response to adjacent muscles at that segmental level
Grade 3	Sustained response and spread of muscle response to distant muscles beyond one segmental level
Grade 4	Sustained response and spread to the contralateral limb and multiple segmental levels

(Phillips and Parks, 1989)

bilaterally of response. In addition, electromyographic signals that are clonic, decremental, incremental, bizarre and high frequency, multiphasic, or sustained beyond the period of stimulation as noted above are considered abnormal by most neurophysiologists, although some also consider decremental responses to be normal (Steinbok *et al.*, 1992b). At our institution, we utilize the grading system derived from Phillips and Parks in 1989 as a good clinical and quantifiable system that can usually be managed well by both the neurophysiologist and the surgical team (Table 15.3).

Utilizing a grading system to identify the response at each level and at each rootlet in an objective fashion is best, and an example of such a grading system is seen in Table 15.3. Since muscle responses need to be evaluated and graded, neuromuscular blockade needs to be avoided, similar to the requirements used during electromyographic recording for pedicle screw monitoring.

In some patients it may be difficult to determine from which level a root or rootlets originate, or they may be difficult to tease apart or separate, owing to abnormalities. In this way, the neurophysiologic interpretation can help determine muscles involved in the response to stimulation very appropriately. In general, 30–50% of dorsal rootlets for each level will typically be sectioned. If the surgeon is unsure whether a given rootlet should be spared, he or she may keep the rootlet during the first trial or pass and go back at a later point during stimulation and retry those rootlets again. In this way information from both the clinical examination at the time of stimulation as well as the interpretation by the neurophysiologist of the electromyographic recording can be used to decide whether to spare or section a given rootlet. There is significant variation in the muscle responses both clinically and electromyographically among different rootlets. It is not unusual to have different rootlets with grade 0 or 1 and grade 3 or 4 in the same root. Grade 3 and 4 responses are typically thought to contribute to the undesirable spastic response. Because of the need to spare rootlets and preserve sensation, grade 2 or lower responses are often spared, at the discretion of the surgeon and depending on the level involved.

Clinical significance and outcome studies

A possible complication of SDR is the development of hyperthermia, possibly due to increased muscle metabolism with spasm as a result of stimulation. Judicious use of anesthetics, as well as lower ambient room temperature and close attention to core temperature are needed. Monitoring the involved muscles ensures that nerve root innervations to these structures will be spared. In a patient still able to ambulate, some degree of extensor tone is needed, so the hip and knee extensors will need to be spared while reducing the degree of adductor spasticity with SDR. Monitoring allows a more sensitive means of accomplishing this selectivity (Engsberg *et al.*, 2006).

In many centers utilizing this procedure and in selected patients, with electromyographic recording, one usually sees functional improvements in the weeks to months following selective dorsal rhizotomy (Engsberg *et al.*, 2007). Improvements in outcome have been described in children after selective dorsal rhizotomy for spasticity and the improvement is long-standing. In some patients, even in the absence of the ability to ambulate, improvements in tone and range of motion allow for improvements in quality of life for activities of daily living. Most notably, these improvements occur in the hip adductors and abductors, knee flexors and extensors, and plantar flexors. Improvement typically occurs several months after surgery and continues over the following year with long-standing physical therapy and rehabilitation (Steinbok and McLeod, 2006). Long-term outcome studies are few in dorsal rhizotomy patients and the published data does not typically include large numbers of patients (Peter and Arens, 1993).

Evaluation of passive range of motion as well as functional scores, such as the Gross Motor Functional Measure or GMFM-88 (Haley *et al.*, 1993) and the

Pediatric Evaluation Disability Inventory, are helpful indicators. The Gross Motor Functional Score is a validated measure developed to assess children with spastic cerebral palsy accurately. This measure has a total score and five subscores of motor function given as a percentage of the maximum score.

In a study by Nordmark *et al.* (2008), 35 children underwent SDR and functional scores were used as a measure of long-term outcome over a five-year period. In these children, 40% of L2 to S2 rootlets were selectively sectioned. Patients had reduction in spasticity postoperatively in the adductors, hamstrings, and dorsiflexors and no recurrence of spasticity over the five year follow-up period. Significant improvements in gross motor function, performance of functional skills, and independence with self care occurred in all groups regardless of the gross motor function prior to SDR.

In another study (van Schie *et al.*, 2005), a group of spastic diplegic but ambulatory cerebral palsy children aged 43–82 months showed functional motor improvements postoperatively at 12 months follow-up after SDR using several rating scales including the Gross Motor Function Measure. Additional functional scores can be obtained with the *Edinburgh Visual Gait Assessment* and in the *Passive Range of Motion Scale* (Mittal *et al.*, 2002), which have been shown to have good interobserver reliability.

With SDR, there is typically improvement in postural stability including in sitting or postural ability (Steinbok *et al.*, 1992a). These observations obtained after electrical stimulation of lumbosacral nerve roots were reported in children with and without lower limb spasticity (Steinbok *et al.*, 1992a). Following the measured range of motion at a joint in a given patient both prior to and after selective dorsal rhizotomy can allow meaningful correlation with functional outcome. In addition, in these children significant physical therapy is needed and functional therapy scores can be utilized to assess the impact of SDR.

Complications and risks associated with SDR

The number of SDR procedures that occur is unknown. However, this procedure is the only treatment currently known to reduce spasticity permanently and is therefore in widespread use for spasticity related to cerebral palsy. Unfortunately, complications can occur in patients undergoing SDR. Urinary retention, back pain, and sensory changes in the lower extremities have been reported, as well as pulmonary complications, including aspiration pneumonia and bronchospasm (Steinbok and Schrag, 1998). The sensory changes included dysesthesias as well as hyperesthesias and may be avoided by limiting the number of dorsal rootlets cut.

Urinary retention requiring catheterization, lasting several days after the procedure, is the most common complication but this complication has decreased in frequency with sparing of the S2 roots and modifying the SDR procedure. The complication of persistent urinary incontinence may occur in up to 5% of patients after SDR. In patients whose S2 roots were spared, urinary and bowel dysfunction does not typically occur. In addition, pudendal nerve monitoring is associated with a significantly lower risk of urinary and fecal incontinence as a consequence of the procedure.

Aspiration pneumonia can occur, although infrequently, particularly in patients with bronchopulmonary dysplasia, gastroesophageal reflux, reactive airway disease, and spastic quadriparesis.

Additionally, significant pain is reported postoperatively in patients after SDR procedures. Postoperative management often requires IV opioids or epidural infusions usually for between several days and a week (Hellelgard *et al.*, 2007). Using an age-appropriate pain scale is important for pain management both in the short and longer term, as the patient undergoes prolonged and often uncomfortable physiotherapy.

There is a great deal of heterogeneity among cerebral palsy patients, whether ambulatory or not, and spasticity in many of these patients contributes in varying degrees to their ability to stand and move (O'Brien *et al.*, 2004). Therefore, SDR can cause deterioration in the ability to ambulate if the procedure is not selective enough or there is a lack of residual voluntary strength. However, with technical expertise, limitations on the surgical expanse, and intraoperative monitoring during dorsal rootlet severance, the risk of deterioration in motor and functional skills is significantly lessened (Engsberg JR, Ross SA, Collins DR, et al., 2007).

Case example for Chapter 15

DM is a 10 year old girl who is wheelchair dependent with spastic cerebral palsy. She has had unsuccessful medication trials to help with caregiver care. On examination, she had significant adductor spasm

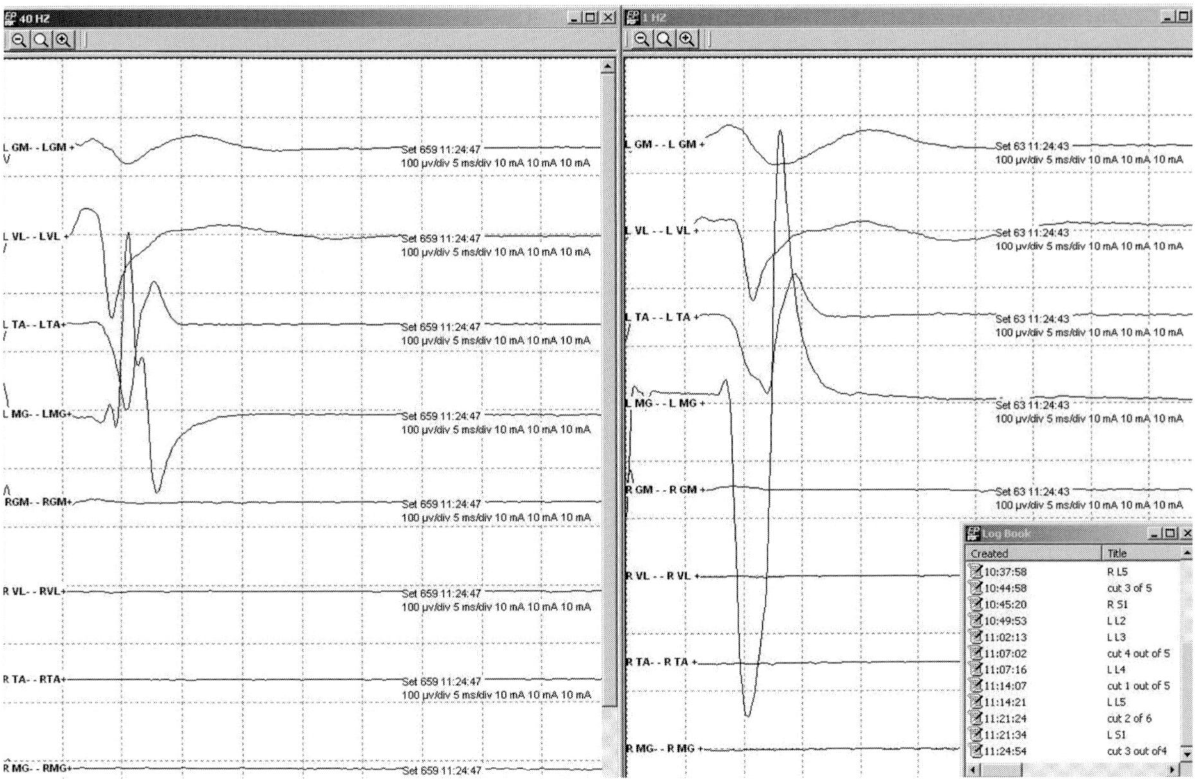

Figure 15.3 Case DM: see text for description. Muscle responses after dorsal rootlet stimulation at the left S1 level. There is spread to myotomes outside of the area of innervation on the same side without spread to the contralateral side. Three of the four rootlets at this level were cut, including this one.

with significant weakness of the hip abductors and contractures at the ankles. Four rootlets were noted on the left at the S1 level. Responses from one of these rootlets are noted in Figure 15.3: with stimulation of that rootlet, there was spread of the response beyond the segmental level but not to the contralateral side. The surgeon cut three out of the four rootlets involved at this level, including the one indicated in Figure 15.3. The patient had an uneventful postoperative course and the surgery was followed by longstanding physical therapy. Improvements in spasticity occurred over the intervening months along with improvements in the ability of the caregiver to assist in daily living.

Questions

1. The type of cerebral palsy most amenable to dorsal rhizotomy is:
   (a) Athetoid
   (b) Ataxic
   (c) Dystonic
   (d) Spastic
2. Which of the following statements is true?
   (a) In the Gross Motor Function test, level 1 is the least functional and level 5 is the most functional
   (b) In the Gross Motor Function test, level 1 is the most functional and level 5 is the least functional
   (c) The Gross Motor Function test uses qualitative, not quantitative, measures of function
3. In general, the percentage of rootlets sectioned in SDR procedures is:
   (a) 10–20%
   (b) 30–50%
   (c) 50–60%
   (d) 60–75%
4. Which of the following statements about SDR is false?
   (a) The frequency of urinary retention has decreased with sparing of the L2 root

(b) The frequency of bowel and bladder dysfunction has decreased with the use of pudendal and urethral monitoring
(c) Dysesthesias and sensory dysfunction are reduced when the number of rootlets severed is limited

5. Which of the following statements regarding stimulation in SDR is correct?
   (a) The stimulation required to generate muscle contraction with rootlet stimulation is significantly less than for stimulation of the entire root
   (b) The stimulation required to generate muscle contraction with rootlet stimulation is significantly more than for stimulation of the entire root
   (c) If a response is not obtained or high levels of stimulation are required, an assessment of the level of neuromuscular blockade and anesthesia should be made immediately
   (d) The stimulation required to generate muscle contraction with rootlet stimulation is the same as that required for stimulation of the entire root
6. Which of the following statements about the external anal sphincter and the external urethral sphincter are true?
   (a) They are both innervated by the tibial nerve and supplied by L4–S1 rootlets
   (b) They are both innervated by the pudendal nerve and supplied by S2–4 roots
   (c) The external anal sphincter is innervated by the posterior tibial nerve and the pudendal nerve is innervated by the superficial peroneal nerve
7. All of the following electromyographic findings are considered abnormal, except for:
   (a) Clonic
   (b) Decremental or incremental
   (c) High-frequency responses
   (d) Non-sustained beyond the period of stimulation
   (e) Sustained beyond the period of stimulation
8. Which of the following is not typically a potential complication of SDR?
   (a) Urinary retention
   (b) Dysesthesias
   (c) Hyperthermia
   (d) Kidney stones
9. Which of the following is true?
   (a) The most common methodology of SDR involves electromyographic recording and clinical muscle examination
   (b) The most common methodology of SDR involves SEP recording and biopsy
   (c) The most common methodology of SDR involves hematologic evaluation and manual muscle testing
10. The expected outcome after SDR is improvement in:
   (a) Visual tracking
   (b) Postural stability
   (c) Incline percent on a treadmill
   (d) Arm span

References

Abbott, R. (1991). Indications for surgery to treat children with spasticity due to cerebral palsy. In *Neurosurgery for Spasticity: A Multidisciplinary Approach*, ed. M. Sindou, R. Abbott, and Y. Keravel. Springer Verlag: Vienna, pp. 215–217.

Abbott, R., Johann-Murphy, M., Shiminsk-Maher, T, *et al.* (1993). Selective dorsal rhizotomy: outcome and complications in treating spastic cerebral palsy. *Neurosurgery*, **33**, 851–857.

Bohannan, R. W. and Smith, M. B. (1987). Interrater reliability of a modified Ashworth scale of muscle spasticity. *Physical Therapy*, **67**, 206–207.

Coster, W., Haley, S., Ni, P., *et al.* (2008). Assessing self-care and social function using a computer adaptive testing version of the pediatric evaluation of disability inventory. *Archives of Physical Medicine and Rehabilitation*, **89**, 622–629.

Deletis, V., Vodusek, D., Abbott, R., Epstein, F., and Turndorf, H. (1992). Intraoperative monitoring of the dorsal sacral roots: minimizing the risks of iatrogenic micturition disorders. *Neurosurgery*, **30**, 72–75.

Engsberg, J. R, Ross, S. A, Collins, D. R., Park, T. S. (2006). Effect of selective dorsal rhizotomy in the treatment of children with cerebral palsy. *Journal of Neurosurgery*, **105**(1), 8–15.

Engsberg, J. R., Ross, S. A., Collins, D. R., Park, T. S. (2007). Predicting functional change from preintervention measures in selective dorsal rhizotomy. *Journal of Neurosurgery*, **106**(4), 282–287.

Fasano, V. A., Broggi, G., Zeme, S., Lo Russo, G., and Squazzi, A. (1980). Long term results of posterior functional rhizotomy. *Acta Neurochirurgica. Supplementum*, **30**, 435–439.

Fasano V. A., Broggi G., and Zeme S. (1988) Intraoperative electrical stimulation for functional posterior rhizotomy. *Scandinavian Journal of Rehabilitation Medicine. Supplement*, **17**, 149–154.

Frigon, C., Sedeek, K., Poulin, C., *et al.* (2008). Does ketamine affect intraoperative electrophysiological monitoring in children undergoing selective posterior rhizotomy? *Pediatric Anesthesia*; **18**, 831–837.

Gregson, J. M., Leathley, M., Moore, A. P., *et al.* (1999). Reliability of the tone assessment scale and the modified Ashworth scale as clinical tools for assessing poststroke spasticity. *Archives of Physical Medicine and Rehabilitation*, **80**, 1013–1016.

Haley, S. M., Ludlow, L. H., and Coster, W. J. (1993). Pediatric evaluation of disability inventory: clinical interpretation of summary scores using Rasch rating scale methodology. *Physical Medicine and Rehabilitation Clinics of North America*, **4**, 529–540.

Hellelgard, K., Reinstrup, P., Stromblad, L. G., *et al.* (2007). Selective dorsal rhizotomy and postoperative pain management. *Pediatric Neurosurgery*, **43**, 107–112.

Katz, K., Rosenthal, A., and Yosipovitch, Z. (1992). Normal ranges of popliteal angle in children. *Journal of Pediatric Orthopedics*, **12**, 229–319.

Kerr, A. M., Hazlewood, M. E., van der Linden, M. I., *et al.* (2002). The Edinburgh Visual Gait Score as an outcome measure after surgical intervention cerebral palsy. *Gait & Posture*, **6**, S116.

Konya, D., Gercek, A., Dagcinar, A., *et al.* (2009). Prevention of brisk hyperactive response during selective dorsal rhizotomy in children with spasticity: isoflurane versus sevoflurane maintenance anesthesia. *Journal of Clinical Neuroscience*, **16**, 241–245.

Maathuis, K., van der Schans, C., van Iperen, A., *et al.* (2005). Gait in children with cerebral palsy: observer reliability of physician rating scale and Edinburgh Visual Gait Analysis interval testing scale. *Journal of Pediatric Orthopaedics*, **25**(3), 268–272.

Mittal, S., Farmer, J. P., Al-Atassi, B., *et al.* (2002). Functional performance following selective posterior rhizotomy: long-term results determined using a validated evaluative measure. *Journal of Neurosurgery*, **97**(3), 510–518.

Nordmark, E., Josenby, A. L., Lagergren, J., *et al.* (2008). Long term outcomes five years after selective dorsal rhizotomy. *BMC Pediatrics*, **8**, 54. www.biomedcentral.com/1471-2431/8/54.

O'Brien, D. F., Park, T. S., Puglisi, J. A., Collins, D. R., and Leuthardt, E. C. (2004). Effect of selective dorsal rhizotomy on need for orthopedic surgery for spastic quadriplegic cerebral palsy: long-term outcome analysis in relation to age. *Journal of Neurosurgery*, **101**(1), 59–63.

Peacock, W. J., Arens, L. J., Berman, B. (1987). Cerebral palsy spasticity. Selective posterior rhizotomy. *Pediatric Neuroscience*, **13**, 61.

Peacock, W. J., Staudt, L. A., and Nuwer, M. R. (1990). A neurological approach to spasticity: selective posterior rhizotomy. In *Neurosurgery Update II*, ed. R. H. Wilkins and S. S. Rengachary. New York: McGraw Hill, pp. 403–407.

Peter, J. C and Arens, L. J. (1993). Selective posterior lumbosacral rhizotomy for the management of cerebral palsy spasticity: a 10 year experience. *South African Medical Journal*, **83**, 745–747.

Phillips, L. and Parks, T. (1989). Electrophysiological studies of selective posterior rhizotomy patients. In *Neurosurgery: State of the Art Reviews: Management of Spasticity in Cerebral Palsy and Spinal Cord Injury*, ed. T. S. Park, L. Phillips, and W. J. Peacock. Philadelphia: Hanley and Belfus, pp. 459–469.

Reed, H. S., Hillmann, S. J., Hazlewood, M. E., *et al.* (1999). The Edinburgh Visual Gait Analysis interval testing (gait) scale. *Gait & Posture*, **10**, 63.

Riegle, E. V., Gunter, J. B., Lagueruela, R. G., *et al.* (1992). Anesthesia for selective dorsal rhizotomy in children. *Journal of Neurosurgery and Anesthesiology*, **4**, 182–187.

Ross, S. A. A. and Engsberg, J. R. (2007). Relationships between spasticity, strength, gait, and the GMFM-66 in persons with spastic diplegia cerebral palsy. *Archives of Physical Medicine and Rehabilitation*, **88**, 1114–1120.

Steinbok, P. and McLeod, K. (2002). Comparison of motor outcomes after selective dorsal rhizotomy with and without preoperative intensified physiotherapy in children with spastic diplegic cerebral palsy. *Pediatric Neurosurgery*, **36**(3), 142–147.

Steinbok, P. and Schrag, C. (1998). Complications after selective posterior rhizotomy for spasticity in children with cerebral palsy. *Pediatric Neurosurgery*, **28**(6), 300–313.

Steinbok, P., Langill, L., Cochrane, D. D., and Keyes, R. (1992a). Observations on electrical stimulation of lumbosacral nerve roots in children with and without lower limb spasticity. *Child's Nervous System*, **8**, 376–382.

Steinbok, P., Reiner, A., Beurchamp, R. D., Cochrane, D. D., and Keyes, R. (1992b). Selective functional rhizotomy for treatment of spastic cerebral palsy in children: review of 50 consecutive cases. *Pediatric Neurosurgery*, **18**, 34–42.

Thompson, N., Jonathan, R., and Saunders, J. (2001). Relevance of the popliteal angle to hamstring length in cerebral palsy crouch gait. *Journal of Pediatric Orthopaedics*, **21**(3), 383–389.

Turner, R. (2009). Neurophysiologic intraoperative monitoring during selective dorsal rhizotomy. *Journal of Clinical Neurophysiology*, **26**(2), 82–84.

van Schie, P. E., Vermeulen, R. J., van Ouwerkerk, W. J., Kwakkel, G., and Becher, J. G. (2005). Selective dorsal rhizotomy in cerebral palsy to improve functional abilities: evaluation of criteria for selection. *Child's Nervous System*, **21**(6), 451–457.

Zaarour, C., Engelhardt, T., Strantzas, S., *et al.* (2007). Effect of low-dose ketamine on voltage requirement for transcranial electrical motor-evoked potentials in children. *Spine*, **32**(22), E627–E630.

16 Interpreting and reporting the neurophysiologic data to the surgical team: how to do it and when it is indicated

Jaime R. López

Introduction

Neurophysiologic intraoperative monitoring (IOM) can be viewed as the application of neurophysiologic, usually electrophysiologic, techniques used in operative procedures to detect changes in the functional state of the nervous system consistent with ischemia or injury. Because of the very nature of operative procedures, this application of neurophysiologic testing is done in a fluid and dynamic fashion. For example, data obtained from recording somatosensory evoked potentials (SEPs) in the operating room are likely to yield different results when compared with the same type of recording, from the same patient, obtained in the outpatient neurodiagnostic laboratory. Furthermore, an intraoperative evoked potential recording is likely to show significant differences during different times and stages of surgery. Frequently, these observed changes are due not to ischemia or injury to the neural pathways being monitored but to other factors.

York *et al.* (1987) found that SEPs obtained during scoliosis surgery showed latency changes of up to 50% and amplitude changes of greater than 50% during the course of the operation. They also reported that patients with these changes did not exhibit any clinical neurologic sequelae. In addition, the variability of SEP monitoring during scoliosis surgery has also been reported to be highest during the early stages of the surgical procedure (Chen *et al.*, 2004). An example scenario would be in comparing SEP results obtained immediately after anesthesia induction during a carotid endarterectomy with those observed two hours later after the administration of thiopental to induce, deliberately, a burst-suppression pattern on the electroencephalogram (EEG) for neuroprotection.

The acquisition of neurophysiologic data in this constantly changing environment requires a clear understanding of the significance of the acquired data and their implication. This has to be taken into consideration prior to raising alarms and altering the surgical management. Proper interpretation of IOM data requires an understanding of several critically important and interrelated properties of each specific monitoring technique. These include:

1. The neural pathway being monitored,
2. The specific neurophysiologic modality monitored,
3. The test sensitivity,
4. The test specificity,
5. The predictive value,
6. Non-neurophysiologic confounding factors, such as temperature, anesthesia, ambient electrical noise, and technical causes.

Data cannot be interpreted accurately, nor can the interpretation be reported, if data are not acquired properly. Improper acquisition of data will yield information that is likely to be erroneous which, in turn, will lead to an incorrect interpretation. Thus, proper acquisition of data is critical to the proper performance and application of IOM.

In this chapter, I will discuss these issues, define some changes that are considered "critical" and require an alarm to be issued, and the method by which this should be done. I will also attempt to demonstrate these concepts using clinical examples.

Intraoperative data acquisition

How the data are acquired in the operating room can have a significant impact on the overall quality and accuracy of the results. The first point in the timeline of a case is when the specific protocol is decided upon for that particular case. Clinical information needs to be attained prior to planning the proper monitoring strategy. The protocol to be used should address several important questions, such as:

1. Will the planned techniques monitor the nervous system territories at risk during surgery?
2. Are the proper neurophysiologic modalities being monitored?
3. Are the sensitivities and specificities of the neurophysiologic tests sufficient to yield proper and accurate results?

To perform proper IOM, we need and should be able to answer the first two questions. Unfortunately, we do not always have clear answers to the last question posed.

Monitoring techniques and nervous system territory at risk during surgery

It may seem obvious that the planned techniques should monitor the neurologic pathways or regions at risk. However, this cannot be planned for if the patient's clinical history and neurologic examination findings are unknown before the surgical procedure. Often, incomplete preoperative clinical information leads to erroneous IOM planning.

For example, in the case where the IOM team are requested to monitor a patient for a planned acoustic neuroma resection, the monitoring strategy would probably consist of monitoring facial and acoustic nerve function. The specific strategy and protocol to use will differ if the patient has non-functional hearing on the side of surgery. This would not require monitoring of brainstem auditory evoked potentials (BAEPs) because of the hearing loss. However, monitoring of BAEPs, SEPs, and transcranial electrical motor-evoked potentials (TcMEPs), in addition to electromyography (EMG) monitoring of the facial nerve, would probably be planned for if it was known that the patient suffered from a large tumor causing brainstem compression.

Another scenario would be using only SEPs to monitor scoliosis surgery. The use of SEPs in such a case may be sufficiently adequate to assess for global spinal cord injury but may be inadequate for focal injury. It is quite conceivable that a patient would wake up after surgery with a pure motor deficit while monitoring revealed no changes in SEPs during surgery. Such cases have been reported in the literature (Minahan *et al.*, 2001). Would this be a false-negative result? No: in this scenario, even though the patient developed a new neurologic deficit, the SEPs accurately predicted the neurologic outcome of the neural pathway (somatosensory) being monitored. Thus, it is a true negative outcome, since the SEPs do not directly monitor motor pathways.

Therefore, it is critical that the nervous system territories and neural pathways at risk for injury are known and understood before planning the protocol to be used. If necessary, and if possible, a preoperative neurologic examination can be performed by the supervising IOM physician, to clarify that the proper nervous system territory is accurately monitored during surgery. This would also serve to identify any preexisting neurologic deficits.

The sensitivity and specificity of the different IOM modalities will vary and are also dependent on their application. The same neurophysiologic test is likely to have different sensitivities and specificities depending on the type of case to which it is applied. An example would be the monitoring of cortical SEPs during scoliosis surgery or during carotid endarterectomy. These two types of cases are vastly different and the nervous system territories at risk are also completely different. However, how the SEPs are obtained and monitored are technically very similar in both instances. A review of some of the carotid endarterectomy literature shows that the sensitivities and specificities of SEPs ranged from 83% to 100% and 83% to 99%, respectively (Lam *et al.*, 1991). Whereas, the sensitivity and specificity of SEPs for scoliosis surgery has been reported to be 92% and 98.8%, respectively (Nuwer, 1999). A review and discussion of the published sensitivities and specificities for all of the different modalities and their different surgical applications is beyond the scope of this chapter. The reader is referred to the excellent chapters in this book where these topics have been reviewed and discussed in detail.

Non-neurophysiologic confounding factors

There are many different factors that are not a result of injury to the nervous system that can influence IOM data and subsequent interpretation. The following list includes some of the common causes of changes in the IOM data which are not a result of nervous system injury.

IOM technical factors

Stimulating and recording electrodes

- Incorrect or improper use of electrodes,
- High impedances,
- Incorrect recording montage,
- Inadequate stimulation parameters,
- Technical errors.

IOM equipment

Equipment and software lacking the ability to perform the chosen IOM protocol.

IOM staff

Inadequately trained and/or inexperienced staff, including:

- Monitoring technologist,
- Clinical neurophysiologist,
- Physician.

Anesthesia

Incorrect anesthetic protocol

In general, most anesthetic agents will have negative effects on the IOM modalities, and the use of neuromuscular blocking agents can make EMG monitoring unreliable. Thus, it is imperative to discuss the plan in advance with the anesthesiologist and together devise an anesthetic protocol that will provide the best results for the planned surgical procedure (see Chapter 5).

Fluctuating or changing anesthetic levels during the case

- It is imperative that the IOM team maintain constant communication with the anesthesiologist throughout the case.
- Blood pressure and mean arterial pressure, inhalational and parenteral levels of anesthetic agents, and temperature, should be reviewed and documented periodically throughout the case. All of these can influence the IOM results.

Operating room environment

Electrical noise contamination of the IOM electrophysiologic potentials

The operating room is a very electrically hostile environment due to the use of a large amount of electrically powered equipment. One or more of these items can produce sufficient electrical artifacts which, in turn, can contaminate the electrophysiologic testing/monitoring being performed by the IOM team.

Special surgical protocols

Induced mild hypothermia

Performed for the purpose of providing neuroprotection in cerebrovascular, cardiac and aortic aneurysm surgery. Causes slowing of peripheral and central nerve conduction velocities resulting in increased SSEP latencies.

Obtaining baseline IOM data prior to final positioning

- Typically, this is done in some cervical spine surgical cases prior to turning the patient onto the prone position.
- This usually requires removing many of the electrodes after the baseline data are obtained and immediate retesting after the patient is in the final surgical position.

Multistage procedures

Different surgical specialties collaborating on the same case

- This may require repositioning of the patient.
- It may also necessitate repositioning and/or removing of electrodes.

Endovascular procedures

The patients are often awake

- The electrophysiologic modalities that can be used for monitoring are limited when the patient is awake,
- EMG, which is a common artifact in an awake patient, can contaminate the EEG and SEPs,
- Electrical noise sources are plentiful but different from those in the operating room.

Identifying confounding factors during surgery

Any one of these factors can have deleterious effects on the IOM modalities being monitored and needs to be either excluded or taken into consideration when interpreting the data. This raises an important question. How much time should be spent in determining whether the changes in the monitored parameters are caused by nervous system injury or dysfunction, or are a result of non-neurophysiologic factors? There are no strict guidelines addressing this issue. Nonetheless, certain things can be concluded, such as that the determination of whether or not the observed change in the IOM data is a result of neural injury should be done as quickly as possible and that it should be given the highest priority. Such a policy would be applicable and appropriate at any time during any surgical procedure. Of course, the urgency is highest during the critical portion of the surgery.

For example, what is the acceptable amount of time to spend determining whether the significant change noted in the cortical SEPs during a scoliosis case occurring at a time when the surgeon is dissecting soft tissue, but has not yet reached the level of spinal column, is a result of neural injury or not? Should anyone be notified of the change before the cause is determined? If so, then who should be notified? The answer to the latter question depends on who is in the operating room acquiring the IOM data. If the IOM technologist is the person in the operating room monitoring and acquiring the data, then the supervising IOM physician should first be notified and together with the IOM technologist attempt to identify the cause. The surgeon need not be notified at that time but would need to be notified if a clear non-neurophysiologic cause is not identified or if it is determined that the change is probably due to spinal cord dysfunction or injury.

Nuwer (1986) states that it is unreasonable to spend five minutes or more attempting to identify the cause of an evoked potential change during spinal cord monitoring. Of course, with the multimodality protocols currently used, in some cases employing up to 32 channels, troubleshooting the cause of an IOM change is more complex and can take more than a few minutes. Regardless, it is advisable to inform the surgeon and anesthesiologist of the difficulty encountered if the cause of the evoked potential change is not identified within five minutes. If necessary, the surgical procedure may be paused to allow the IOM team to identify the cause of the observed IOM changes.

When the IOM change coincides with the critical period of the operative procedure, the determination as to the cause should be done within a few minutes. Again, no specific guidelines are published on this matter but it stands to reason that the longer it takes to intervene, in an attempt to reverse changes consistent with neural injury, the higher the probability of sustaining permanent neural damage.

Critical IOM changes

Once it has been determined that the observed IOM changes are not a result of non-neurophysiologic factors but probably due to impending neural injury, then it has to be determined whether the change is significant. Unfortunately, there is no clear consensus as to what constitutes a "critical" change in most of the IOM modalities used. Furthermore, what is considered to be "critical" may differ according to the type of procedure being monitored, such as the use of TcMEPs in surgical procedures of the spine or their use in cerebrovascular surgery.

Whatever criteria are used to define a "critical" change, it is required that the degree of change be of such magnitude that there is a high probability of neural dysfunction or injury. Unfortunately, regardless of the criteria used, there will always be a probability that the intraoperative critical change will occur in patients with and without new postoperative neurologic deficits. If the warning criteria are too conservative, then cases with new deficits will be missed but the criteria will probably be more specific. Conversely, if the criteria are too liberal then more cases without new deficits will fall into this category, resulting in more false-positive alarms, making the criteria less specific but perhaps more sensitive. Before issuing an alarm, several principles must be fulfilled:

1. Non-neurophysiologic factors, as possible contributors to the observed changes, must first be eliminated.
2. Critical changes must be reproduced. Some authors state that the changes should be recorded for a sufficient period of time to ensure that the change is different from spontaneous fluctuations (Schramm, 1985).
3. Baseline parameters should be established after anesthesia induction (Schramm, 1985; Chen *et al.*, 2004). In addition, new baseline values should be re-assessed and reestablished as necessary during the case. Good examples of this include procedures where induced hypothermia is used and cases where the anesthetic regimen is altered.
4. Different alarm criteria for different nervous system regions, such as the different SEP criteria used for spinal versus cortical recordings (Schramm, 1985), should be used.
5. Different alarm criteria are required for patients with impaired nervous system function, such

Table 16.1 Alarm or alert criteria for SEP changes in spine surgery as reported in a variety of publications between 1985 and 2009

	Amplitude reduction	Latency increase	Reproducibility
Breitner and Matzen, 1985	Instability or loss of SEP	N/A	N/A
Brown and Nash, 1985	>50%	>3 ms	Persistent change over 15 minutes
Koht *et al.*, 1985	≥50%	≥10%	Over two successive recordings
Young and Berenstein, 1985	>Baseline variation	N/A	Over three recordings or 5 minutes
Nuwer, 1986	50%	N/A	N/A
Dawson *et al.*, 1991	50%	10%	N/A
Nuwer, 1999	50%	>5%	N/A
Seyal and Mull, 2002	>50%	>5–10%	N/A
MacDonald *et al.*, 2003	"amplitude alteration occurring abruptly or as a trend"	N/A	"clearly exceeding trial-to-trial variability"
Chen *et al.*, 2004	50%	10%	Compared with baseline recordings obtained after anesthesia induction
Schwartz *et al.*, 2007	> or equal to	Specifically not used to issue an alert	N/A
Strommen, 2009	50%	5%	N/A

as those with impaired spinal cord function (Tamaki *et al.*, 1985a).

6. It is important to insure that the proper neurophysiologic modality corresponding to the neural territory at risk is being monitored. This is particularly important since most cases use multimodality monitoring. For example, it would be important to concentrate on monitoring of posterior tibial nerve SEPs during temporary occlusion of the anterior cerebral artery (ACA) during aneurysm surgery, even though median nerve SEPs might also be monitored during the procedure.

Thus, in the following section it would be wise to spend some time discussing what criteria are used to define a "critical" change in some of the different IOM modalities commonly being used. I will also attempt to demonstrate these concepts using clinical examples. However, I do not intend to review all of the modalities.

Somatosensory evoked potentials (SEPs)

Somatosensory evoked potentials have the highest level of consensus for what constitutes a "critical" change. In addition, most practitioners in IOM have a good understanding of the degree of change from the baseline SEP values that is required to be considered significant enough to issue an alarm. The criteria are typically applied to SEPs obtained after upper or lower extremity stimulation, usually from median, ulnar, or posterior tibial nerves. The pathophysiologic rationale for using SEPs as surrogate markers of spinal cord injury is based on strong experimental evidence in animal studies (Schramm *et al.*, 1983; Tamaki *et al.*, 1985b; Arai *et al.*, 2000; Carlson *et al.*, 2003,).

Table 16.1 shows the alert or alarm criteria reported in different articles and published studies on spinal cord monitoring. All of the studies recorded conventional SEPs, stimulating peripheral nerves, usually the posterior tibial nerve, and recording cortical-generated SEPs using scalp electrodes. Even though the alarm criteria were set arbitrarily (Schramm, 1985), they appear to have stood the test of time. It is now generally accepted that a 50% amplitude drop in the recorded potentials is considered sufficient to raise an alarm (Nuwer, 1986; 1999; Seyal and Mull, 2002; Strommen, 2009). In regard to latency changes, a greater than 5–10% increase in latency has also been used to indicate a critical change and raise an alarm (Nuwer 1999; Seyal and Mull, 2002; Strommen, 2009). However, others report that latency change rarely occurs (Jones, 1994). Schwartz *et al.* (2007) report that they do not use latency to define an alarm, since they have never observed a relevant latency shift as being the principal indicator of neurologic injury in spine surgery.

Table 16.2 Alarm or alert criteria for TcMEPs in spine surgery

	Type of change	Reproducibility
Calancie *et al.*, 1998	"Threshold-level"	N/A
Schwartz *et al.*, 2007	Unilateral or bilateral amplitude reduction ≥65%	A minimum of three test trials
Calancie and Molano, 2008	"Threshold-level" increase of 100 volts	N/A

The same criteria to raise an alarm are used when the SEPs are used for other non-spine surgical procedures, specifically IOM of cerebrovascular disorders. In these cases, there is a well-established physiologic basis for the changes in amplitude, latency, and central conduction time associated with cerebral ischemia (Branston *et al.*, 1974; 1977; Hargadin *et al.*, 1980; Symon, 1980; Jones *et al.*, 1981; Branston *et al.*,1984).

Transcranial magnetic evoked potentials (TcMEPs)

The alarm criteria for TcMEPs are not as straightforward as those for SEPs (Table 16.2). When used to monitor disorders of the spine and spinal cord, different alarm criteria are used, according to the type of motor-evoked potential recorded.

These include D wave potentials recorded directly from the spinal cord and myogenic evoked potentials recorded from muscle. The generally accepted criterion for a critical change in D wave amplitudes is a 50% reduction (Deletis and Sala, 2008). However, when it comes to myogenic TcMEP alarm criteria, there is still an ongoing debate without clear consensus. Alarm criteria have been reported based on "presence or absence" of myogenic TcMEPs, amplitude reduction, and "threshold-level" change needed to trigger TcMEPs (Calancie *et al.*, 1998; Schwartz *et al.*, 2007; Calancie and Molano, 2008; Deletis and Sala, 2008).

Brainstem auditory evoked potentials (BAEPs)

The generally accepted alarm criteria for changes in the BAEPs are (Burke *et al.*, 1999):

1. Persistent loss of waves I, III, and V,
2. Increase in the I–V interpeak interval by >1 ms,
3. Reduction of wave V amplitude, inverting the I/V ratio.

EMG (electromyography)

When monitoring continuous, free-EMG activity, a situation that warrants issuing an alarm is a neurotonic discharge, because these occur in response to mechanical, metabolic, or thermal irritation to the nerve innervating the corresponding muscle (Daube and Harper, 1989; Crum, 2009; Strommen, 2009). Unfortunately, there is no clear consensus in terms of degree, frequency, or size of the discharges seen during a surgical procedure. A reasonable conclusion is that the more neurotonic discharges (not resulting from irrigation) noted during a surgical procedure the higher the probability of nerve injury causing a postoperative weakness.

Interpreting the data

Interpretation of IOM data is a continuous, dynamic process, which starts at the time that the first neurophysiologic data are obtained, in order to begin establishing the baseline values. This is in contrast with laboratory and diagnostic evoked potential testing, or standard outpatient EEG studies where the data are only interpreted after the procedure has already been completed. Continuous interpretation of the data occurs throughout the operative procedure until the recording electrodes are removed. If this is the case, then who should be interpreting the data? Before answering the question, the roles of the different members of the IOM team should be defined.

Fortunately, these issues have been addressed by several different neurology specialty and electroneurodiagnostic technologists' societies. I will review the position statements of the American Association of Neuromuscular and Electrodiagnostic Medicine (AANEM) and The American Society of Electroneurodiagnostic Technologists (ASET) that cover these issues. There are also the published guidelines from the American Electroencephalographic Society (now the American Clinical Neurophysiology Society), *Guideline Eleven: Guidelines for Intraoperative Monitoring of Sensory Evoked Potentials* (Evoked Potentials Committee, 1987; 1994).

The AANEM's position statement on "*The Role of the Intraoperative Monitoring Team*," approved September 2008, states that, "The team should be composed of the surgeon, clinical neurophysiologist, monitoring technologist, and anesthesiologist." It also states that, "A trained clinical neurophysiologist (MD or DO) must be a member of the IOM team," and that the clinical neurophysiologist should have, "completed additional training in neurophysiologic monitoring." In regard to intraoperative monitoring technologists, the position statement indicates that their role

"... is to assist the surgical team. As such, the monitoring technologist should be under the direct supervision of the clinical neurophysiologist (MD or DO) who has training in electrophysiologic monitoring and the fundamentals of clinical neurophysiology."

It further states that, depending on the clinical problem, the clinical neurophysiologist may be in one of the following locations:

1. In the operating room,
2. In the same building,
3. Real-time monitoring from a remote site, when a monitoring technologist is continuously present in the operating room, or
4. At a location where the operating room is accessible within minutes to view the recording procedure.

The ASET's Position Statement on *Electroneurodiagnostic Technologists in the Operating Room* (revised 1998) states:

> "Interpretation of the intraoperative recording, recommendations, and decisions regarding action or consequence are the responsibility of a qualified physician or clinical neurophysiologist (MD, DO, PhD). It is recommended that an interpreting physician or neurophysiologist be physically present or present by means of a real-time, remote mechanism for all END monitoring situations.
>
> When electrical activity is recorded directly from the surface of the brain, a physician neurophysiologist *must* be physically present or present by means of real-time, remote mechanism to interpret the recording and to advise the surgeon.
>
> For other types of monitoring, when the physician or neurophysiologist is not present physically or by means of remote mechanism and when the technologist is providing the waveform description, the surgeon or anesthesiologist is responsible for the interpretation, diagnosis, and course of action; with the medical director of the electroneurodiagnostic department or a physician neurophysiologist bearing the responsibility for educating the surgeon regarding possible interpretations and potential consequences of the waveforms identified by the technologist."

Given the defined roles as described above, the principal responsibility of the technologist, once the electrodes have been placed on the patient, is to access and continuously monitor the IOM data and report when the IOM results exceed a predetermined range. For example, in cerebral aneurysm surgery, a critical change in the cortical SEP would correlate with an amplitude reduction of ≥50% or an increase in the latency of 10%.

Once an alarm has been triggered, the technologist must exclude the possibility that the observed findings are due to non-neurophysiologic factors. Simultaneously, the technologist must reproduce the observed changes by immediately repeating the test to see if the findings are reproducible. If the change is reproduced and non-neurophysiologic factors are excluded as causing or significantly contributing to the change, then the technologist should simultaneously inform the IOM physician, surgeon, and anesthesiologist of the observed findings.

The IOM physician should immediately review the data and render an interpretation of the findings. The IOM physician's opinion should be relayed to the surgeon and the rest of the surgical team. This can be done directly by the IOM physician, if already present, or close by the operating room. If the IOM physician is monitoring the case remotely, then the information can be provided to the surgical team by the technologist who has already reviewed the findings with the IOM physician. The IOM physician must use whatever available means to communicate with the surgeon and the other members of the surgical team.

It is not the role or responsibility of the IOM technologist to interpret the data or render an opinion based on the data independently; neither should the surgeon or anesthesiologist attempt to interpret the IOM data independently. This is especially true if they do not have the required training, expertise, and experience. Furthermore, even if they were experienced clinical IOM neurophysiologists, they would have to review the baseline data, background variability, and non-neurophysiologic factors. This could significantly divert their attention away from their primary responsibility. Ultimately, the NIOM technologist and supervising clinical neurophysiologist must work seamlessly as a team. To do this, they must have the proper skills, training, knowledge, abilities, and experience with intraoperative clinical neurophysiology, the surgical procedures, and the different clinical situations (Nuwer, 2008).

Finally, it is necessary to generate a final IOM report after the case is completed, which should be included in the patient's chart or electronic medical record. In addition, copies of the final report should be provided to the surgeon and anesthesiologist.

As the field of IOM has grown and its application has expanded, there has developed a need to establish and maintain high levels of excellence in the delivery of this medical service. Fortunately, this is now being addressed by the recent development of the ability to obtain accreditation specifically in IOM. The American Board of Registration of Electroencephalographic and Evoked Potential Technologists (ABRET), in collaboration with the AANEM, the American Academy of Neurology (AAN), and the ACNS, has developed an accreditation for IOM laboratories. The purpose of this accreditation is to recognize through an objective peer review process those laboratories performing IOM that,

"... comply with established standards and guidelines, have policies and procedures in place that meet its standards of care; and support professionalism and quality patient care."

Conclusions

Accurate interpretation of IOM data is the culmination of a long process, which begins after the request for IOM is received by the IOM service. Several things have to be known and understood before an IOM protocol is decided upon for each patient. This is the first step in a long process, which must be done correctly, to obtain accurate and valid data.

Thereafter, in the operating room, data analysis can only be done by ensuring that the data are valid and not overly influenced by one or more of the many confounding factors commonly encountered. Once satisfied of the validity of the data, then ultimately it is the responsibility of the IOM physician to interpret the data, render an opinion, and inform the surgeon and any other members of the surgical team as soon as possible.

Of course, this process is not possible without understanding the limitations imposed by the currently available electrophysiologic techniques in assessing different regions of the nervous system.

Finally, it is of great importance to acknowledge that this whole process would not be possible without the highly trained and dedicated IOM technologists, who are indispensable in setting up cases, acquiring the data, and doing the minute-by-minute data assessment and analysis that allow physicians to be in more than one place at the same time.

Case examples for Chapter 16

Case example 1

A 68-year-old male with a history of TIAS who was found to have a high-grade left internal carotid artery stenosis. The patient underwent carotid endarterectomy with the aid of IOM. Six-channel EEG and bilateral median nerve SSEPs were monitored during the surgical procedure. In this example, the pre-clamp EEG tracings (Figure 16.1a) show a burst-suppression pattern, the result of thiopental administration for neuroprotection prior to cross-clamping of the left internal carotid artery (ICA). This shows the effects of a non-neurophysiologic confounding factor (thiopental) on an IOM modality, the EEG. In addition, the morphology of cortical SEPs after bilateral independent median nerve stimulation was altered from the original baseline (not shown). One minute after the carotid clamp was placed, there were no significant changes in either EEG or the cortical SEPs. However, at five minutes after the carotid clamp was placed, there appeared to be a subtle difference in the EEG, showing reduced voltage on the left. No corresponding right median nerve cortical SEP change was seen (trace 8). Is this change significant to warrant an alarm? How much of this change is due to the administration of thiopental?

In Figure 16.1b, the EEG asymmetry is more pronounced and an alert is given. The SEPs now also indicate a mild difference in morphology, with the left median nerve cortical SEP (trace 7) showing a later peak, which is absent after right median nerve stimulation. There is slight improvement in the EEG one minute after the placement of an indwelling carotid shunt with resolution of all EEG and SEP asymmetries five minutes later. This example demonstrates how non-neurophysiologic factors can affect the IOM data, making the data more difficult to interpret and possibly delay any alarms.

In this case, the supervising IOM physician was present in the operating room during the period of carotid cross-clamping, and the alarm was provided by the IOM supervising physician directly to the surgeon and anesthesiologist. No further significant changes were noted in the EEG or SEPs during the remainder of the operative procedure. The final interpretation provided to the surgeon before the patient awoke was that the changes noted were consistent with transient left cerebral ischemia, which resolved after placement of the left carotid indwelling shunt, and that there was

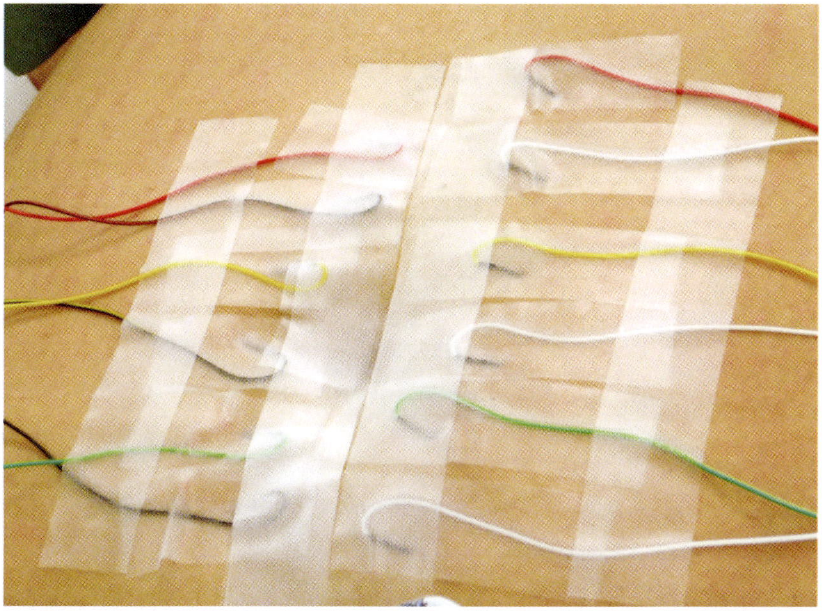

Figure 4.1 A set of needle electrodes placed by the NIOM technologist in the abdominal muscles to record triggered and spontaneous electromyographic responses during scoliosis surgery. The electrode wires are color coded to identify and label the different muscle groups easily.

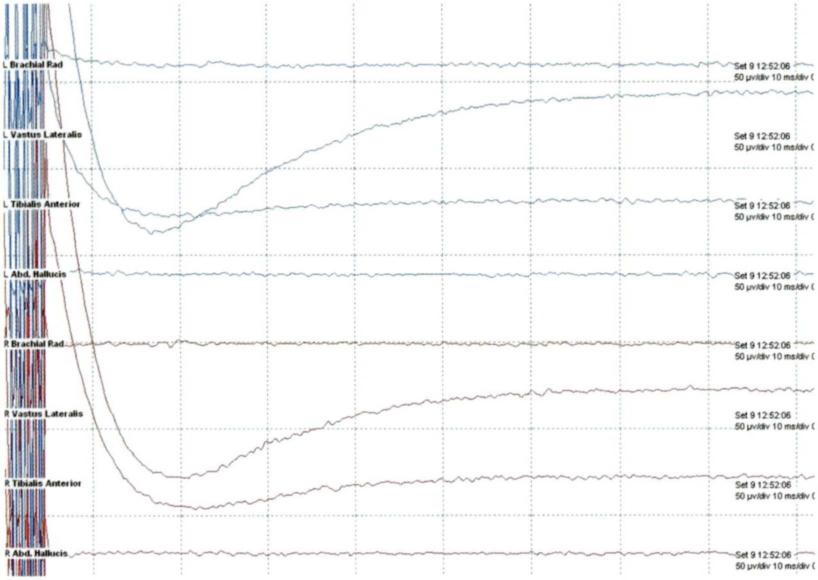

Figure 5.2 Effect of increasing the sevoflurane level to 2.5%, showing loss of all previously recorded MEPs.

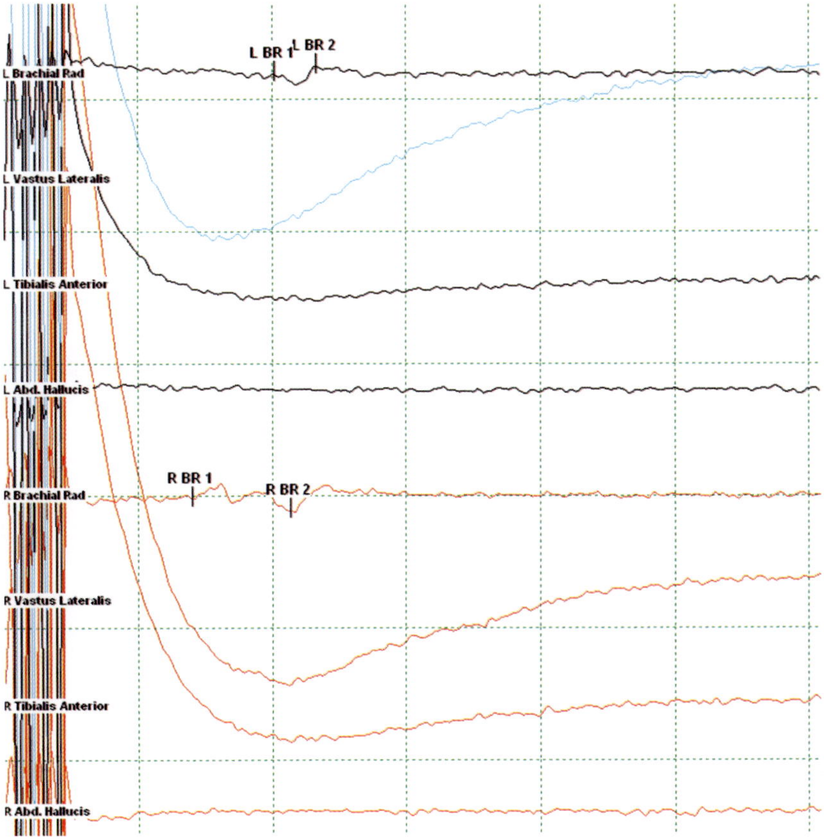

Figure 5.3 Loss of lower extremity MEPs with decrease of MAP during scoliosis repair.

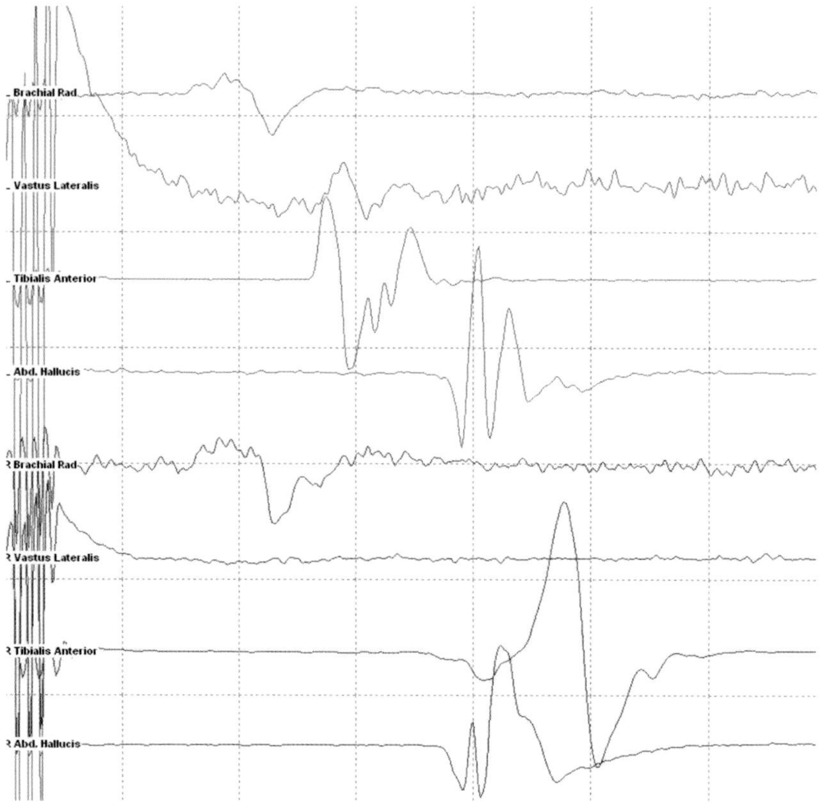

Figure 5.4 Return of lower extremity MEPs, although small in amplitude, after the MAP was increased in the patient in Figure 5.3.

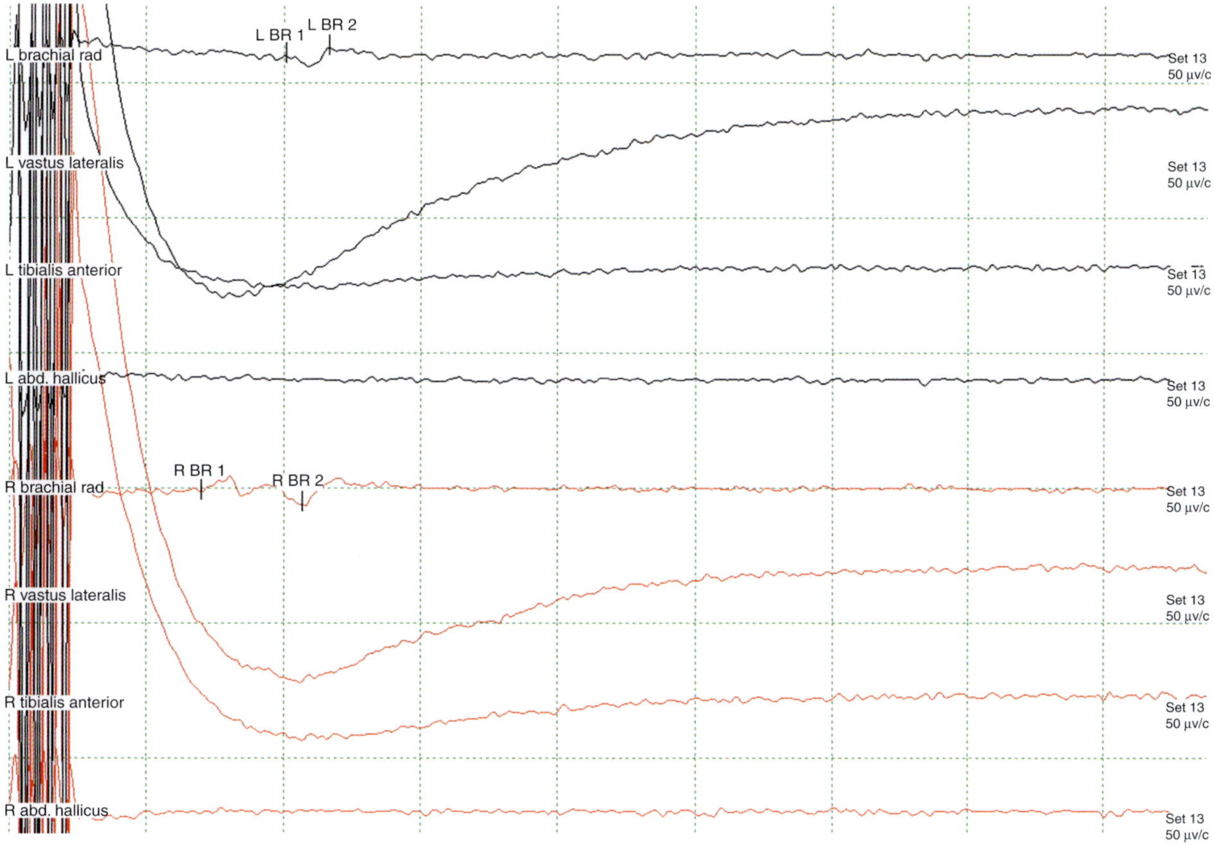

Figure 7.1 Loss of MEPs from the lower extremities following reduction in the mean arterial pressure.

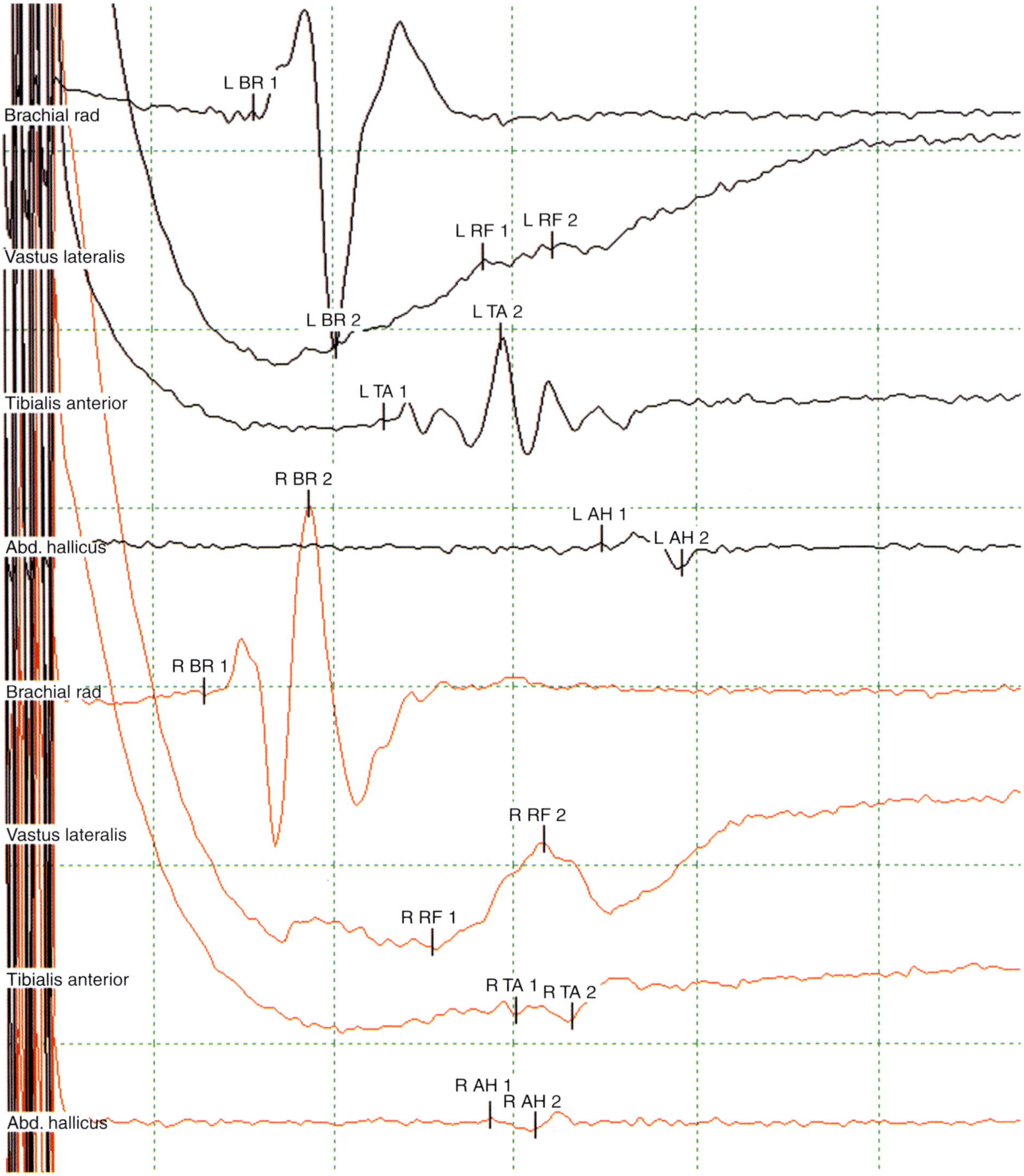

Figure 7.2 Return of MEPs in the lower extremities following an increase in the mean arterial pressure.

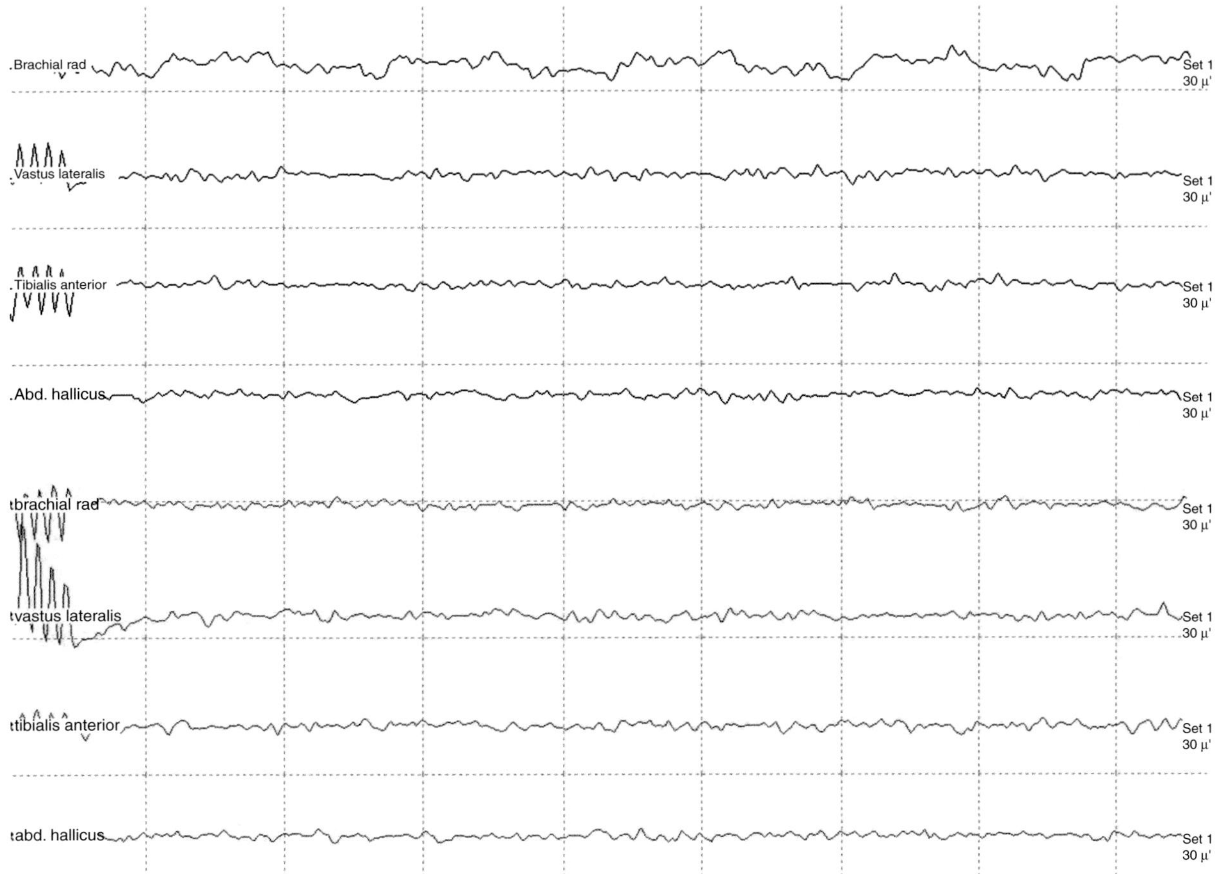

Figure 7.3 Increasing the level of sevoflurane anesthesia resulted in a loss of MEP potentials from legs and arms.

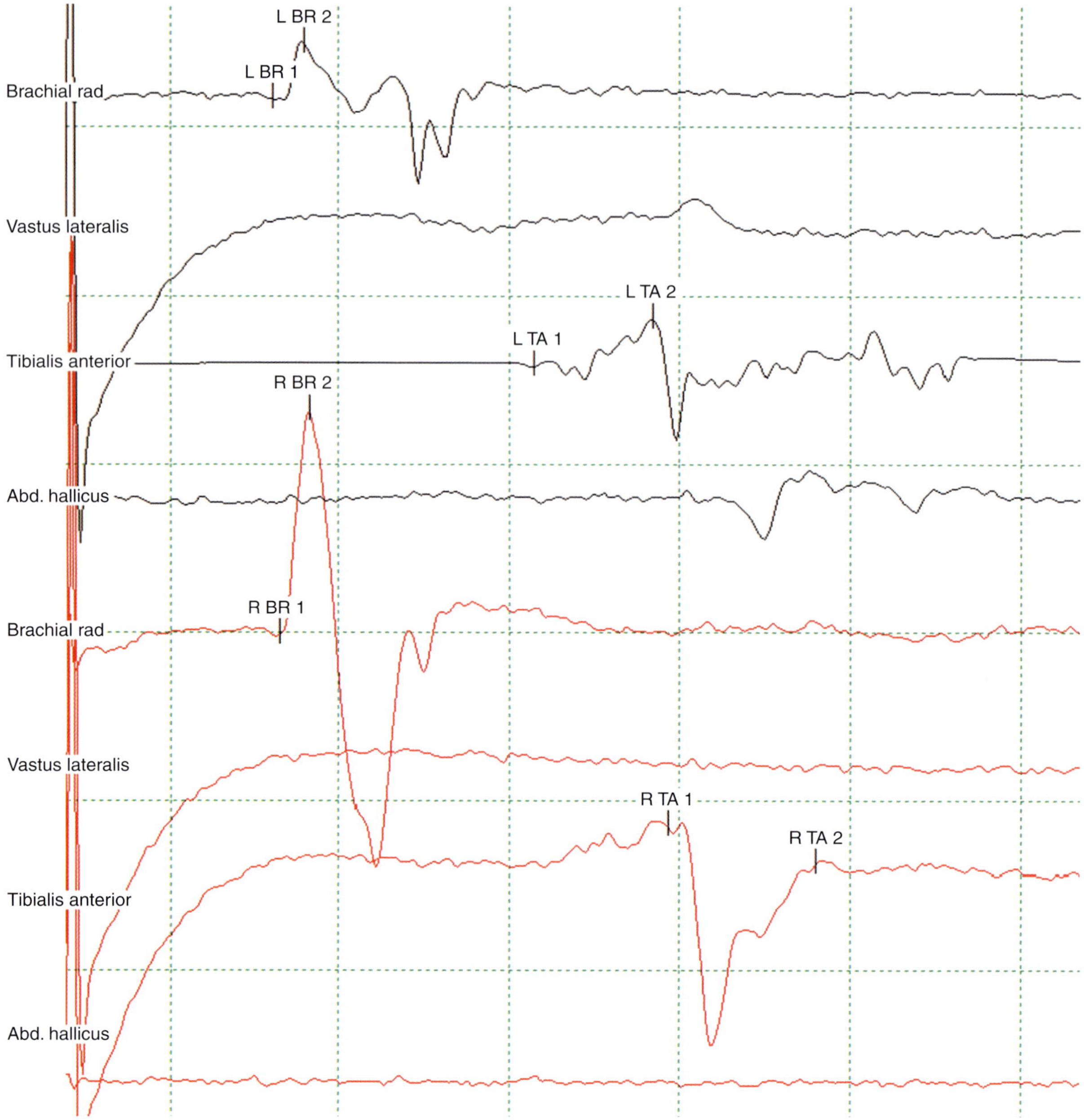

Figure 7.4 With a reduction in sevoflurane level, the MEP responses returned in all limbs.

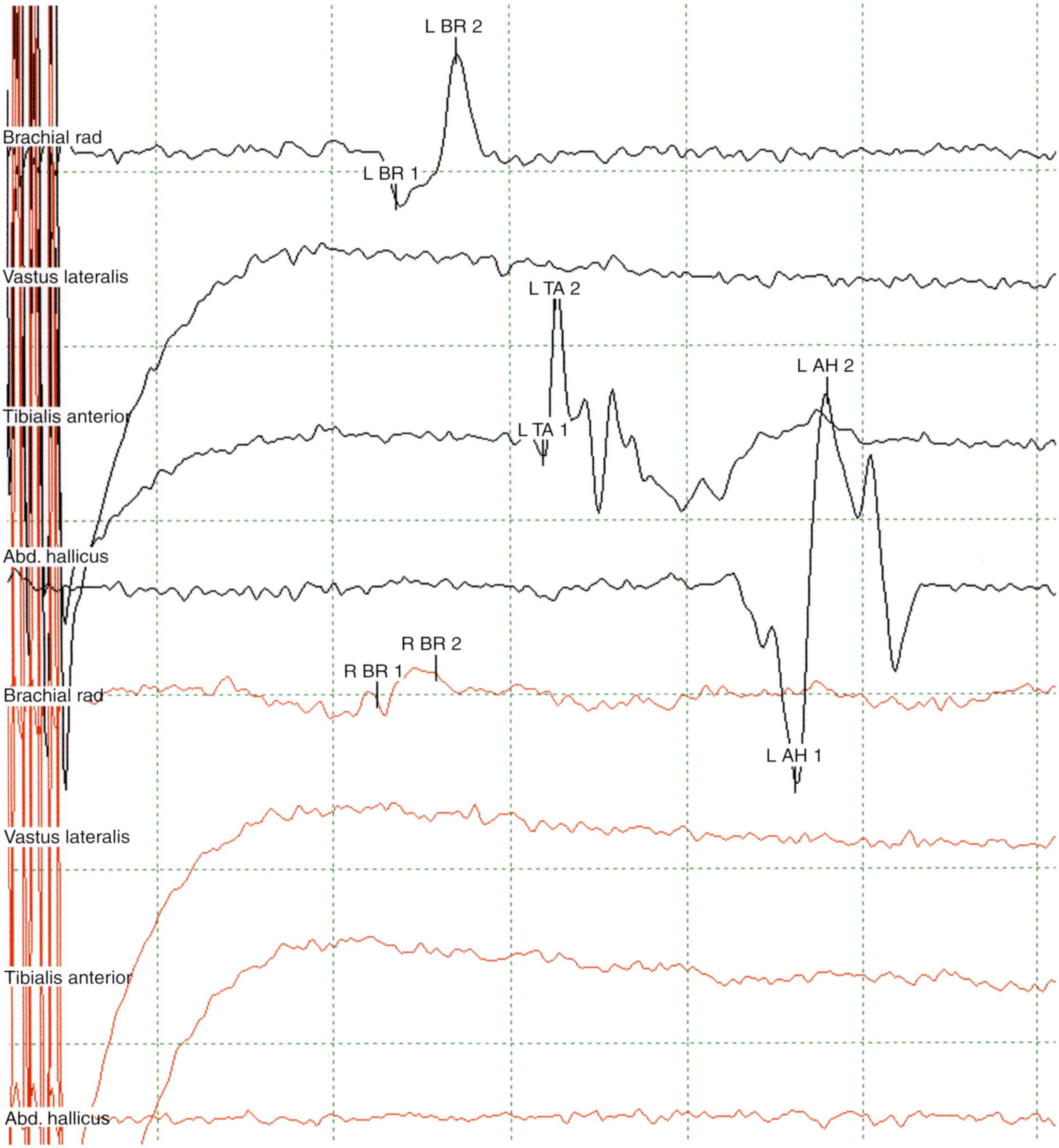

Figure 7.5 Loss of MEPs from the right leg following tightening of a rod during scoliosis surgery.

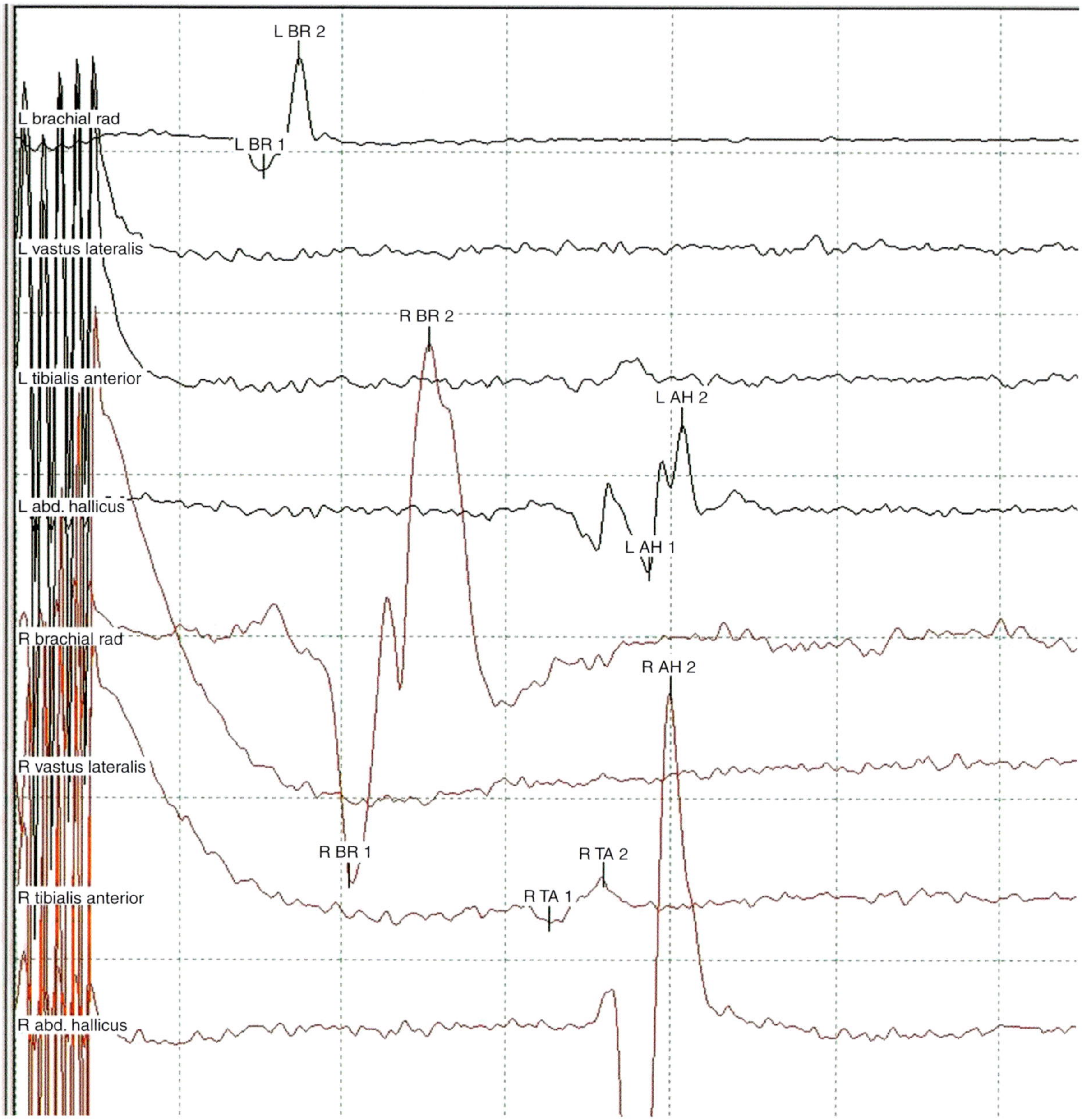

Figure 7.6 Return of MEPs from the right leg with loosening of rod during scoliosis surgery.

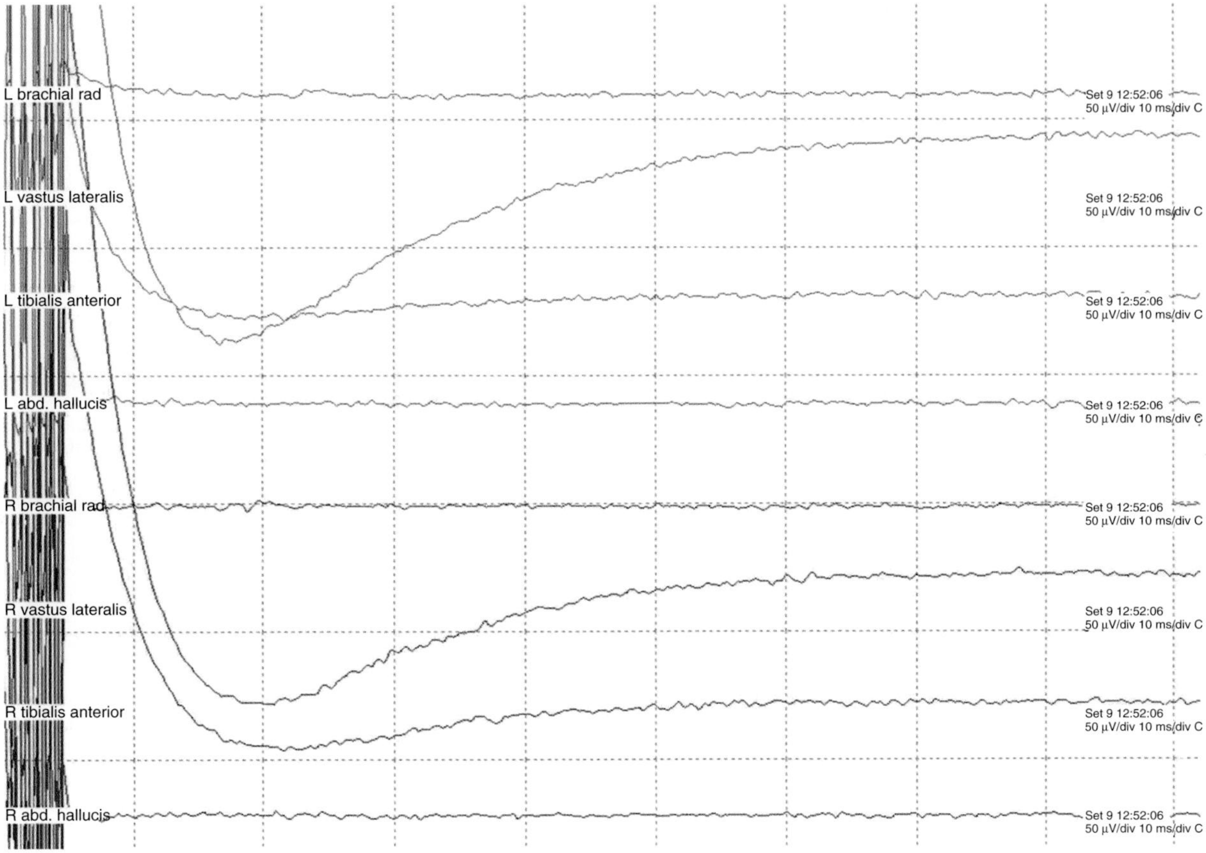

Figure 18.1 Absence of transcranial electric motor stimulation responses from all limbs due to elevated sevoflurane levels during spinal surgery.

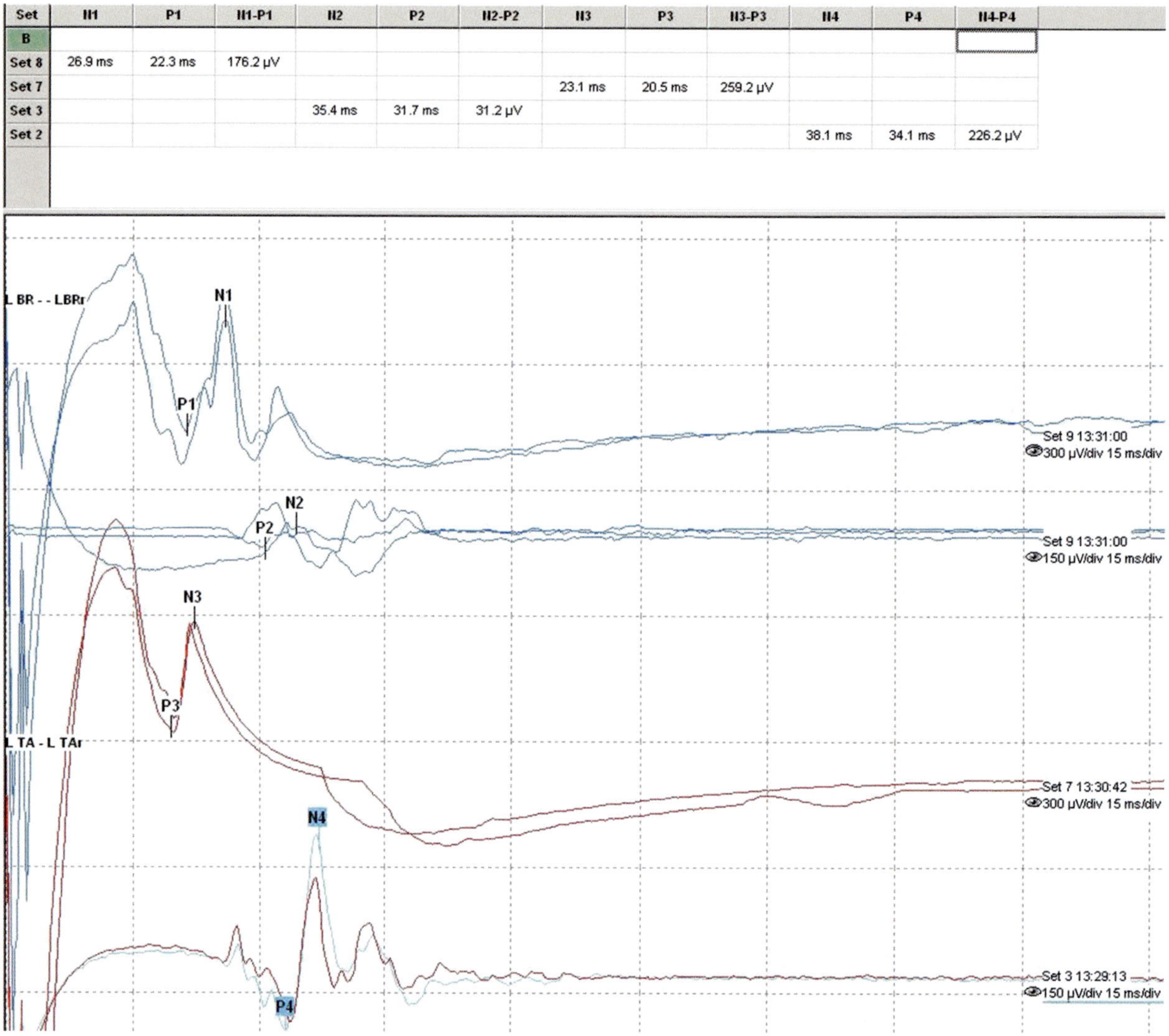

Set	N1	P1	N1-P1	N2	P2	N2-P2	N3	P3	N3-P3	N4	P4	N4-P4
B												
Set 8	26.9 ms	22.3 ms	176.2 µV									
Set 7							23.1 ms	20.5 ms	259.2 µV			
Set 3				35.4 ms	31.7 ms	31.2 µV						
Set 2										38.1 ms	34.1 ms	226.2 µV

Figure 18.2 Responses from all limbs seen on transcranial magnetic stimulation on postoperative study from patient in Figure 18.1.

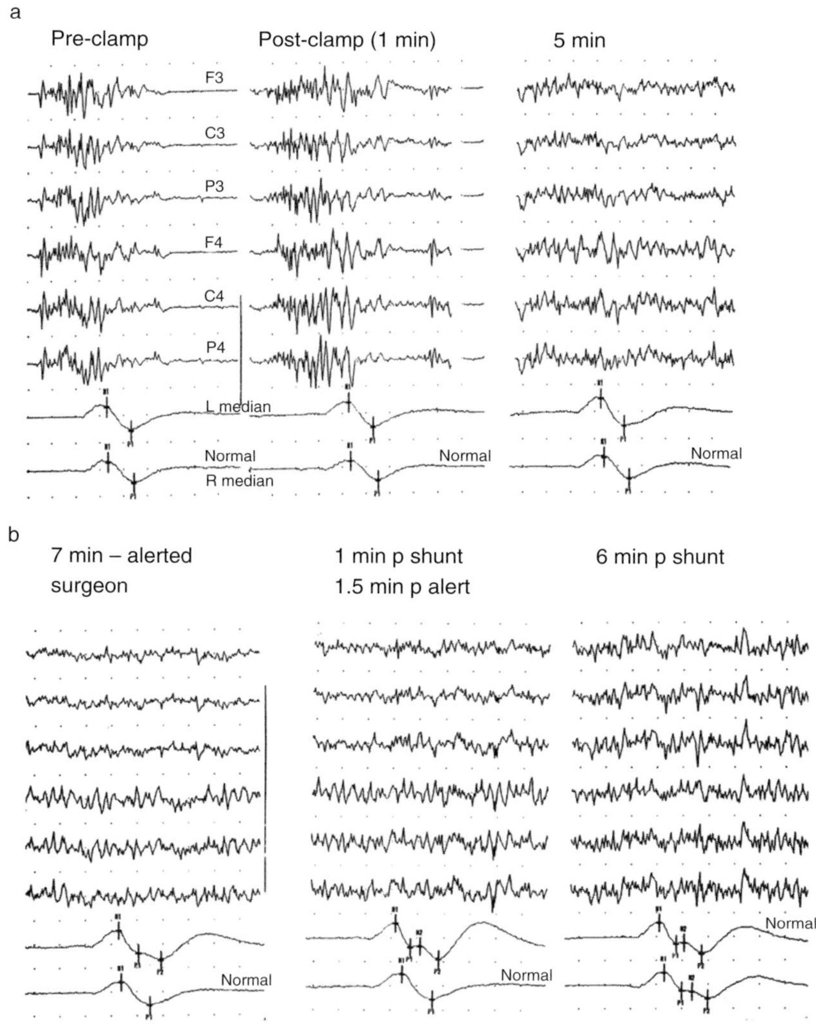

Figure 16.1 (a) EEG tracings recorded after thiopental administration but before cross-clamping of the left internal carotid artery. (b) This EEG recording shows a pronounced asymmetry, sufficient to cause alarm.

low probability of the patient developing a new neurological deficit. The patient experienced no new postoperative neurological deficits.

Case example 2

On neurologic evaluation, a 58-year-old female who presented with progressive ascending lower extremity numbness and paresthesias was found to have a T8–T9 intradural tumor. The plan was for the patient to undergo surgical resection of her thoracic spinal tumor. The IOM physician discussed the case with the attending neurosurgeon prior to surgery in order to plan the proper IOM protocol and to learn about the patient's preoperative neurologic examination findings. The patient's physical examination findings were remarkable for mild lower extremity numbness to touch but normal lower extremity strength.

The IOM protocol used to monitor this patient consisted of bilateral median and posterior tibial nerve SEPs, TcMEPs with myogenic recordings from bilateral hands, anterior tibialis (AT), and abductor hallucis (AH) muscles. During the surgical procedure the anesthetic regimen included stable low levels of isoflurane and nitrous oxide, and a remifentanyl infusion. The mean arterial pressures were maintained at 70 mmHg and above, temperature was kept constant at 37 °C; and neuromuscular blocking agents were not used.

Figure 16.2a shows well-developed and reproducible cortical and cervical spinal (C7) baseline SEPs and myogenic TcMEPs. Figure 16.2b shows loss of the TcMEP from the right anterior tibialis, slightly smaller right abductor hallucis, and no changes from the right or left upper and lower extremities. In addition, there is only a mild amplitude reduction in the cortical

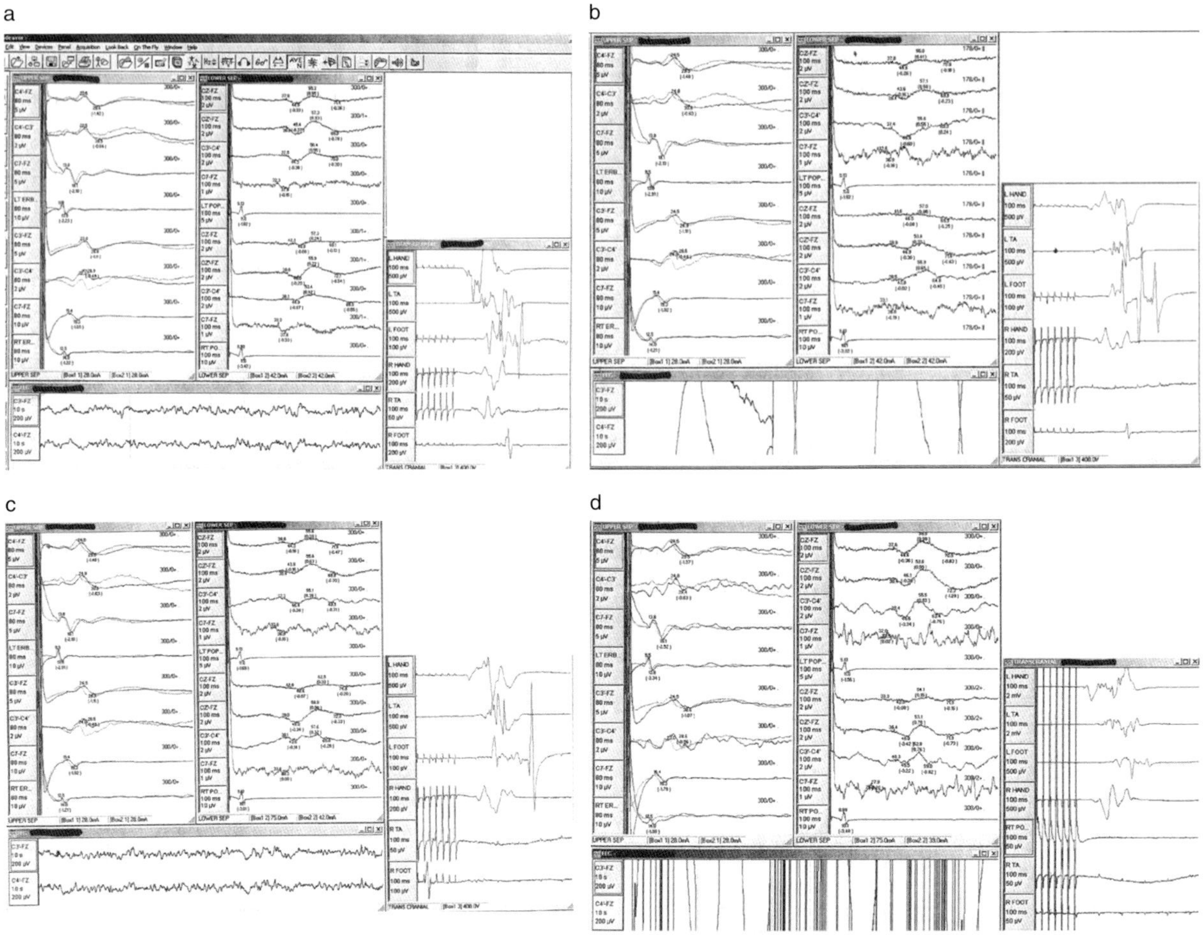

Figure 16.2 For (a), (b), (c), (d): Median nerve SSEPs are depicted in the left upper panel. Recordings are from C4–Fz, C4–C3, C7–Fz, and left Erb's–right Erb's (traces 1–4, respectively), after left median nerve stimulation. Traces 5–8 show recording from C3–Fz, C3–C4, C7–Fz, and right Erb's–left Erb's, after right median nerve stimulation. Middle panel shows SSEPs recording after left posterior tibial nerve stimulation from Cz–FZ, Cz′–Fz, C3′–C4′, C7–Fz and left popliteal fossa (traces 1–5, respectively). Traces 6–10 are recorded after right posterior tibial nerve stimulation from Cz–Fz, Cz′–Fz, C4′–C3′, C7–Fz and right popliteal fossa. TcMEP results are shown in the panel on the right. Recordings were obtained from the left hand, left TA, left AH, right hand, right TA, and right AH for traces 1–6 respectively. Two channels of C3–Fz and C4–Fz EEG recordings are shown in the left lower panel (traces 1 and 2, respectively). (a) Baseline. (b) Changes in the TcMEP and SEP recordings. (c) Further losses in TcMEPs. (d) Recording at time of closing, no recovery of TcMEPs from right leg, but some improvement in cortical SEPs.

posterior tibial SEPs. Are these changes sufficient to raise an alarm? If we apply the "presence or absence" rule for myogenic TcMEPs, then an alarm should not be raised since there is still a TcMEP from the right abductor hallucis. However, if we apply criteria of amplitude reduction, then an alert is warranted given the changes noted in the right leg. This shows the dilemma that can exist when the IOM results are not completely clear. In this case, the IOM physician was in the operating room at the time of change and issued an alert to the surgeon. The anesthesiologist was also informed and the mean arterial pressure was raised.

Figure 16.2c now shows loss of the TcMEPs from both AT and AH muscles. The lower extremity cortical SEPs show only slight amplitude reduction. Figure 16.2d was obtained at time of closing and shows that there was no recovery of the TcMEPs from the right leg. However, there was improvement in the amplitudes of the cortical SEPs after posterior tibial nerve stimulation.

The interpretation of these findings, made in the operating room and before the patient awoke, was that there was a high probability of mild-to-moderate spinal cord injury and that she would most likely develop right leg weakness. She was noted to have right leg weakness in the immediate postoperative period and worsening of her lower extremity numbness. Two months later, the patient's leg numbness had nearly resolved but she continued to have right leg weakness, although it was significantly improved. This case shows the importance of having the IOM physician available to review the data quickly, make an interpretation, discuss the case with the other members of the surgical team (surgeon and anesthesiologist), alter management, and make a prospective prediction of the patient's lik\ely outcome.

Case example 3

A 24-year-old female with a diagnosis of left hemifacial spasm was noted to have a prominent vessel adjacent to her left facial nerve, after undergoing brain MRI. Other than her left hemifacial spasm, the patient exhibited no other neurologic abnormalities on neurologic examination. As a result of her MRI findings she opted for surgical treatment of her hemifacial spasm. She subsequently underwent left microvascular decompression of her facial nerve. The IOM protocol consisted of continuous, free-running EMG recorded from the left masseter, orbicularis oculi, orbicularis oris, and trapezius and, in addition, bilateral BAEP monitoring and left lateral-spread testing. The anesthetic regimen included low-level isoflurane and nitrous oxide combination with remifentanyl. Neuromuscular blocking agents were not used.

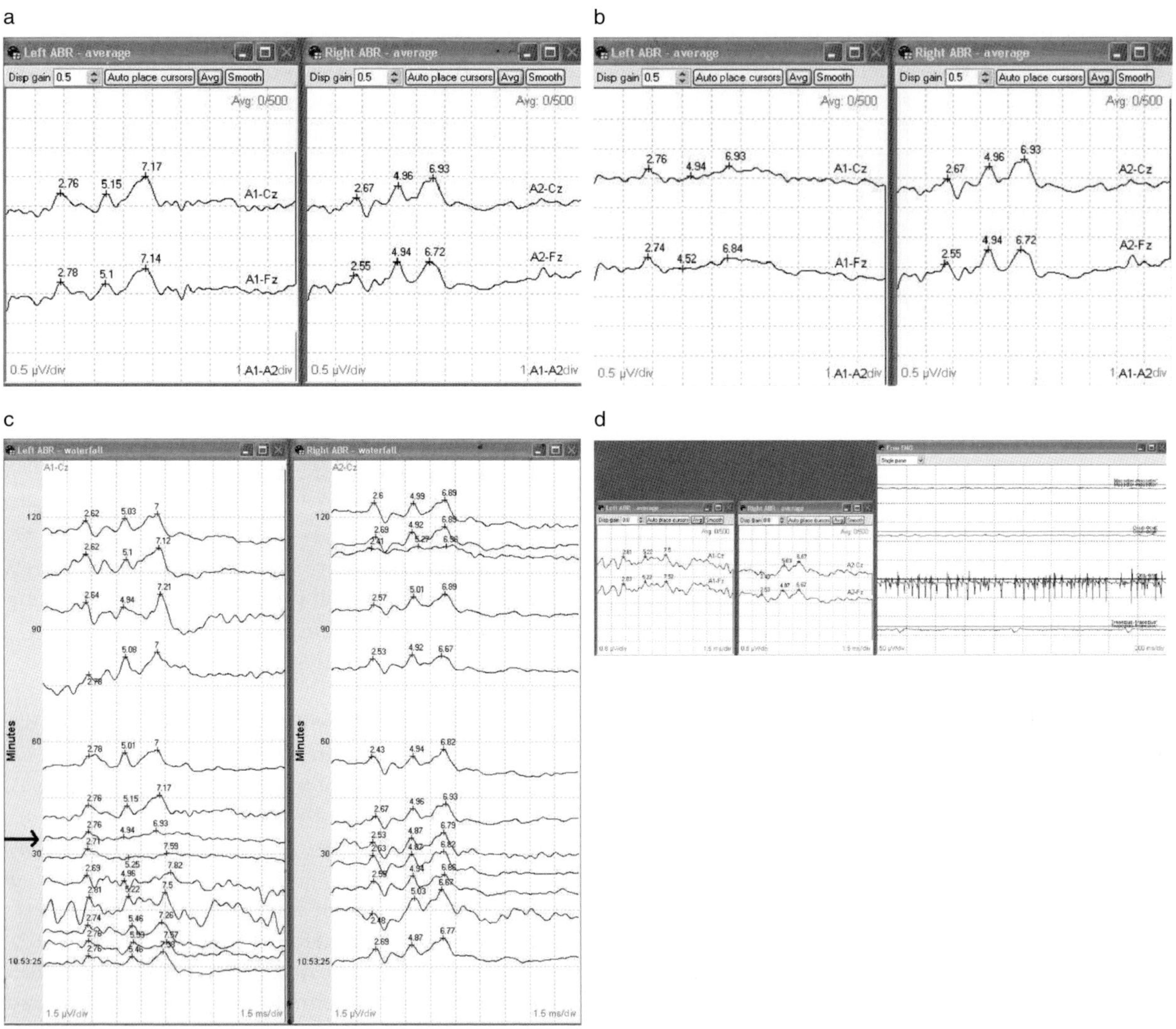

Figure 16.3 (a) Baseline BAEPS. Left panel: BAEPS after left ear stimulation, recorded from A1–Cz and A2–Cz, respectively. Right panel: BAEP after right ear stimulation, recorded from A2–Cz and A1–Cz, respectively. (b) Significant amplitude reduction of left BAEP waves III and V. (c) Note change of the left BAEPs (arrow) with subsequent recovery at 10:53:25. (d) Neurotonic discharges emanating from the left orbicularis oris muscles.

When compared with the baseline traces seen in Figure 16.3a, Figure 16.3b shows a significant amplitude reduction of waves III and V of the Left BAEP. Note that there are no significant amplitude or latency changes in wave I from the baseline values. This change occurred relatively quickly and required that the surgeon be notified immediately after the change was reproduced on the subsequent trace (arrow), as seen in Figure 16.3c. The change occurred after the placement of retractors, which were removed after the BAEP change was noted. There was a gradual improvement of the left BAEP back to baseline levels after approximately 20 minutes.

The change in BAEPs occurred while the IOM physician was reviewing another IOM case in a different operating room. However, the IOM physician was able to view the case remotely using real-time remote monitoring software, confirm the left BAEP change, and arrive at this patient's operating room within minutes to discuss the findings with the neurosurgeon. There were no further changes in the BAEPs during the remainder of the surgical procedure. The transient left BAEP changes were interpreted to indicate that there was transient impairment of impulse conduction along the intrinsic brainstem auditory pathways, proximal to the cochlear nucleus and most likely at the level of the pons. Furthermore, the IOM physician opined that there was a low probability of the patient developing a new postoperative neurologic deficit, including left-sided hearing loss, since the changes were transient. This interpretation and opinion was rendered by the IOM physician before the patient awoke from surgery. The patient experienced no new postoperative neurologic deficits.

Figure 16.3d shows an example of neurotonic discharges emanating from the left orbicularis oris muscles, which occurred during manipulation of the facial nerve. The surgeon is informed of the presence of neurotonic discharges, and can also listen to via a loudspeaker. He/she is also told which muscle is producing the EMG and/or neurotonic discharges.

Questions

1. Which of the following is considered to be a critical change when monitoring cortical generated SEPs after posterior tibial stimulation during scoliosis surgery?
   (a) An amplitude reduction of 10%
   (b) An amplitude reduction of 20%
   (c) An amplitude reduction of 30%
   (d) An amplitude reduction of 40%
   (e) An amplitude reduction of 50%
2. What should the IOM technologist do next when a critical change is noted in the posterior tibial nerve scalp-recorded SEPs during scoliosis surgery?
   (a) Immediately alert the surgeon
   (b) Ask the anesthesiologist to reduce the concentration of anesthetic agents
   (c) Turn off and reboot the IOM equipment
   (d) Perform another posterior tibial nerve SEP
   (e) Replace the posterior tibial nerve stimulating electrode
3. How much time is considered reasonable to spend attempting to identify the cause of an evoked potential change during spinal cord monitoring?
   (a) 5 minutes
   (b) 10 minutes
   (c) 15 minutes
   (d) 20 minutes
   (e) None of the above
4. Once a critical change has been established as valid and accurate, what is the next step that the IOM technologist should take?
   (a) Alert the surgeon
   (b) Alert the anaesthesiologist
   (c) Alert the IOM physician
   (d) All of the above
   (e) None of the above
5. If warning criteria are too conservative, then cases with new deficits will probably be missed, making the criteria less specific
   (a) True
   (b) False
6. Which of the following BAEP changes fulfill the criteria for issuing an alarm?
   (a) Transient loss of wave IV only
   (b) Persistent loss of wave IV
   (c) Decrease in the I–V interpeak interval by >1 ms
   (d) Persistent loss of wave I
   (e) Increased amplitude of wave III
7. Once changes sufficient to raise an alarm have been identified by the technologist, the IOM physician should do which of the following first?
   (a) Ask the anesthesiologist to raise the patient's mean arterial pressure

(b) Review the data and confirm the changes
(c) Immediately alert the surgeon
(d) Check electrode impedances
(e) All of the above

8. Which of the following is more likely to change as a result of spinal cord ischemia during scoliosis surgery?
(a) Posterior tibial nerve SEP central conduction time
(b) Posterior tibial nerve SEP cortical latency
(c) Posterior tibial nerve SEP amplitude
(d) All occur equally in frequency
(e) None of the above

9. Who of the following is responsible for interpreting the IOM data and writing the final report
(a) Surgeon
(b) IOM physician
(c) IOM technologist
(d) The surgeon and the IOM physician
(e) All of the above

10. Which of the following is probably the most important aspect of accurate IOM
(a) Reporting the data
(b) Alerting the IOM physician
(c) Alerting the surgeon
(d) Interpreting the data
(e) Acquiring the data

References

American Association of Neuromuscular and Electrodiagnostic Medicine (AANEM) (2008). *The Role of the Intraoperative Monitoring Team.* AANEM position statement.

American Society of Electroneurodiagnostic Technologists (ASET) (1998). *Electroneurodiagnostic Technologists in the Operating Room.* ASET position statement. www.aset.org.

Arai, M., Goto, T., Seichi, A., Miura, T., and Nakamura, K. (2000). Comparison of spinal cord evoked potentials and peripheral nerve evoked potentials by electric stimulation of the spinal cord under acute spinal cord compression in cats. *Spinal Cord*, **38**, 403–408.

Branston, N. M., Symon, L., Crockard, H. A., and Pasztor, E. (1974). Relationship between the cortical evoked potential and local cortical blood flow following acute middle cerebral artery occlusion in the baboon. *Experimental Neurology*, **45**, 195–208.

Branston, N. M., Strong, A. J., and Symon, L. (1977). Extracellular potassium activity, evoked potential and tissue blood flow: relationships during progressive ischaemia in baboon cerebral cortex. *Journal of the Neurological Sciences*, **32**, 305–321.

Branston, N. M., Ladds, A., Symon, L., and Wang, A. D. (1984). Comparison of the effects of ischaemia on early components of the somatosensory evoked potential in brainstem, thalamus, and cerebral cortex. *Journal of Cerebral Blood Flow and Metabolism*, **4**, 68–81.

Breitner, S. and Matzen, K. A. (1985). Scalp recorded somatosensory evoked potentials during spinal surgery. In *Spinal Cord Monitoring*, ed. J. Schramm and S. J. Jones. Berlin: Springer, pp. 173–178.

Brown, R. H. and Nash, C. L. (1985). The "grey zone" in intraoperative SCEP monitoring. In *Spinal Cord Monitoring*, ed. J. Schramm and S. J. Jones. Berlin: Springer, pp. 179–185.

Burke, D., Nuwer, M. R., Daube, J., *et al.* (1999). Recommendations for the practice of clinical neurophysiology guidelines of the International Federation of Clinical Physiology. *Electroencephalography and Clinical Neurophysiology*, **52***(suppl.)*, 133–148.

Calancie, B. and Molano, M. R. (2008). Alarm criteria for motor evoked potentials. What's wrong with the "presence-or-absence" approach? *Spine*, **33**, 406–414.

Calancie, B., Harris, W., Broton, J. G., Alexeeva, N., and Green, B. A. (1998). "Threshold-level" multipulse transcranial electrical stimulation of motor cortex for intraoperative monitoring of spinal motor tracts: description of method and comparison to somatosensory evoked potential monitoring. *Journal of Neurosurgery*, **88**, 457–470.

Carlson, G. D., Gorden, C. D., Oliff, H. S., Pillai, J. J., and LaManna, J. C. (2003). Sustained spinal-cord compression. Part 1: time-dependent effect on long-term pathophysiology. *Journal of Bone and Joint Surgery*, **85**, 86–94.

Chen, Z.-Y., Wong, H.-K., and Chan, Y.-H. (2004). Variability of somatosensory evoked potential monitoring during scoliosis surgery. *Journal of Spinal Disorders and Techniques*, **17**, 470–476.

Crum, B. A. (2009). Brain stem and cranial nerve monitoring. In *Clinical Neurophysiology*, ed. J. R. Daube and D. I. Devon. New York: Oxford University Press. pp. 739–749.

Daube, J. R. and Harper, C. M. (1989). Surgical monitoring of cranial and peripheral nerves. In *Neuromonitoring in Surgery*, ed. J. E. Desmedt. Amsterdam: Elsevier, pp. 115–138.

Dawson, E. G., Sherman, J. E., Kanim, L. E. A., and Nuwer, M. R. (1991). Spinal cord monitoring results of the Scoliosis Research Society and the European Spinal Deformity Society survey. *Spine*, **16**, S361–S364.

Deletis, V. and Sala, F. (2008). Intraoperative neurophysiological monitoring of the spinal cord during spinal-cord and spine surgery: a review focusing on the corticospinal tracts. *Clinical Neurophysiology*, **119**, 248–264.

Evoked Potentials Committee (1987). American Electroencephalographic Society guidelines for intraoperative monitoring of sensory evoked potentials. *Journal of Clinical Neurophysiology*, **4**, 397–416.

Evoked Potentials Committee (1994). Guideline 11: guidelines for intraoperative monitoring of sensory evoked potentials. *Journal of Clinical Neurophysiology*, **11**, 77–87.

Hargadine, J. R., Branston, N. M., and Symon, L. (1980). Central conduction time in primate brain ischemia – a study in baboons. *Stroke*, **11**, 637–642.

Jones, S. J. (1994). Somatosensory evoked potentials and their use for spinal-cord monitoring. In *Handbook of Spinal Cord Monitoring. Proceedings of the Fifth International Symposium on Spinal Cord Monitoring*, ed. S. J. Jones, S. Boyd, M. Hetreed, and N. J. Smith. Boston: Kluwer Academic.

Jones, T. H., Morawetz, R. B., Crowell, R. M., *et al.* (1981). Thresholds of focal ischemia in awake monkeys. *Journal of Neurosurgery*, **54**, 773–782.

Koht, A., Sloat, T., Ronai, A., and Toleikis, J. R. (1985). Intraoperative deterioration of evoked potentials during spinal surgery. In *Spinal Cord Monitoring*, ed. J. Schramm and S. J. Jones. Berlin: Springer, pp. 161–166.

Lam, A. M., Manninen, P. H., Ferguson, G. G., and Nantua, W. (1991). Monitoring electrophysiologic function during carotid endarterectomy: a comparison of somatosensory evoked potentials and conventional electroencephalogram. *Anesthesiology*, **75**, 15–21.

MacDonald, D. B., Al Zayed, A., Khoudeir, I., and Stigsby, B. (2003). Monitoring scoliosis surgery with combined multiple pulse transcranial electric motor and cortical somatosensory evoked potentials from the lower and upper extremities. *Spine*, **28**, 194–203.

Minahan, R. E., Sepkuty, J. P., Lesser, R. P., Sponseller, P. D., and Kostuik, J. P. (2001). Anterior spinal-cord injury with preserved neurogenic "motor" evoked potentials. *Clinical Neurophysiology*, **112**, 1442–1450.

Nuwer, M. R. (1986). Spinal cord monitoring. In *Evoked Potential Monitoring in the Operating Room*, ed. M. R. Nuwer. New York: Raven Press, pp. 49–101.

Nuwer, M. R. (1999). Spinal cord monitoring. *Muscle & Nerve*, **22**, 1620–1630.

Nuwer, M. R. (2008). Intraoperative monitoring of spinal cord. *Clinical Neurophysiology*, **119**, 247.

Schramm, J. (1985). Spinal cord monitoring: current status and new developments. *Central Nervous System Trauma*, **2**, 207–227.

Schramm, J., Shigeno, T., and Brock, M. (1983). Clinical signs and evoked response alterations associated with chronic experimental cord compression. *Journal of Neurosurgery*, **58**, 734–741.

Schwartz, D. M., Auerbach, J. D., Dormans, *et al.* (2007). Neurophysiological detection of impending spinal-cord injury during scoliosis surgery. *Journal of Bone and Joint Surgery (America)*, **89**, 2440–2449.

Seyal, M. and Mull, B. (2002). Mechanisms of signal change during intraoperative somatosensory potential monitoring of the spinal cord. *Journal of Clinical Neurophysiology*, **19**, 409–415.

Strommen, J. A. (2009). Spinal-cord monitoring. In *Clinical Neurophysiology*, ed. J. R. Daube and D. I. Devon. New York: Oxford University Press, pp. 751–776.

Symon, L. (1980). The relationship between CBF, evoked potentials and the clinical features in cerebral ischaemia. *Proceedings of the 23rd Scandinavian Neurological Congress. Acta Neurologica Scandinavica*, **62**(suppl.), 175–190.

Tamaki, T., Takano, H., Takakuwa, K., *et al.* (1985a). An assessment of the use of spinal-cord evoked potentials in prognosis estimation of injured spinal cord. In *Spinal Cord Monitoring*, ed. J. Schramm and S. J. Jones. Berlin: Springer, pp. 221–226.

Tamaki, T., Takano, H., and Takakuwa, K. (1985b). Spinal-cord monitoring: basic principles and experimental aspects. *Central Nervous System Trauma*, **2**, 137–149.

York, D. H., Chabot, R. J., and Gaines, R. W. (1987). Response variability of somatosensory evoked potentials during scoliosis surgery. *Spine*, **12**, 864–876.

Young, W. and Berenstein, A. (1985). Somatosensory evoked potentials monitoring or intraoperative procedures. In *Spinal Cord Monitoring*, ed. J. Schramm and S. J. Jones. Berlin: Springer, pp. 197–203.

17 Intraoperative wake-up test

Marc R. Nuwer

Before intraoperative neurophysiologic monitoring (IOM) became widely used, the test for spinal-cord integrity was the wake-up test. This remains a useful test for resolving uncertainty in IOM or when IOM is not possible.

As described by Stagnara and colleagues (Vauzelle *et al.*, 1973), the procedure is typically performed in spinal deformity surgery. The patient is awakened during surgery after the main part of the surgery is completed, e.g., shortly after instrumentation is placed and the spine is distracted. The patient remains on the table and intubated with the surgical site still open. Anesthetic is discontinued or greatly lightened. Neuromuscular blockade is avoided after opening early in the case. Once the patient is sufficiently awake, he or she is asked to move his or her feet. This demonstrates intact spinal-cord motor function. When the test is complete, anesthesia is reinstituted, the surgical procedure is completed, and the back is closed.

This was once the only practical way to test spinal-cord function during scoliosis and other spinal deformity surgery. It had significant risks, including excessive movement on the operating room table, accidental endotracheal tube extubation, and dislodgement of intravenous lines.

Disadvantages

The wake-up test is usually performed only once during surgery. As such it is not really a continuous monitoring technique. It may identify problems too late to intervene. Once the wake-up test is performed, it is not usually performed again. Therefore, the remainder of the case proceeds without any testing until the patient is awakened again at the end of the case.

The desire for continuous monitoring eventually led to spinal-cord monitoring with somatosensory evoked potentials (SEPs). Bradshaw and colleagues (1984) recommended that SEP monitoring should replace wake-up as the test of choice. Padberg and colleagues (1998) recommended that monitoring SEPs and spinal stimulation motor EPs should be the standard of care for scoliosis surgery, and that these tests were good enough to replace the wake-up test. Eggspuehler and colleagues (2007) reported that the wake-up test was obsolete in their institution.

Yet, even with more or less continuous evoked potential monitoring, there are times when monitoring EPs could not be used to answer the relevant clinical questions, e.g., when EPs could not be obtained. Sometimes the recorded EPs changed, but questions remain about whether the patient's motor status was sufficiently changed that the surgeon should remove the spinal instrumentation.

Advantages

Many users see the wake-up test as an additional tool to be used if necessary to test directly the patient's motor function. Wake-up testing may confirm the motor loss predicted by IOM (Glassman *et al.*, 1993; Wilson-Holden *et al.*, 2002). Wake-up testing can demonstrate that the clinical interventions undertaken when the monitored signals changed, e.g., raising the blood pressure, were sufficient to preserve motor function (Othman *et al.*, 2004). The wake-up test may show that the motor examination remains intact, so that the recorded changes may have been falsely positive (Wilson-Holden *et al.*, 1999; Kim *et al.*, 2007; Cheh *et al.*, 2008). Intraoperative monitoring may show real deterioration, while the wake-up test itself can be falsely normal (Ben-David *et al.*, 1987).

Sometimes the signals cannot be obtained, such as in patients with substantial preexisting impairment (Thuet *et al.*, 2005). Those patients with unobtainable signals are at higher risk of neurological postoperative complications, and several authors

Table 17.1 Consider a wake-up test under certain circumstances

When the IOM results are unreliable
When the IOM results show changes but questions remain about the patient's motor status
When the patient is sufficiently cooperative to participate safely
When the benefits of awakening with the back still opened outweigh the risks

Table 17.2 Role of the intraoperative monitoring team in the wake-up test

Go under the surgical drapes at the foot of the table to assess the foot and leg movement
Ask anesthesia to instruct the patient loudly to move the feet up and down
Report accurately the amount of foot or leg movement
Do not mistake small leg movements that are due to patient's trunk movement
More than an inch up and down foot movements are a positive test
Knee flexion while raising the lower leg up off the table is a positive test

recommended considering wake-up tests when EPs could not be obtained. Patients with some neurodegenerative disorders e.g., Friedreich's ataxia have spinal disorders as well as no recordable IOM signals, so the wake-up test was used instead (Milbrandt *et al.*, 2008). For some patients, motor-evoked potentials (MEPs) are avoided because of a history of seizures, and wake-up testing may be added to SEPs to monitor those patients (Master *et al.*, 2008).

As commonly performed, the wake-up test requires the patient to cooperate in moving the legs. Pediatric patients may be unable to cooperate sufficiently, so to compensate for that, a stimulator may be placed on the plantar foot surface or on the posterior tibial nerve that innervates the foot. Stimulation can reliably prompt even young children or neonates to demonstrated flexion and extension of hips and knees (Polly *et al.*, 2000; Govindarajan *et al.*, 2006).

Overall, many authors have concluded that the wake-up test remains a valuable tool that should be used in certain circumstances (Table 17.1). The IOM team members may participate in the wake-up test by observing under the drapes to assess whether or not leg or foot movements occurred (Table 17.2). Optimally, the movements should be substantial, not just slight twitches.

Anesthesia

Anesthesia options have been reviewed elsewhere (Gambrall, 2007). Midazolam may allow for a quicker wake-up test when the benzodiazepine reversal agent flumazenil is used to awaken the patient (Eldar *et al.*, 1992). This benzodiazepine tactic may be quicker than when using propofol anesthesia (Koscielniak-Nielsen *et al.*, 1998). Others have noted that desfluane allows faster awakening than propofol (Grottke *et al.*, 2004; Ting *et al.*, 2004). Lidocaine around the endotracheal tube can allow for better endotracheal tube tolerance (Govindarajan *et al.*, 2006).

Some anesthesiologists use the bispectral index (BIS) to predict when the patient is lightly enough anesthetized to undergo the wake-up procedure (McCann *et al.*, 2002, Govindarajan *et al.*, 2006). Predicting movement is helpful because abrupt unexpected movement can dislodge intravenous lines or endotracheal intubation, or cause other unwanted problems. Pneumothorax with soft tissue emphysema has been reported as a complication of abrupt wake-up testing (Maestrello *et al.*, 2001). Even otherwise uneventful wake-up tests can produce substantial catecholamine stress release despite virtually unchanged mean arterial pressure and heart rate (Eroglu *et al.*, 2003).

Conclusions

No test is perfect in predicting postoperative outcomes. Most notably, both IOM and wake-up testing may give normal results in cases of delayed-onset anterior spinal artery syndromes – those with onset of weakness hours after surgery (Stöckl *et al.*, 2005). Neither test predicts future complications.

In summary, the wake-up test remains a viable back-up intraoperative test for use in certain circumstances. The IOM team itself may participate directly in the test by observing under the drapes to ascertain whether the legs or feet are moving. Intraoperative monitoring teams should be aware of this test as an option to discuss or recommend.

Questions

1. The spinal-cord evaluation surgical wake-up test is sometimes referred to as the:
   (a) Vauzelle wake-up test
   (b) Stagnara wake-up test
   (c) Tamaki wake-up test
   (d) Brown wake-up test
2. To prepare for the wake-up test, the following is performed:
   (a) The patient is extubated
   (b) The surgical site first is closed
   (c) The neuromuscular blockade is discontinued
   (d) The patient is turned supine
3. The wake-up test usually is performed:
   (a) Every hour
   (b) At baseline and again at the end of the case
   (c) Once late after distraction is accomplished
   (d) Repeatedly, like motor-evoked potentials
4. The wake-up test is used sometimes when:
   (a) EPs cannot be performed on that patient
   (b) Inhalation anesthetics must be used in the case
   (c) The patient is too young for EP testing
   (d) EPs remained stable during the case
5. To coax pediatric patients to move their feet or legs, some teams:
   (a) Play a tape recording of the parents telling the patient to move the legs
   (b) Tickle the feet
   (c) Use electrical stimulation of the plantar foot surface
   (d) Use bright flashing light to startle the patient
6. In a positive wake-up test, the patient should be able to:
   (a) Move the knees off the operating table
   (b) Move the lower legs off the table against resistance
   (c) Demonstrate inches of bilateral plantar and dorsiflexion foot movement
   (d) Wiggle the toes on both feet
7. Known complications of the wake-up test are all of the following, *except:*
   (a) Dislodgement of the intravenous line
   (b) Endotracheal tube extubation
   (c) Pneumothorax
   (d) Anterior spinal artery occlusion

References

Ben-David, B., Taylor, P. D., and Haller, G. S. (1987). Posterior spinal fusion complicated by posterior column injury. A case report of a false-negative wake-up test. *Spine*, **12**(6), 540–543.

Bradshaw, K., Webb, J. K., and Fraser, A. M. (1984). Clinical evaluation of spinal-cord monitoring in scoliosis surgery. *Spine*, **9**(6), 636–643.

Cheh, G., Lenke, L. G., Padberg, A. M., *et al.* (2008). Loss of spinal-cord monitoring signals in children during thoracic kyphosis correction with spinal osteotomy: why does it occur and what should you do? *Spine*, **33**(10), 1093–1099.

Eggspuehler, A., Sutter, M. A., Grob, D., Jeszenszky, D., and Dvorak, J. (2007). Multimodal intraoperative monitoring during surgery of spinal deformities in 217 patients. *European Spine Journal*, **2**, S188–196.

Eldar, I., Lieberman, N., Shiber, R., Nerubay, J., and Perel, A. (1992). Use of flumazenil for intraoperative arousal during spine fusion. *Anesthesia and Analgesia*. **75**(4), 580–583.

Eroglu, A., Solak, M., Ozen, I., and Aynaci, O. (2003). Stress hormones during the wake-up test in scoliosis surgery. *Journal of Clinical Anesthesia*, **15**(1), 15–18.

Gambrall, M. A. (2007). Anesthetic implications for surgical correction of scoliosis. *AANA Journal*, **75**(4), 277–285.

Glassman, S. D., Johnson, J. R., Shields, C. B., *et al.* (1993). Correlation of motor evoked potentials, somatosensory-evoked potentials, and the wake-up test in a case of kyphoscoliosis. *Journal of Spinal Disorders*, **6**(3), 194–198.

Govindarajan, R., Babalola, O., Gad-El-Kareem, M., *et al.* (2006). Intraoperative wake-up test in neonatal neurosurgery. *Paediatric Anaesthesia*, **16**(4), 451–453.

Grottke, O., Dietrich, P. J., Wiegels, S., and Wappler, F. (2004). Intraoperative wake-up test and postoperative emergence in patients undergoing spinal surgery: a comparison of intravenous and inhaled anesthetic techniques using short-acting anesthetics. *Anesthesia and Analgesia*, **99**(5), 1521–1527.

Kim, D. H., Zaremski, J., Kwon, B., *et al.* (2007). Risk factors for false positive transcranial motor-evoked potential monitoring alerts during surgical treatment of cervical myelopathy. *Spine*, **32**(26), 3041–3046.

Koscielniak-Nielsen, Z. J., Stens-Pedersen, H. L., and Hesselbjerg, L. (1998). Midazolam-flumazenil versus propofol anaesthesia for scoliosis surgery with wake-up tests. *Acta Anaesthesiologica Scandinavica*, **42**(1), 111–116.

Maestrello, C. L., Campbell, R. L., and Campbell, J. R. (2001). Pneumothorax with soft tissue emphysema following abrupt wake-up and self-extubation. *Anesthesia Progress*, **48**(1), 27–31.

Master, D. L., Thompson, G. H., Poe-Kochert, C., Biro, C. (2008). Spinal-cord monitoring for scoliosis surgery in Rett syndrome: can these patients be accurately monitored? *Journal of Pediatric Orthopedics*, **28**(3), 342–346.

McCann, M. E., Brustowicz, R. M., Bacsik, J., *et al.* (2002). The bispectral index and explicit recall during the intraoperative wake-up test for scoliosis surgery. *Anesthesia and Analgesia*, **94**(6), 1474–1478.

Milbrandt, T. A., Kunes, J. R., and Karol, L. A. (2008). Friedreich's ataxia and scoliosis: the experience at two institutions. *Journal of Pediatric Orthopedics*, **28**(2), 234–238.

Othman, Z., Lenke, L. G., Bolon, S. M., and Padberg, A. (2004). Hypotension-induced loss of intraoperative monitoring data during surgical correction of Scheuermann kyphosis: a case report. *Spine*, **29**(12), E258–E265.

Padberg, A. M., Wilson-Holden, T. J., Lenke, L. G., and Bridwell, K. H. (1998). Somatosensory and motor evoked potential monitoring without a wake-up test during idiopathic scoliosis surgery. An accepted standard of care. *Spine*, **23**(12), 1392–1400.

Polly, D. W., Jr., Klemme, W. R., Fontana, J. L, and Sterbis, M. D. (2000). A modified wake-up test for use in very young children undergoing spinal surgery. *Journal of Pediatric Orthopedics*, **20**(1), 64–65.

Stöckl, B., Wimmer, C., Innerhofer, P., Kofler, M., and Behensky, H. (2005). Delayed anterior spinal artery syndrome following posterior scoliosis correction. *European Spine Journal*, **14**(9), 906–909.

Thuet, E. D., Padberg, A. M., Raynor, B. L., *et al.* (2005). Increased risk of postoperative neurologic deficit for spinal surgery patients with unobtainable intraoperative evoked potential data. *Spine*, **30**(18), 2094–2103.

Ting, C. K., Hu, J. S., Teng, Y. H., *et al.* (2004). Desflurane accelerates patient response during the wake-up test for scoliosis surgery. *Canadian Journal of Anaesthesia*, **51**(4), 393–397.

Vauzelle, C., Stagnara, P., and Jouvinroux, P. (1973). Functional monitoring of spinal-cord activity during spinal surgery. *Clinical Orthopaedics and Related Research*, (93), 173–178.

Wilson-Holden, T. J., Padberg, A. M., Lenke, L. G., *et al.* (1999). Efficacy of intraoperative monitoring for pediatric patients with spinal-cord pathology undergoing spinal deformity surgery. *Spine*, **24**(16), 1685–1692.

Wilson-Holden, T. J., VanSickle, D., and Lenke, L. G. (2002). The benefit of neurogenic mixed evoked potentials for intraoperative spinal-cord monitoring during correction of severe scoliosis: a case study. *Spine*, **27**(10), E258–E265.

18 Postoperative studies and outcomes: clinical indications and usefulness of a postoperative study

Gloria M. Galloway

Background and the use of multimodality monitoring (MIOM)

Multimodality intraoperative monitoring has become the standard of care in many surgeries in which neural structures may be at risk of injury. Increasingly, it is also being utilized in cardiac and systemic vascular surgeries, owing to the risk of cerebral ischemia with these surgeries (Razumovsky *et al.*, 2006).

In the 1980s, monitoring somatosensory evoked potentials and brainstem monitoring were typical during complicated surgeries with risk of neural injury. Since then, monitoring has become increasingly more complex, with numerous modalities now being employed during surgical cases. In addition to somatosensory evoked potentials and motor potentials, electromyographic potentials for cranial nerve monitoring, stimulated electromyographic recording in pedicle screw surgeries, peripheral nerve monitoring, dorsal rhizotomy monitoring with low frequency and high frequency stimulation, sphincter and pudendal monitoring are all encountered.

Multiple muscle recordings are commonplace and anesthetic regimens have become increasingly more complex, all creating an environment that is complex to monitor and increases the risk of postoperative injury. Fortunately, with the advent of increasing technologies, postoperative outcomes have improved and the risk of injury has lessened when monitoring is supervised by those skilled and experienced in MIOM (Krassioukov *et al.*, 2004).

Unfortunately, the field is relatively new and as such it is becoming increasingly clear that proper guidelines are required for administration, to ensure that patient care is maintained. Therefore, regulations aimed at limiting the number of cases that can be monitored simultaneously, and that specify the education and training that the supervising physician overseeing the monitoring should have, and how it should proceed, are being developed. Many of these regulations are a response to the fact that there are relatively few physicians trained appropriately and willing to be involved in IOM. The recent creation of a neurology fellowship specifically dedicated to IOM as part of a clinical neurophysiology program of study may increase the numbers of those qualified in intraoperative monitoring.

When neurophysiologic intraoperative monitoring results in changes in any of the monitoring modalities employed, particularly changes that persist beyond the period of surgical closing, a postoperative study can be very useful (Weinzierl *et al.*, 2007). This postoperative study may show improvements not yet indicated on clinical examination or may suggest that neurologic abnormalities may be persistent or longstanding.

In the absence of a clear etiology of a change observed during intraoperative monitoring and particularly if this change is not attributed to systemic variables – further information on the etiology may become apparent during a postoperative monitoring study involving the modalities that underwent the change. Typically, employing the modality that had undergone the change during the surgical procedure along with at least one other complementary modality will provide additional useful information to the clinical team. A description of some of the modalities that may be employed during surgical procedures along with a description of those modalities likely to be considered useful postoperatively in certain clinical situations will be discussed in this chapter.

Transcranial magnetic motor and somatosensory postoperative studies

Transcranial electric motor stimulation (TES) is commonly used in numerous intramedullary spinal-cord surgeries, spinal corrective surgeries, spinal vascular and aortic repair surgeries, and tethered cord repair, in addition to numerous other surgeries in which the motor pathways may be compromised. Most often, TES is combined with evoked somatosensory potentials to evaluate both the anterior and the posterior components of the spinal cord.

A young otherwise healthy female, aged 15 years, underwent insertion of growing rods for scoliosis with IOM. Our protocol involves somatosensory evoked potential and transcranial electric monitoring in the upper and lower extremities. A preoperative study had not been done-just "prior" to this surgery but the patient had had a preoperative study with transcranial magnetic stimulation and somatosensory evoked potentials one year before, as part of her original scoliosis repair, and these studies were normal.

In the operating room, scouting evoked potentials were obtained but motor-evoked potentials could not be obtained despite changes to a variety of stimulation parameters. The patient's blood pressure was elevated, and in an attempt to lower the blood pressure, the levels of halogenated anesthetic had been raised to 2.3% MAC sevoflurane. In this case, using blood pressure lowering agents rather than raising the anesthetic level may have been more useful for IOM.

Baseline somatosensory potentials were then obtained and revealed significantly longer latencies than had been obtained in the patient's prior preoperative study and than normal data, although the patient had had prior surgery. Motor-evoked potentials could not be obtained from any of the extremities. All this suggested that the anesthetic level was probably causing the inability to record motor-evoked potentials in the operating room (Figure 18.1).

The surgery was relatively short but it was very difficult to awaken the patient after surgery and during recovery. The patient vomited twice in recovery and

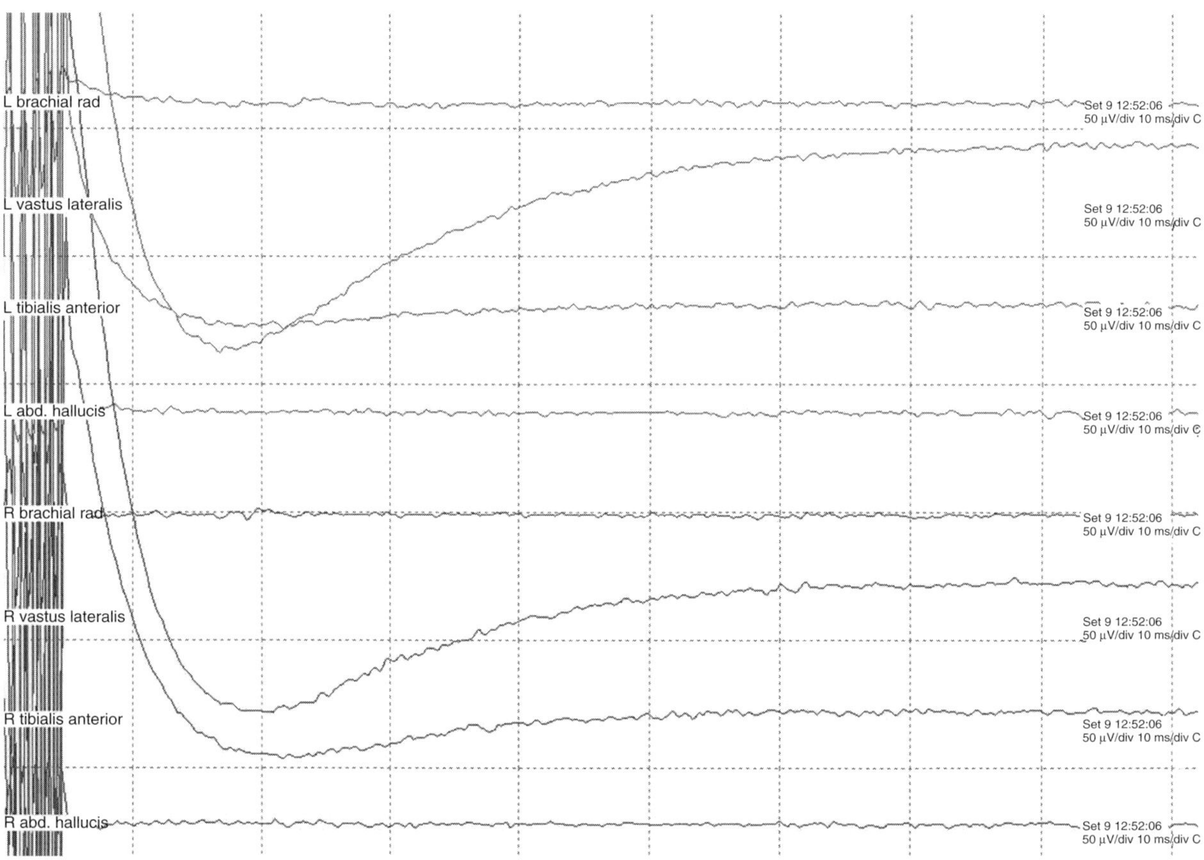

Figure 18.1 Absence of transcranial electric motor stimulation responses from all limbs due to elevated sevoflurane levels during spinal surgery. For colour version, see plate section.

was not cooperative in tests of her motor strength. It was not clear whether the patient was moving all of her limbs over the next day and a postoperative motor study was arranged (Figure 18.2), in which motor-evoked potentials were obtained from all limbs postoperatively using transcranial magnetic stimulation: this suggested that a motor deficit in the motor pathways was unlikely. One could argue that careful attention to anesthetic levels may have improved this patient's postoperative experience.

This is an example of the usefulness of a postoperative study in a patient in which intraooperative changes occurred in transcranial electrical motor monitoring, where information was not adequately obtained in the operating room and the patient had a postoperative complication.

Unfortunately, because this patient had difficulties postoperatively, not having the intraoperative IOM information complicated the situation. Evaluation of motor pathway conduction could be obtained without much patient effort through a postoperative study.

If available, postoperative testing with transcortical magnetic motor potentials in addition to

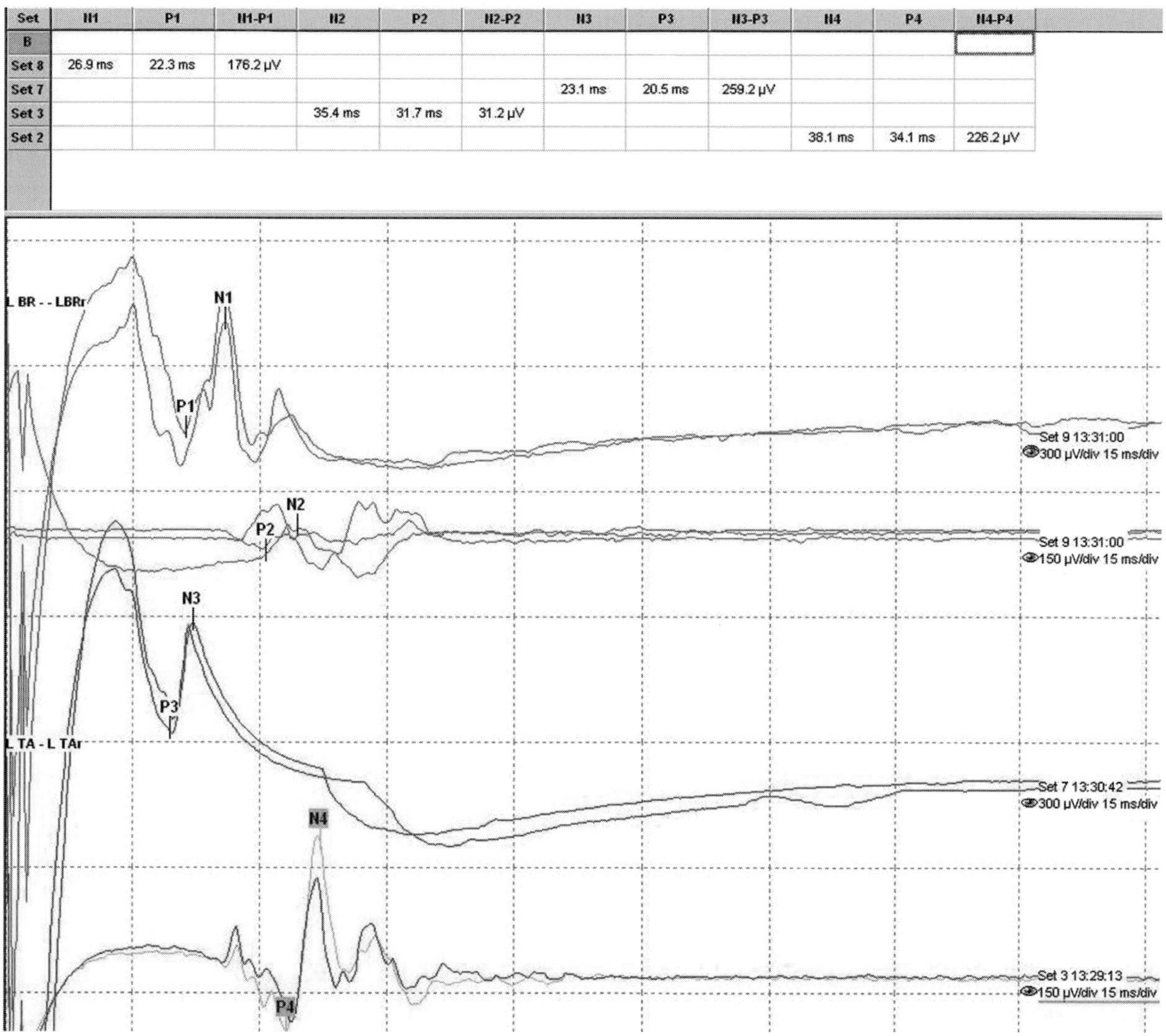

Figure 18.2 Responses from all limbs seen on transcranial magnetic stimulation on postoperative study from patient in Figure 18.1. For color version, see plate section.

somatosensory evoked potential testing will allow assessment of the spinal motor and sensory pathways reliably. In this case, delays were noted on conduction in the somatosensory study intraoperatively, which complemented and supported the findings seen in the motor monitoring. When changes are found using transcranial electric stimulation during intraoperative monitoring, somatosensory evoked potentials may or may not show changes intraoperatively. Somatosensory potential changes may occur at a delayed time or to a smaller degree than the changes seen using TES. If transcranial magnetic stimulation is not available, somatosensory potential testing can be done postoperatively to look for a delay or improvement to the intraoperative study. Unfortunately the somatosensory postoperative study, although helpful, would not directly give information on the spinal motor pathways.

Brainstem auditory evoked potentials

In cases of surgery to the posterior fossa, it is common that brainstem auditory evoked potential monitoring will be used (Rizvi *et al.*, 1999). This may be complemented by cranial nerve monitoring, depending on the surgery involved. In surgery to the posterior fossa, in which changes occurred intraoperatively in brainstem auditory evoked responses (BAERs), monitoring BAERs postoperatively to evaluate recovery neurophysiologically will provide useful prognostic information.

Because of the interrelationship of various neurological structures, more than one modality will often be affected by a change during the intraoperative procedure. Therefore, multimodality monitoring (MIOM) has become the standard method of monitoring complicated surgeries in which monitoring modalities are utilized (Wiedemayer *et al.*, 2002).

Pedicle screw stimulation with electromyographic monitoring

Pedicle screw stimulation with electromyographic monitoring was described in a study reported by Sutter *et al.* (2007), in which MIOM was evaluated during 409 lumbosacral surgical procedures. The monitoring modalities included motor and sensory evoked potentials, as well as electromyographic recording during pedicle screw stimulation. This study reported a sensitivity of MIOM at 90% and a specificity of 99.7% for procedures of decompression and fusion of the lumbosacral spine. The complication rate was 4.9%. Seventeen patients presented with positive MIOM findings and mild or moderate postoperative complications, which included parasthesias or radiculopathy. Five of these 17 patients had partial recovery and the remainder had complete recovery of symptoms.

A patient underwent spinal corrective surgery for scoliosis and a compound muscle action potential (CMAP) was seen on pedicle screw stimulation in the thoracic area. No radicular complaints were noted postoperatively but the patient awoke with a unilateral facial palsy. The palsy resolved completely two weeks later, as suggested by a delayed but present facial nerve response on a postoperative study made 10 days after onset. Bell's palsy has been reported to occur after surgical procedures, although it was not directly related to the surgical procedure, and in this case obtaining a postoperative nerve conduction and EMG study at least seven days after the onset assured the family of a good prognostic outcome.

Selective dorsal rhizotomy (SDR) with dorsal root stimulation and electromyographic monitoring

The procedure involves surgical lumbar laminectomy with an incision that exposes both the lumbar and sacral spinal nerves from the L2 to S2 levels. The S1 root is easily identified, since it is the largest nerve root in the lumbosacral area, and serves as a useful landmark. At each spinal level, the surgeon severs several selected dorsal rootlets, to decrease the facilitory afferent input (see Chapter 15). Threshold stimulation is followed by stimulation at 30–50 Hz, referred to as tetanic stimulation, for one second at each rootlet level. Tetanic stimulation will involve a significantly greater intensity of muscle response than that seen with threshold testing and allows interpretation of the range of abnormality and muscle involvement, as well as the spread of response in that patient.

After selective dorsal rhizotomy (SDR), one often sees functional improvement in the weeks to months after the procedure. Postoperative studies are not usual in these patients to evaluate spasticity or outcome but may be requested if loss of motor function occurs in the postoperative period. Clinical improvements in outcome in children after selective dorsal rhizotomy for spasticity are typical and are well described and the improvement continues for a long time on follow-up examinations in these patients.

In a study by van Schie et al. (2005), a group of spastic diplegic but ambulatory children with cerebral palsy aged 43–82 months showed functional motor improvements postoperatively at 12 months follow-up after selective dorsal rhizotomy using several rating scales, including the gross motor function measure. With SDR, there is typically improvement in postural stability including stability for sitting (Steinbok *et al.*, 1992).

Complications have been reported in patients undergoing SDR. These include urinary retention, back pain, and sensory changes in the lower extremities, as well as pulmonary complications, including aspiration pneumonia, and bronchospasm (Steinbok and Schrag, 1998). The sensory changes included both dysesthesias and hyperesthesias. These sensory changes may be avoided by limiting the number of dorsal rootlets cut.

In some patients, a postoperative *nerve conduction and electromyography test* may be requested to evaluate these sensory complaints further. Possible findings include radiculopathies, peripheral nerve entrapments, and polyneuropathies that are probably unrelated to the surgical procedure. Urinary retention requiring catheterization lasting several days after the procedure is the most common complication and has decreased with sparing of S2 roots and modification of the SDR procedure. The complication of persistent urinary incontinence may occur in up to 5% of patients after SDR. In patients with S2 sparing, urinary and bowel dysfunction does not typically occur.

Pudendal nerve monitoring and postoperative nerve conduction and electromyography

When *pudendal nerve monitoring* is included as part of the selective rhizotomy procedure, the risks of postoperative complication can be further reduced. Pudendal nerve monitoring is associated with significantly less risk of urinary and fecal incontinence as a complication of SDR. Since the presence of spasticity in many patients with cerebral palsy contributes to varying degrees to their ability to stand and ambulate, SDR can cause deterioration in the ability to ambulate if it is not selective enough.

Deterioration in the ability to ambulate can also occur as a consequence of lack of residual voluntary strength and deconditioning. In these cases, a postoperative nerve conduction and electromyography test may be requested to look for an etiology and explanation of the increased difficulty. However, with technical expertise, and limitations of the surgical expanse using intraoperative monitoring during dorsal rootlet severance, the risk of deterioration in motor and functional skills is significantly lessened (Nordmark *et al.*, 2008).

Complex spinal surgeries with pedicle screw stimulation may be associated with sensory or motor symptoms postoperatively. Radicular pain or radicular palsy may occur after surgery in the lumbosacral area, including surgeries for stenosis, scoliosis, spondylosis, spondylolisthesis, or any spinal decompressive surgery (Hu and Beren, 2006). During the procedure, free-run mode EMG recordings are used, as well as stimulated electromyography during stimulation of the pedicle screws. Pedicle screw stimulation will help to decrease the risk of root injury during complicated spinal corrective surgeries. If complaints suggestive of radiculitis or radiculopathy persist postoperatively, a postoperative nerve conduction and electromyography test (NC/EMG) can be useful in determining accurately the level involved electrophysiologically in explaining the area of complaint.

It is not uncommon for a CMAP to occur on stimulation of a pedicle screw (see Figure 18.3). Upon palpation a small breach may be discovered in the medial wall of the pedicle. In this case the surgeon can apply bone wax to the breach area and restimulate the screw, to reevaluate for the presence of a CMAP. Many studies have shown that manual palpation of the pedicle wall is of limited value in identifying a breach of the medial or lateral pedicle wall. A study by Donohue *et al.* (2007) used a high-frequency pulse train to activate the adjacent thoracic spinal-cord axons by way of the thoracic pedicle screw track. The effect of stimulation was monitored by seeing if there were muscle responses from the lower limbs. This study showed that pedicle screw stimulation was an effective means of identifying a medial wall breach since lower-limb stimulation could not be explained otherwise.

Postoperatively, if symptoms of radicular injury occur, EMG and nerve conduction testing can be diagnostic but will typically remain negative for the first 7–10 days post injury. Somatosensory testing postoperatively will probably be negative because the

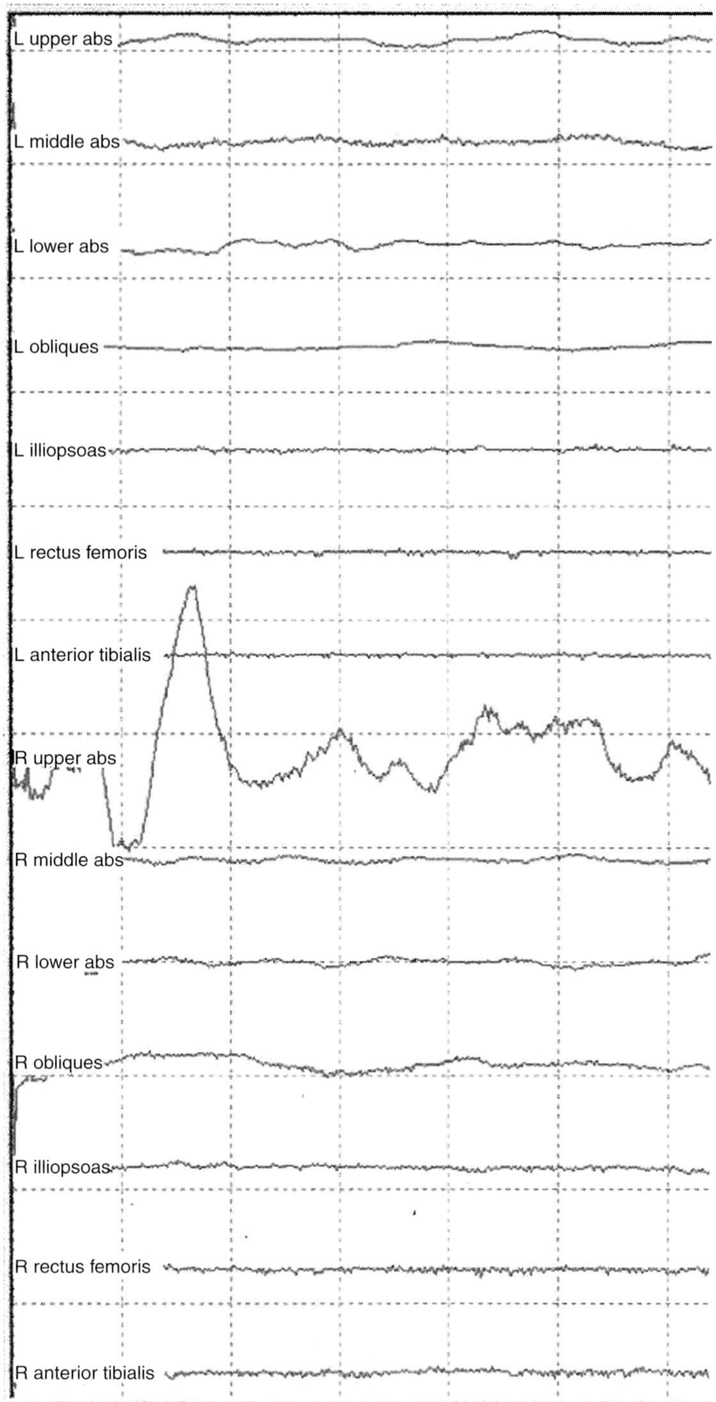

Figure 18.3 A compound muscle action potential (CMAP) is seen during pedicle screw stimulation in the thoracic area.
(100 uV/div)

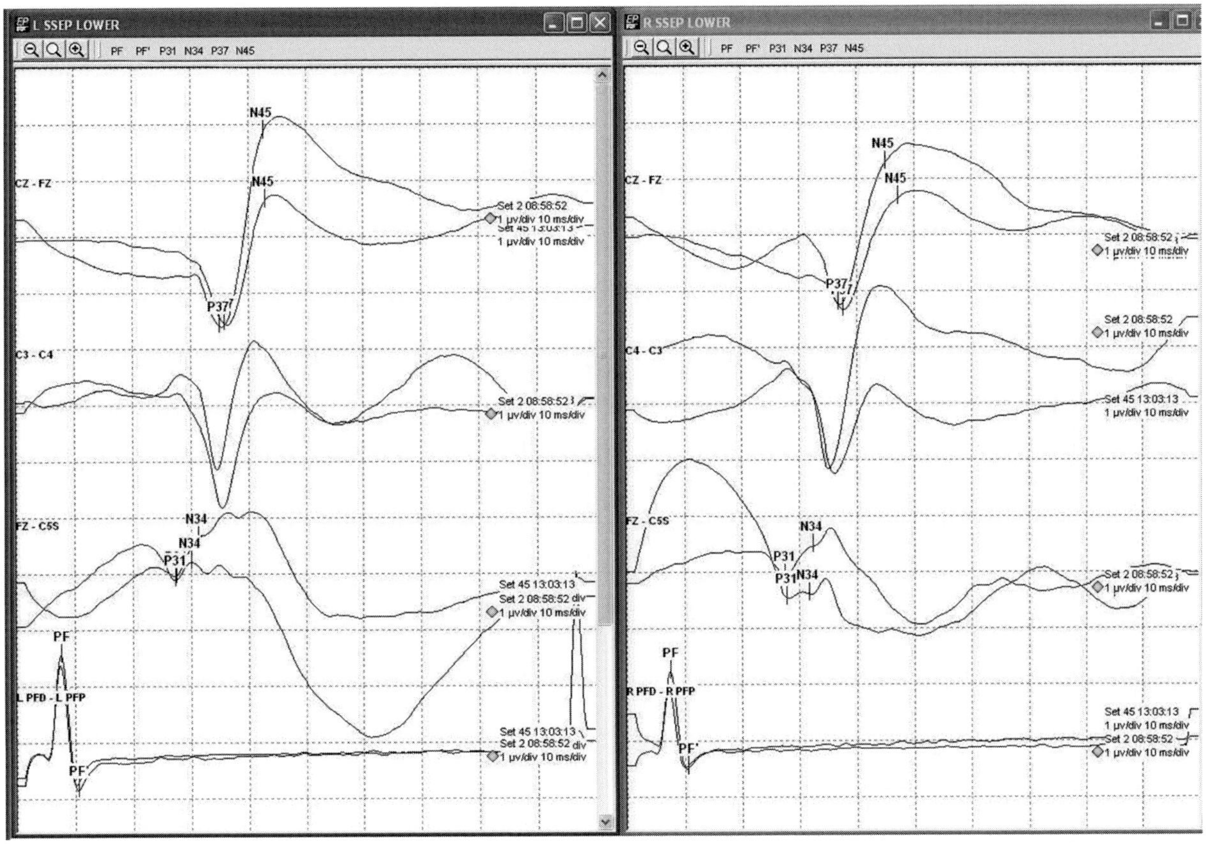

Figure 18.4 Posterior tibial SEP, showing intraoperative attenuation in the amplitude of the subcortical and cortical responses in response to a significant drop in mean arterial pressure.

diagnostic utility of SEPs is limited to dorsal column function. So although somatosensory evoked potentials may be affected during a spinal-cord lesion that predominately involves the sensory tract, these potentials should not be abnormal with root injuries or compression.

Monitoring with somatosensory and transcranial motor stimulation in spinal corrective surgeries

Ischemic or mechanical spinal-cord injury can be followed and assessed postoperatively with somatosensory evoked potentials. Somatosensory evoked potentials should be sensitive to injury or ischemia to the dorsal columns or posterior portion of the spinal cord. To assess motor pathways, information from the anterior portion of the spinal cord will probably be needed. Figure 18.4 shows a posterior tibial SEP recording for a patient undergoing spinal corrective surgery. In this case, there was a significant loss of amplitude at the subcortical and cortical levels bilaterally during loss of mean arterial pressure. The mean arterial pressure was raised and anesthesia levels lowered. There was improvement in the posterior tibial SEP, although this remained significantly decreased compared with the baseline. Transcranial motor potentials did not significantly change. A postoperative study the next morning (Figure 18.5) revealed improvement in the posterior tibial SEP recording and the patient did not develop symptoms of spinal-cord ischemia.

If available, transcranial magnetic stimulation would be useful to test motor pathways, which can be injured from lesions affecting the anterior portion of the spinal cord. In some cases, SEPs will be sensitive in detecting anterior cord insult, although in most anterior spinal-cord insults, either no change in SEPs or a delayed change would be expected. Alternatively, in some cases, TMS may be sensitive

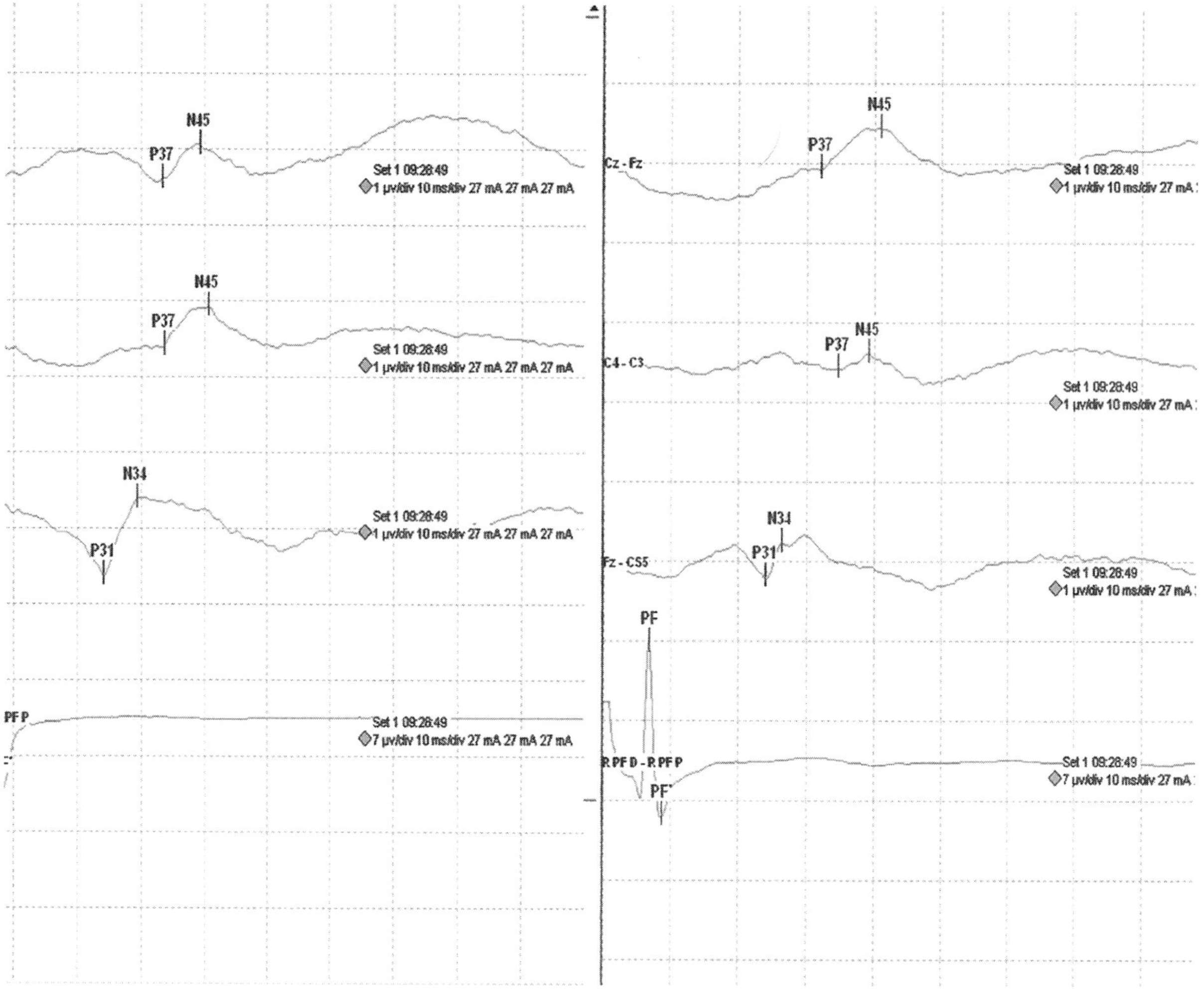

Figure 18.5 Improvement in posterior tibial SEP with increase in mean arterial pressure.

for evaluation of posterior cord insults although in a similar way, this would not be expected to be as sensitive in detecting posterior cord insults as somatosensory evoked potentials.

A study by Takaso *et al.* (2010) described posterior spinal fusion for severe scoliosis in patients with Duchenne muscular dystrophy (DMD) and a mean preoperative scoliosis of 98° (range 81°–130°). In many such cases, surgeons may be hesitant to operate on patients with DMD, owing to the risk of complications, and these patients may not undergo treatment until late in the scoliotic curve progression. Forced vital capacity (FVC) was less than 30% in 13 of the 14 patients reported in this study. All patients underwent a six-week preoperative period with a pulmonary trainer device for respiratory muscle training and improvement of FVC prior to surgical correction. The authors reported good outcomes in the patients studied, with improvements in FVC and extubation on the operative day without respiratory complications, suggesting that with improvement in FVC and careful attention, even patients with Duchenne dystrophy and severe scoliosis can have a good postoperative outcome after scoliosis correction.

Vascular surgeries with transcranial motor and somatosensory evoked potential monitoring

Aortic artery repair can place the patient at risk of spinal-cord ischemia. Multimodality intraoperative monitoring has been shown to improve neurological outcome following aortic clamp procedures (Sloan, 2004). Similarly, multimodality monitoring has been useful in patients during cardiopulmonary bypass procedures in which spinal-cord ischemia is possible (Rao *et al.*, 2008). Any loss or change in MEPs during this procedure suggests anterior spinal-cord ischemia. Improvement can be seen in a postoperative transcranial magnetic study. Somatosensory evoked potentials may not change intraoperatively. Chapter 14 gives examples of monitoring cases involving vascular surgeries and the need for postoperative studies, depending on the patient's intraoperative course and postoperative clinical outcome.

Tethered cord surgery and transcranial electric motor and somatosensory evoked potential monitoring

Surgery for repair of tethered cord is common in patients of all ages, from the very young to adults. Symptoms include low back and leg pain as well as urological and bowel dysfunction. Often the anatomy of that area is variable and there may be longstanding developmental anomalies. Complication rates have been reported to be as high as 13% (Huttmann *et al.*, 2001). Spina bifida may be present as well as cauda equina tumors and lipomas in association with a tethered cord.

Intraoperative monitoring in these cases often involves motor and sensory monitoring as well as free run or continuous electromyography (EMG) in the lower limbs to record from muscles in the L2 to S2 myotomes. In addition recordings are often also taken from the anal and urethral sphincters because of the risk of bowel and bladder impairment (Quinones-Hinojosa *et al.*, 2004) Limb muscles may include the quadriceps femoris, medial or lateral gastrocnemius, and tibialis anterior. The most common neurophysiologic monitoring modality used in tethered cord surgery is continuous EMG recording. This is because of the risk of inadvertently sectioning functional neural tissue within the filum terminale, has been reported in up to 36% of cases (Legatt *et al.*, 1992).

The recordings obtained at baseline vary considerably depending on the extent of involvement. Abnormalities can be present in either the motor-evoked or sensory-evoked potentials and can be bilateral or asymmetric in nature. Often there is baseline irritability on free-run EMG or there may be no muscle activity at all. Electromyographic activity, if not present prior to surgical manipulation, may develop after manipulation or may increase or decrease afterwards. The EMG activity is considered significant if it is burst-like in nature and prolonged. If the EMG activity is transient in nature, it may be seen as a consequence of manipulation without any pathological significance. In a study by Paradiso *et al.* (2006) on 44 adult patients undergoing surgical repair of tethered cord syndrome, continuous EMG had a sensitivity of 100% for identifying new postoperative neurological deficits, although there was a low specificity.

Needless to say, preparation and limitation of perioperative complications will decrease the risk of postoperative complications and the need for postoperative studies. Perioperative management includes optimizing pulmonary, cardiac, and anesthetic conditions so that the patient will have the lowest surgical risk (Glassman *et al.*, 2007). Limitation in the expanse of the surgical procedure and the use of intraoperative monitoring together create a state in which postoperative complications can be potentially reduced and patient care improved.

Cranial nerve monitoring and brainstem auditory evoked potential monitoring

Cranial nerves VII, IX, X are most often monitored, with cranial nerves V and XI not uncommonly monitored as well, depending on the surgical procedure. Patients with acoustic neuromas typically do not need a postoperative study, as they often do well. However, if there is a complicated postoperative course, a hearing deficit does not improve, or facial weakness occurs, a postoperative brainstem evoked potential study can be very useful. If facial weakness is the only abnormality, a postoperative nerve conduction and electromyography study may be useful. Intraoperative removal of the acoustic neuroma is often guided by the use of BAER recording as well as cranial nerve monitoring of the seventh or facial nerve, because of its close proximity to the area of resection and the risk of mechanical injury to this nerve. In addition, vital brainstem structures may be at risk of ischemic injury.

Questions

1. The most common neurophysiologic monitoring modality reported to be used in tethered cord surgery is:
   (a) Recording SEPs from the upper and lower limbs
   (b) Transcranial electric motor stimulation
   (c) Continuous EMG recording from limbs and anal or urethral sphincters
   (d) BAER recording
2. Aortic artery repair can place the patient at risk of spinal-cord ischemia. Which of the following combinations of neurological structures and monitoring modality is most useful for this type of surgical procedure?
   (a) Posterior spinal cord and SEPs of lower limbs
   (b) Anterior spinal cord and MEPs of lower limbs
   (c) Brainstem and BAER
3. Which monitoring modality can be useful if available to test the motor pathways in an awake unsedated patient?
   (a) Monitoring SEPs in all four limbs
   (b) BAER monitoring with increasing levels of intensity testing
   (c) Transcranial electric motor stimulation
   (d) Transcranial magnetic motor stimulation
4. The symptoms of tethered cord may include:
   (a) Tingling over the face and tongue
   (b) Inability to close eyelids
   (c) Urinary and bowel dysfunction
5. The myotomes that should be recorded from for tethered cord repair include:
   (a) T5–10
   (b) C5–8
   (c) L2–S2
   (d) C1–C8
6. A surgeon deciding to palpate the pedicle during spinal repair surgery is most probably looking for:
   (a) A lateral wall breach
   (b) The pedicle screw
   (c) The spinal cord
   (d) A medial wall breach
7. The likelihood of inadvertently severing neural tissue in the filum terminale may occur during which surgical procedure?
   (a) Spinal decompression
   (b) Posterior fossa decompression
   (c) Tethered cord repair
   (d) Pedicle screw placement
8. Which two modalities are most useful in spinal surgery to evaluate for spinal cord ischemia?
   (a) Transcranial motor stimulation and BAER recording
   (b) Cranial nerve monitoring and transcranial motor stimulation
   (c) Somatosensory recording and transcranial motor stimulation
   (d) Somatosensory recording and continuous EMG recording
9. Selective dorsal rhizotomies are done to improve or reduce which of the following?
   (a) Fasciculations
   (b) Radicular complaints
   (c) Spasticity
   (d) Kyphosis
10. Patients with acoustic neuromas often do well postoperatively. What monitoring modalities are typically utilized during the surgery?
   (a) BAER and CN 7 monitoring
   (b) SEPs and TcMEPs
   (c) Pedicle screw stimulation
   (d) Dermatomal SEPs

References

Donohue, M., Murtagh-Schaffer, C., Basta J., *et al.* (2008). Pulse train stimulation for detecting medial malposition of thoracic pedicle screws. *Spine*, **33**(12), E378–E385.

Glassman, S., Hamill, C., Bridwell, K., *et al.* (2007). The impact of perioperative complications on clinical outcome in adult deformity surgery. *Spine*, **32**(24), 2764–2770.

Hu, S. and Beren, S. (2006). Preparing the adult deformity patient for spinal surgery. *Spine*, **31**(195), 5126–5131.

Huttmann, S., Krauss J., Collmann H., Sorensen, N., and Roosen, K. (2001). Surgical management of tethered spinal cord in adults: report of 54 cases. *Journal of Neurosurgery*, **95**, 173–178.

Krassioukov, A. V., Sarjeant, R., Arkia, H., and Fehlings, M. G. (2004). Multimodality intraoperative monitoring during complex lumbosacral procedures; indications, techniques, and long- term follow-up review of 61 consecutive cases. *Journal of Neurosurgery: Spine*, **1**(3), 243–253.

Legatt, A. D., Schroeder, C. E., Gill, B., and Goodrich, J. T. (1992). Electrical stimulation and multichannel EMG recording for identification of functional neural tissue during cauda equina surgery. *Child's Nervous System*, **8**, 185–189.

Nordmark, E., Josenby, A. I., Lagergren, J., *et al.* (2008). Long-term outcomes five years after selective dorsal rhizotomy. *BMC Pediatrics*, **8**, 54.

Paradiso, G., Lee, G., Sarjeant R., *et al.* (2006). Multimodality intraoperative neurophysiologic monitoring findings during surgery for tethered cord syndrome: analysis of a series of 44 patients with long-term follow-up. *Spine*, **31**(18), 2095–2102.

Quinones-Hinojosa, A., Gadkary, C. A., Gulati, M., *et al.* (2004). Neurophysiological monitoring for safe surgical tethered cord syndrome release in adults. *Surgical Neurology*, **62**(2), 127–133. Discussion 133–135.

Rao, S. L., Bundy, J. T., and Stephenson, E. (2008). Neurophysiologic monitoring during cardiopulmonary bypass in a patient with critical aortic stenosis and cervical spine stenosis. *Journal of Cardiothoracic and Vascular Anesthesia*, **22**(1), 115–118.

Razumovsky, A. Y., Gugino, L. D., and Owen, J. H. (2006). Advanced neurologic monitoring for cardiac surgery. *Current Cardiology Reports*, **8**(1), 17–22.

Rizvi, S. S., Goyal, R. N., and Calder, H. B. (1999). Hearing preservation in microvascular decompression for trigeminal neuralgia. *Laryngoscope*, **109**(4), 591–594.

Sloan, T. B. (2004). Electrophysiologic monitoring during surgery to repair the thoracoabdominal aorta. *Seminars in Cardiothoracic and Vascular Anesthesia*, **8**(2), 113–125.

Steinbok, P. and Schrag, C. (1998). Complications after selective posterior rhizotomy for spasticity in children with cerebral palsy. *Pediatric Neurosurgery*, **28**, 300–313.

Steinbok, P., Langill, L., Cochrane, D. D., and Keyes, R. (1992). Observations on electrical stimulation of lumbosacral nerve roots in children with and without lower limb spasticity. *Child's Nervous System*, **8**, 376–382.

Sutter, M., Eggspuehler, A., Grob, D., *et al.* (2007). Multimodal intraooperative monitoring (MIOM) during 409 lumbosacral surgical procedures in 409 patients. *European Spine Journal*, **16**(suppl. 2), S221–S228.

Takaso, M., Nakazawa, T., Imura, T., *et al.* (2010). Surgical management of severe scoliosis with high-risk pulmonary dysfunction in Duchenne muscular dystrophy. *International Orthopedics*. **34**, 401–406.

van Schie, P. E. M., Vermeulen, R. J., Van Ouwerkerk, W. J., Kwakkel G., and Becher, J. G. (2005). Selective dorsal rhizotomy in cerebral palsy to improve functional abilities: evaluation of criteria for selection. *Child's Nervous System*, **21**(6), 451–457.

Weinzierl, M. R., Reinacher, P., Gilsbach, J. M., and Rohde, V. (2007). Combined motor and somatosensory evoked potentials for intraoperative monitoring: intra- and postoperative data in a series of 69 operations. *Neurosurgical Review*, **30**(2), 109–116.

Wiedemayer, H., Fauser, B., Sandalcioglu, I. E., Schafer, H., and Stolke, D. (2002). The impact of neurophysiological intraoperative monitoring on surgical decisions: a critical analysis of 423 cases. *Journal of Neurosurgery*, **96**, 255–262.

Index